RILEM State-of-the-Art Reports

RILEM STATE-OF-THE-ART REPORTS

Volume 23

RILEM, The International Union of Laboratories and Experts in Construction Materials, Systems and Structures, founded in 1947, is a non-governmental scientific association whose goal is to contribute to progress in the construction sciences, techniques and industries, essentially by means of the communication it fosters between research and practice. RILEM's focus is on construction materials and their use in building and civil engineering structures, covering all phases of the building process from manufacture to use and recycling of materials. More information on RILEM and its previous publications can be found on www.RILEM.net.

The RILEM State-of-the-Art Reports (STAR) are produced by the Technical Committees. They represent one of the most important outputs that RILEM generates—high level scientific and engineering reports that provide cutting edge knowledge in a given field. The work of the TCs is one of RILEM's key functions.

Members of a TC are experts in their field and give their time freely to share their expertise. As a result, the broader scientific community benefits greatly from RILEM's activities.

RILEM's stated objective is to disseminate this information as widely as possible to the scientific community. RILEM therefore considers the STAR reports of its TCs as of highest importance, and encourages their publication whenever possible.

The information in this and similar reports is mostly pre-normative in the sense that it provides the underlying scientific fundamentals on which standards and codes of practice are based. Without such a solid scientific basis, construction practice will be less than efficient or economical.

It is RILEM's hope that this information will be of wide use to the scientific community.

More information about this series at http://www.springer.com/series/8780

Sofiane Amziane · Florence Collet
Editors

Bio-aggregates Based Building Materials

State-of-the-Art Report of the RILEM Technical Committee 236-BBM

Springer

Editors
Sofiane Amziane
Polytech Clermont-Ferrand
Clermont-Ferrand
France

Florence Collet
Université de Rennes 1
Rennes
France

ISSN 2213-204X ISSN 2213-2031 (electronic)
RILEM State-of-the-Art Reports
ISBN 978-94-024-1030-3 ISBN 978-94-024-1031-0 (eBook)
DOI 10.1007/978-94-024-1031-0

Library of Congress Control Number: 2016961667

Printed on acid-free paper

This Springer imprint is published by Springer Nature
The registered company is Springer Science+Business Media B.V.
The registered company address is: Van Godewijckstraat 30, 3311 GX Dordrecht, The Netherlands

Preface

Global warming, energy savings, and life cycle analysis issues are the factors that have contributed to the rapid expansion of plant-based materials for buildings, which can be qualified as environmental-friendly, sustainable and efficient multi-functional materials. These materials are obtained from the processing of hemp, flax, miscanthus, pine, maize, sunflower and bamboo.

The work of the Technical Committee (TC 236-BBM) was dedicated to the study of construction materials made from plant particles. Are concerned building materials containing as main raw material renewable, recyclable and easily available plant particles. However, the work was relatively centred on hemp because hemp shiv is the bio-aggregate that is the most widely used in building materials and the most studied in the literature.

This state-of-the-art report reflects the current knowledge on the assessment of the chemical, physical and mechanical properties of bio-aggregate and vegetal concrete. It presents an overview on the several possibilities developed worldwide about the use of plant aggregate to design bio-based building materials. The first five chapters relate to the description of the vegetal aggregate. Then, hygrothermal properties, fire resistance, durability and finally the impact of the variability of the method of production of bio-based concrete are assessed on Chaps. 7–9.

Vegetal aggregates are generally highly porous with a low apparent density and a complex architecture marked by a multi-scale porosity. Chapters 1–5 give an overview on the physical properties of the vegetal aggregate and the methods to assess these characteristics.

These geometrical characteristics result in noteworthy hygrothermal performances. This is one of the essential characteristics, which differ from vegetal concrete compared to the tradition mineral-based concretes. Chapter 6 gives the state of the art of previous studies on hygric and thermal properties of bio-aggregate based building materials. Hygric characteristics such as sorption isotherms, water vapor permeability and moisture diffusivity are given. The ability of bio-aggregate-based building materials to moderate ambient relative humidity may be valued using moisture buffer value. Thermal properties (thermal conductivity, thermal diffusivity conductivity and specific heat capacity) are then reported. Finally, concluding remarks on hygrothermal behavior with simultaneous heat and mass transfer are provided.

Chapter 7 concerns behaviour of bio-aggregate based building materials exposed to fire. Discrepancies between fire reaction and fire resistance are highlighted in this chapter. Various results of fire reaction test performed on bio-based materials are reported. Bio-aggregates are often in Class F, while concretes range in class B1. In some of the presented case studies, render and plaster play a key role in the fire resistance. EI 90 fire resistance appeared to be accessible with conventional technologies.

Chapter 8 deals with the impact of biological and environmental ageing on the durability of the multiphysical performances of vegetal concretes. It is important to note that in the case of hemp concretes, hygrothermal changes can also lead to specific disorders, such as variations in dimensional stability, of microstructure and of functional properties (mechanical, acoustical and hygrothermal properties). Moreover, the growth of microorganisms can also be observed and further aggravates the pathologies mentioned above. All these evolutions may as well lead to variations in the functional properties of the materials during their use.

Chapter 9 investigates the effect of production parameters including curing conditions (65% vs. >95% RH), time of demoulding and specimen geometry (cylinder vs. cube) on the concrete's strength which relates to density and therefore to thermal and hygric properties. It studies hydration in the concrete's microstructure and measures the compressive strength development at intervals between 1 day and 1 month.

Moulding time and curing conditions influence drying and therefore may impact binder hydration and consequently strength evolution. Specimen geometry may affect drying and can also determine how strain builds up in the concrete and thus when failure occurs.

The state-of-the-art report is followed by an appendix containing a TC report which presents the experience of a working group within the RILEM Technical Commitee 236-BBM 'Bio-aggregate based building Materials'. The work of the group was dedicated to the study of construction materials made from plant particles. Are concerned building materials containing as main raw material renewable, recyclable and easily available plant particles. These materials are obtained from the processing of hemp, flax, miscanthus, pine, maize, sunflower, bamboo and others. However, the work of the round robin test has been centred on hemp because hemp shiv is the bio-aggregate that is the most widely used in building materials and the most studied in the literature.

The first round robin test of the TC 236-BBM was carried out to compare the protocols in use by different laboratories (labs) to measure initial water content, bulk density, water absorption, particle grading and thermal conductivity. The aim is to define a characterisation protocol derived from those used by different laboratories.

This first round robin test was carried out on one variety of hemp shiv. Nine laboratories from European universities and research centers were involved (Table 1).

Table 1 Participating Labs

Letter	City	Labs
A	Bath (UK)	BRE Centre for Innovative Construction Materials/University of Bath
B	Clermont Ferrand (France)	Institut Pascal
C	Lorient (France)	LIMatB/Université de Bretagne Sud
D	Lyon (France)	DGCB/ENTPE
E	Paris (France)	IFSTTAR
F	Rennes (France)	LGCGM/Rennes 1
G	Toulouse (France)	LMDC/Université de Toulouse/UPS/INSA
I	Combloux (Belgium)	Combloux-Agro ressource—Université de Liège

The test results of 7 laboratories constitute a set of statistically representative data in order to propose recommendations to characterise hemp shiv after analysing the different methodologies in use in these laboratories. In addition to 7 laboratories, in the last round-robin test on "Variability of mechanical properties of hemp concrete", Vicat (France), Queen's University Belfast (UK), and Trinity college Dublin (Ireland) were involved in the comparison of the results between different laboratories. The results were published elsewhere (2016)[1].

Finally, this state of the art on bio-based materials allows to show both advantages and limitations that can be expected in plant-based building materials.

Ongoing research will certainly provide more insights that are not included in this book. But in terms of performance already demonstrated, we already have peace of mind about the potential for the benefit of biomaterials compared to conventional solutions that have a strong impact on our environmental.

Clermont-Ferrand, France

Prof. Sofiane Amziane
Chair of TC 236-BBM

[1]Niyigena, C., Amziane, S., Chateauneuf, A., Arnaud, L., Bessette, L., Collet, F., Lanos, C., Escadeillas, G., Lawrence, M., Magniont, C., Marceau, S., Pavia, S., Peter, U., Picandet, V., Sonebi, M., Walker, P., Variability of the mechanical properties of hemp concrete, (2016) Materials Today Communications, 7, pp. 122–133.

Contents

TC 236-BBM Members

Chairman:
Sofiane Amziane, Institut Pascal, Clermont Université, France.

Secretary:
Florence Collet, Université de Rennes 1, LGCGM, France.

TC Members:
Laurent Arnaud, Ecole Nationale Supérieure d'Arts et Métiers, ENSAM Cluny, France.
Laëtitia Bessette, Centre Technique Louis VICAT, L'Isle d'Abeau, France.
Paulien De Bruijn, Lund University, Faculty of Engineering (LTH), Dept of Building Materials, Lund, Sweden.
Samuel Dubois, Université de Liège (Ulg), Gembloux Agro-Bio Tech, Belgique.
Gilles Escadeillas, Université de Toulouse, UPS, INSA, LMDC, France.
Harald Garrecht, Universität Stuttgart, Institut für Werkstoffe im Bauwesen (IWB), Stuttgart, Deutschland.
Etienne Gourlay, CETE de l'Est—Laboratoire Régional de Strasbourg, France.
André Klatt, Universität Stuttgart, Institut für Werkstoffe im Bauwesen (IWB), Stuttgart, Deutschland.
Christophe Lanos, Université de Rennes 1, LGCGM, France.
Mike Lawrence, University of Bath, BRE Centre for Innovative Construction Materials, UK.
Camille Magniont, Université de Toulouse, UPS, INSA, LMDC, France.
Sandrine Marceau, Université Paris-Est, IFSTTAR, France.
Sara Pavía, Department of Civil Engineering, Trinity College, Ireland.
Ulrike Peter, Lhoist, Lhoist Recherche et Développement S.A., Belgium.
Vincent Picandet, IRDL, Université de Bretagne Sud, France.
Mohammed Sonebi, School of Planning, Architecture and Civil Engineering, Queen's University Belfast, UK.
Peter Walker, University of Bath, BRE Centre for Innovative Construction Materials, UK.

Other Contributors:

Thibaut Colinart, IRDL, Université de Bretagne Sud, France.
Guillaume Delannoy, Université Paris-Est, IFSTTAR, France.
Yunhong Jiang, University of Bath, BRE Centre for Innovative Construction Materials, UK.
César Niyigena, Institut Pascal, Clermont Université, France.
Vincent Nozahic, Institut Pascal, Clermont Université, France.
Sylvie Pretot, Université de Rennes 1, LGCGM, France.
Pierre Tronet, IRDL, Université de Bretagne Sud, France.

RILEM Publications

The following list is presenting the global offer of RILEM Publications, sorted by series. Each publication is available in printed version and/or in online version.

RILEM Proceedings (PRO)

PRO 1: Durability of High Performance Concrete (ISBN: 2-912143-03-9); *Ed. H. Sommer*

PRO 2: Chloride Penetration into Concrete (ISBN: 2-912143-00-04); *Eds. L.-O. Nilsson and J.-P. Ollivier*

PRO 3: Evaluation and Strengthening of Existing Masonry Structures (ISBN: 2-912143-02-0); *Eds. L. Binda and C. Modena*

PRO 4: Concrete: From Material to Structure (ISBN: 2-912143-04-7); *Eds. J.-P. Bournazel and Y. Malier*

PRO 5: The Role of Admixtures in High Performance Concrete (ISBN: 2-912143-05-5); *Eds. J. G. Cabrera and R. Rivera-Villarreal*

PRO 6: High Performance Fiber Reinforced Cement Composites—HPFRCC 3 (ISBN: 2-912143-06-3); *Eds. H. W. Reinhardt and A. E. Naaman*

PRO 7: 1st International RILEM Symposium on Self-Compacting Concrete (ISBN: 2-912143-09-8); *Eds. Å. Skarendahl and Ö. Petersson*

PRO 8: International RILEM Symposium on Timber Engineering (ISBN: 2-912143-10-1); *Ed. L. Boström*

PRO 9: 2nd International RILEM Symposium on Adhesion between Polymers and Concrete ISAP'99 (ISBN: 2-912143-11-X); *Eds. Y. Ohama and M. Puterman*

PRO 10: 3rd International RILEM Symposium on Durability of Building and Construction Sealants (ISBN: 2-912143-13-6); *Eds. A. T. Wolf*

PRO 11: 4th International RILEM Conference on Reflective Cracking in Pavements (ISBN: 2-912143-14-4); *Eds. A. O. Abd El Halim, D. A. Taylor and El H. H. Mohamed*

PRO 12: International RILEM Workshop on Historic Mortars: Characteristics and Tests (ISBN: 2-912143-15-2); *Eds. P. Bartos, C. Groot and J. J. Hughes*

PRO 13: 2nd International RILEM Symposium on Hydration and Setting (ISBN: 2-912143-16-0); *Ed. A. Nonat*

PRO 14: Integrated Life-Cycle Design of Materials and Structures—ILCDES 2000 (ISBN: 951-758-408-3); (ISSN: 0356-9403); *Ed. S. Sarja*

PRO 15: Fifth RILEM Symposium on Fibre-Reinforced Concretes (FRC)—BEFIB'2000 (ISBN: 2-912143-18-7); *Eds. P. Rossi and G. Chanvillard*

PRO 16: Life Prediction and Management of Concrete Structures (ISBN: 2-912143-19-5); *Ed. D. Naus*

PRO 17: Shrinkage of Concrete—Shrinkage 2000 (ISBN: 2-912143-20-9); *Eds. V. Baroghel-Bouny and P.-C. Aïtcin*

PRO 18: Measurement and Interpretation of the On-Site Corrosion Rate (ISBN: 2-912143-21-7); *Eds. C. Andrade, C. Alonso, J. Fullea, J. Polimon and J. Rodriguez*

PRO 19: Testing and Modelling the Chloride Ingress into Concrete (ISBN: 2-912143-22-5); *Eds. C. Andrade and J. Kropp*

PRO 20: 1st International RILEM Workshop on Microbial Impacts on Building Materials (CD 02) (e-ISBN 978-2-35158-013-4); *Ed. M. Ribas Silva*

PRO 21: International RILEM Symposium on Connections between Steel and Concrete (ISBN: 2-912143-25-X); *Ed. R. Eligehausen*

PRO 22: International RILEM Symposium on Joints in Timber Structures (ISBN: 2-912143-28-4); *Eds. S. Aicher and H.-W. Reinhardt*

PRO 23: International RILEM Conference on Early Age Cracking in Cementitious Systems (ISBN: 2-912143-29-2); *Eds. K. Kovler and A. Bentur*

PRO 24: 2nd International RILEM Workshop on Frost Resistance of Concrete (ISBN: 2-912143-30-6); *Eds. M. J. Setzer, R. Auberg and H.-J. Keck*

PRO 25: International RILEM Workshop on Frost Damage in Concrete (ISBN: 2-912143-31-4); *Eds. D. J. Janssen, M. J. Setzer and M. B. Snyder*

PRO 26: International RILEM Workshop on On-Site Control and Evaluation of Masonry Structures (ISBN: 2-912143-34-9); *Eds. L. Binda and R. C. de Vekey*

PRO 27: International RILEM Symposium on Building Joint Sealants (CD03); *Ed. A. T. Wolf*

PRO 28: 6th International RILEM Symposium on Performance Testing and Evaluation of Bituminous Materials—PTEBM'03 (ISBN: 2-912143-35-7; e-ISBN: 978-2-912143-77-8); *Ed. M. N. Partl*

PRO 29: 2nd International RILEM Workshop on Life Prediction and Ageing Management of Concrete Structures (ISBN: 2-912143-36-5); *Ed. D. J. Naus*

PRO 30: 4th International RILEM Workshop on High Performance Fiber Reinforced Cement Composites—HPFRCC 4 (ISBN: 2-912143-37-3); *Eds. A. E. Naaman and H. W. Reinhardt*

PRO 31: International RILEM Workshop on Test and Design Methods for Steel Fibre Reinforced Concrete: Background and Experiences (ISBN: 2-912143-38-1); *Eds. B. Schnütgen and L. Vandewalle*

PRO 32: International Conference on Advances in Concrete and Structures 2 vol. (ISBN (set): 2-912143-41-1); *Eds. Ying-shu Yuan, Surendra P. Shah and Heng-lin Lü*

PRO 33: 3rd International Symposium on Self-Compacting Concrete (ISBN: 2-912143-42-X); *Eds. Ó. Wallevik and I. Níelsson*

PRO 34: International RILEM Conference on Microbial Impact on Building Materials (ISBN: 2-912143-43-8); *Ed. M. Ribas Silva*
PRO 35: International RILEM TC 186-ISA on Internal Sulfate Attack and Delayed Ettringite Formation (ISBN: 2-912143-44-6); *Eds. K. Scrivener and J. Skalny*
PRO 36: International RILEM Symposium on Concrete Science and Engineering—A Tribute to Arnon Bentur (ISBN: 2-912143-46-2); *Eds. K. Kovler, J. Marchand, S. Mindess and J. Weiss*
PRO 37: 5th International RILEM Conference on Cracking in Pavements—Mitigation, Risk Assessment and Prevention (ISBN: 2-912143-47-0); *Eds. C. Petit, I. Al-Qadi and A. Millien*
PRO 38: 3rd International RILEM Workshop on Testing and Modelling the Chloride Ingress into Concrete (ISBN: 2-912143-48-9); *Eds. C. Andrade and J. Kropp*
PRO 39: 6th International RILEM Symposium on Fibre-Reinforced Concretes—BEFIB 2004 (ISBN: 2-912143-51-9); *Eds. M. Di Prisco, R. Felicetti and G. A. Plizzari*
PRO 40: International RILEM Conference on the Use of Recycled Materials in Buildings and Structures (ISBN: 2-912143-52-7); *Eds. E. Vázquez, Ch. F. Hendriks and G. M. T. Janssen*
PRO 41: RILEM International Symposium on Environment-Conscious Materials and Systems for Sustainable Development (ISBN: 2-912143-55-1); *Eds. N. Kashino and Y. Ohama*
PRO 42: SCC'2005—China: 1st International Symposium on Design, Performance and Use of Self-Consolidating Concrete (ISBN: 2-912143-61-6); *Eds. Zhiwu Yu, Caijun Shi, Kamal Henri Khayat and Youjun Xie*
PRO 43: International RILEM Workshop on Bonded Concrete Overlays (e-ISBN: 2-912143-83-7); *Eds. J. L. Granju and J. Silfwerbrand*
PRO 44: 2nd International RILEM Workshop on Microbial Impacts on Building Materials (CD11) (e-ISBN: 2-912143-84-5); *Ed. M. Ribas Silva*
PRO 45: 2nd International Symposium on Nanotechnology in Construction, Bilbao (ISBN: 2-912143-87-X); *Eds. Peter J. M. Bartos, Yolanda de Miguel and Antonio Porro*
PRO 46: ConcreteLife'06—International RILEM-JCI Seminar on Concrete Durability and Service Life Planning: Curing, Crack Control, Performance in Harsh Environments (ISBN: 2-912143-89-6); *Ed. K. Kovler*
PRO 47: International RILEM Workshop on Performance Based Evaluation and Indicators for Concrete Durability (ISBN: 978-2-912143-95-2); *Eds. V. Baroghel-Bouny, C. Andrade, R. Torrent and K. Scrivener*
PRO 48: 1st International RILEM Symposium on Advances in Concrete through Science and Engineering (e-ISBN: 2-912143-92-6); *Eds. J. Weiss, K. Kovler, J. Marchand, and S. Mindess*
PRO 49: International RILEM Workshop on High Performance Fiber Reinforced Cementitious Composites in Structural Applications (ISBN: 2-912143-93-4); *Eds. G. Fischer and V.C. Li*

PRO 50: 1st International RILEM Symposium on Textile Reinforced Concrete (ISBN: 2-912143-97-7); *Eds. Josef Hegger, Wolfgang Brameshuber and Norbert Will*

PRO 51: 2nd International Symposium on Advances in Concrete through Science and Engineering (ISBN: 2-35158-003-6; e-ISBN: 2-35158-002-8); *Eds. J. Marchand, B. Bissonnette, R. Gagné, M. Jolin and F. Paradis*

PRO 52: Volume Changes of Hardening Concrete: Testing and Mitigation (ISBN: 2-35158-004-4; e-ISBN: 2-35158-005-2); *Eds. O. M. Jensen, P. Lura and K. Kovler*

PRO 53: High Performance Fiber Reinforced Cement Composites—HPFRCC5 (ISBN: 978-2-35158-046-2); *Eds. H. W. Reinhardt and A. E. Naaman*

PRO 54: 5th International RILEM Symposium on Self-Compacting Concrete (ISBN: 978-2-35158-047-9); *Eds. G. De Schutter and V. Boel*

PRO 55: International RILEM Symposium Photocatalysis, Environment and Construction Materials (ISBN: 978-2-35158-056-1); *Eds. P. Baglioni and L. Cassar*

PRO 56: International RILEM Workshop on Integral Service Life Modelling of Concrete Structures (ISBN 978-2-35158-058-5); *Eds. R. M. Ferreira, J. Gulikers and C. Andrade*

PRO 57: RILEM Workshop on Performance of cement-based materials in aggressive aqueous environments (e-ISBN: 978-2-35158-059-2); *Ed. N. De Belie*

PRO 58: International RILEM Symposium on Concrete Modelling—CONMOD'08 (ISBN: 978-2-35158-060-8); *Eds. E. Schlangen and G. De Schutter*

PRO 59: International RILEM Conference on On Site Assessment of Concrete, Masonry and Timber Structures—SACoMaTiS 2008 (ISBN set: 978-2-35158-061-5); *Eds. L. Binda, M. di Prisco and R. Felicetti*

PRO 60: Seventh RILEM International Symposium on Fibre Reinforced Concrete: Design and Applications—BEFIB 2008 (ISBN: 978-2-35158-064-6); *Ed. R. Gettu*

PRO 61: 1st International Conference on Microstructure Related Durability of Cementitious Composites 2 vol., (ISBN: 978-2-35158-065-3); *Eds. W. Sun, K. van Breugel, C. Miao, G. Ye and H. Chen*

PRO 62: NSF/RILEM Workshop: In-situ Evaluation of Historic Wood and Masonry Structures (e-ISBN: 978-2-35158-068-4); *Eds. B. Kasal, R. Anthony and M. Drdácký*

PRO 63: Concrete in Aggressive Aqueous Environments: Performance, Testing and Modelling, 2 vol., (ISBN: 978-2-35158-071-4); *Eds. M. G. Alexander and A. Bertron*

PRO 64: Long Term Performance of Cementitious Barriers and Reinforced Concrete in Nuclear Power Plants and Waste Management—NUCPERF 2009 (ISBN: 978-2-35158-072-1); *Eds. V. L'Hostis, R. Gens, C. Gallé*

PRO 65: Design Performance and Use of Self-consolidating Concrete—SCC'2009 (ISBN: 978-2-35158-073-8); *Eds. C. Shi, Z. Yu, K. H. Khayat and P. Yan*

PRO 66: 2nd International RILEM Workshop on Concrete Durability and Service Life Planning—ConcreteLife'09 (ISBN: 978-2-35158-074-5); *Ed. K. Kovler*

PRO 67: Repairs Mortars for Historic Masonry (e-ISBN: 978-2-35158-083-7); *Ed. C. Groot*

PRO 68: Proceedings of the 3rd International RILEM Symposium on 'Rheology of Cement Suspensions such as Fresh Concrete (ISBN 978-2-35158-091-2); *Eds. O. H. Wallevik, S. Kubens and S. Oesterheld*

PRO 69: 3rd International PhD Student Workshop on 'Modelling the Durability of Reinforced Concrete (ISBN: 978-2-35158-095-0); *Eds. R. M. Ferreira, J. Gulikers and C. Andrade*

PRO 70: 2nd International Conference on 'Service Life Design for Infrastructure' (ISBN set: 978-2-35158-096-7, e-ISBN: 978-2-35158-097-4); *Ed. K. van Breugel, G. Ye and Y. Yuan*

PRO 71: Advances in Civil Engineering Materials—The 50-year Teaching Anniversary of Prof. Sun Wei' (ISBN: 978-2-35158-098-1; e-ISBN: 978-2-35158-099-8); *Eds. C. Miao, G. Ye, and H. Chen*

PRO 72: First International Conference on 'Advances in Chemically-Activated Materials—CAM'2010' (2010), 264 pp, ISBN: 978-2-35158-101-8; e-ISBN: 978-2-35158-115-5, *Eds. Caijun Shi and Xiaodong Shen*

PRO 73: 2nd International Conference on 'Waste Engineering and Management—ICWEM 2010' (2010), 894 pp, ISBN: 978-2-35158-102-5; e-ISBN: 978-2-35158-103-2, *Eds. J. Zh. Xiao, Y. Zhang, M. S. Cheung and R. Chu*

PRO 74: International RILEM Conference on 'Use of Superabsorsorbent Polymers and Other New Additives in Concrete' (2010) 374 pp., ISBN: 978-2-35158-104-9; e-ISBN: 978-2-35158-105-6; *Eds. O.M. Jensen, M.T. Hasholt, and S. Laustsen*

PRO 75: International Conference on 'Material Science—2nd ICTRC—Textile Reinforced Concrete—Theme 1' (2010) 436 pp., ISBN: 978-2-35158-106-3; e-ISBN: 978-2-35158-107-0; *Ed. W. Brameshuber*

PRO 76: International Conference on 'Material Science—HetMat—Modelling of Heterogeneous Materials—Theme 2' (2010) 255 pp., ISBN: 978-2-35158-108-7; e-ISBN: 978-2-35158-109-4; *Ed. W. Brameshuber*

PRO 77: International Conference on 'Material Science—AdIPoC—Additions Improving Properties of Concrete—Theme 3' (2010) 459 pp., ISBN: 978-2-35158-110-0; e-ISBN: 978-2-35158-111-7; *Ed. W. Brameshuber*

PRO 78: 2nd Historic Mortars Conference and RILEM TC 203-RHM Final Workshop—HMC2010 (2010) 1416 pp., e-ISBN: 978-2-35158-112-4; *Eds J. Válek, C. Groot, and J. J. Hughes*

PRO 79: International RILEM Conference on Advances in Construction Materials Through Science and Engineering (2011) 213 pp., ISBN: 978-2-35158-116-2, e-ISBN: 978-2-35158-117-9; *Eds Christopher Leung and K.T. Wan*

PRO 80: 2nd International RILEM Conference on Concrete Spalling due to Fire Exposure (2011) 453 pp., ISBN: 978-2-35158-118-6, e-ISBN: 978-2-35158-119-3; *Eds E.A.B. Koenders and F. Dehn*

PRO 81: 2nd International RILEM Conference on Strain Hardening Cementitious Composites (SHCC2-Rio) (2011) 451 pp., ISBN: 978-2-35158-120-9, e-ISBN: 978-2-35158-121-6; *Eds R.D. Toledo Filho, F.A. Silva, E.A.B. Koenders and E.M.R. Fairbairn*

PRO 82: 2nd International RILEM Conference on Progress of Recycling in the Built Environment (2011) 507 pp., e-ISBN: 978-2-35158-122-3; *Eds V.M. John, E. Vazquez, S.C. Angulo and C. Ulsen*

PRO 83: 2nd International Conference on Microstructural-related Durability of Cementitious Composites (2012) 250 pp., ISBN: 978-2-35158-129-2; e-ISBN: 978-2-35158-123-0; *Eds G. Ye, K. van Breugel, W. Sun and C. Miao*

PRO 84: CONSEC13—Seventh International Conference on Concrete under Severe Conditions—Environment and Loading (2013) 1930 pp., ISBN: 978-2-35158-124-7; e-ISBN: 978-2-35158-134-6; *Eds Z.J. Li, W. Sun, C.W. Miao, K. Sakai, O.E. Gjorv and N. Banthia*

PRO 85: RILEM-JCI International Workshop on Crack Control of Mass Concrete and Related issues concerning Early-Age of Concrete Structures—ConCrack 3—Control of Cracking in Concrete Structures 3 (2012) 237 pp., ISBN: 978-2-35158-125-4; e-ISBN: 978-2-35158-126-1; *Eds F. Toutlemonde and J.-M. Torrenti*

PRO 86: International Symposium on Life Cycle Assessment and Construction (2012) 414 pp., ISBN: 978-2-35158-127-8, e-ISBN: 978-2-35158-128-5; *Eds A. Ventura and C. de la Roche*

PRO 87: UHPFRC 2013—RILEM-fib-AFGC International Symposium on Ultra-High Performance Fibre-Reinforced Concrete (2013), ISBN: 978-2-35158-130-8, e-ISBN: 978-2-35158-131-5; *Eds F. Toutlemonde*

PRO 88: 8th RILEM International Symposium on Fibre Reinforced Concrete (2012) 344 pp., ISBN: 978-2-35158-132-2, e-ISBN: 978-2-35158-133-9; *Eds Joaquim A.O. Barros*

PRO 89: RILEM International workshop on performance-based specification and control of concrete durability (2014) 678 pp, ISBN: 978-2-35158-135-3, e-ISBN: 978-2-35158-136-0; *Eds. D. Bjegović, H. Beushausen and M. Serdar*

PRO 90: 7th RILEM International Conference on Self-Compacting Concrete and of the 1st RILEM International Conference on Rheology and Processing of Construction Materials (2013) 396 pp, ISBN: 978-2-35158-137-7, e-ISBN: 978-2-35158-138-4; *Eds. Nicolas Roussel and Hela Bessaies-Bey*

PRO 91: CONMOD 2014—RILEM International Symposium on Concrete Modelling (2014), ISBN: 978-2-35158-139-1; e-ISBN: 978-2-35158-140-7; *Eds. Kefei Li, Peiyu Yan and Rongwei Yang*

PRO 92: CAM 2014—2nd International Conference on advances in chemically-activated materials (2014) 392 pp., ISBN: 978-2-35158-141-4; e-ISBN: 978-2-35158-142-1; *Eds. Caijun Shi and Xiadong Shen*

PRO 93: SCC 2014—3rd International Symposium on Design, Performance and Use of Self-Consolidating Concrete (2014) 438 pp., ISBN: 978-2-35158-143-8; e-ISBN: 978-2-35158-144-5; *Eds. Caijun Shi, Zhihua Ou, Kamal H. Khayat*

PRO 94 (online version): HPFRCC-7—7th RILEM conference on High performance fiber reinforced cement composites (2015), e-ISBN: 978-2-35158-146-9; *Eds. H.W. Reinhardt, G.J. Parra-Montesinos, H. Garrecht*

PRO 95: International RILEM Conference on Application of superabsorbent polymers and other new admixtures in concrete construction (2014), ISBN: 978-2-35158-147-6; e-ISBN: 978-2-35158-148-3; *Eds. Viktor Mechtcherine, Christof Schroefl*

PRO 96 (online version): XIII DBMC: XIII International Conference on Durability of Building Materials and Components(2015), e-ISBN: 978-2-35158-149-0; *Eds. M. Quattrone, V.M. John*

PRO 97: SHCC3—3rd International RILEM Conference on Strain Hardening Cementitious Composites (2014), ISBN: 978-2-35158-150-6; e-ISBN: 978-2-35158-151-3; *Eds. E. Schlangen, M.G. Sierra Beltran, M. Lukovic, G. Ye*

PRO 98: FERRO-11—11th International Symposium on Ferrocement and 3rd ICTRC—International Conference on Textile Reinforced Concrete (2015), ISBN: 978-2-35158-152-0; e-ISBN: 978-2-35158-153-7; *Ed. W. Brameshuber*

PRO 99 (online version): ICBBM 2015—1st International Conference on Bio-Based Building Materials (2015), e-ISBN: 978-2-35158-154-4; *Eds. S. Amziane, M. Sonebi*

PRO 100: SCC16—RILEM Self-Consolidating Concrete Conference (2016), ISBN: 978-2-35158-156-8; e-ISBN: 978-2-35158-157-5; *Ed. Kamal H. Kayat*

PRO 101 (online version): III Progress of Recycling in the Built Environment (2015), e-ISBN: 978-2-35158-158-2; *Eds I. Martins, C. Ulsen and S. C. Angulo*

PRO 102 (online version): RILEM Conference on Microorganisms-Cementitious Materials Interactions (2016), e-ISBN: 978-2-35158-160-5; *Eds. Alexandra Bertron, Henk Jonkers, Virginie Wiktor*

PRO 103 (online version): ACESC'16—Advances in Civil Engineering and Sustainable Construction (2016), e-ISBN: 978-2-35158-161-2

PRO 104 (online version): SSCS'2015—Numerical Modeling—Strategies for Sustainable Concrete Structures (2015), e-ISBN: 978-2-35158-162-9

PRO 105: 1st International Conference on UHPC Materials and Structures (2016), ISBN: 978-2-35158-164-3, e-ISBN: 978-2-35158-165-0

PRO 106: AFGC-ACI-fib-RILEM International Conference on Ultra-High-Performance Fibre-Reinforced Concrete—UHPFRC 2017 (2017), ISBN: 978-2-35158-166-7, e-ISBN: 978-2-35158-167-4; *Eds. François Toutlemonde & Jacques Resplendino*

PRO 107 (online version): XIV DBMC—14th International Conference on Durability of Building Materials and Components (2017), e-ISBN: 978-2-35158-159-9; *Eds. Geert De Schutter, Nele De Belie, Arnold Janssens, Nathan Van Den Bossche*

PRO 108: MSSCE 2016—Innovation of Teaching in Materials and Structures (2016), ISBN: 978-2-35158-178-0, e-ISBN: 978-2-35158-179-7; *Ed. Per Goltermann*

PRO 109 (two volumes): MSSCE 2016—Service Life of Cement-Based Materials and Structures (2016), ISBN Vol. 1: 978-2-35158-170-4, Vol. 2: 978-2-35158-171-4, Set Vol. 1&2: 978-2-35158-172-8, e-ISBN: 978-2-35158-173-5; *Eds. Miguel Azenha, Ivan Gabrijel, Dirk Schlicke, Terje Kanstad and Ole Mejlhede Jensen*

PRO 110: MSSCE 2016—Historical Masonry (2016), ISBN: 978-2-35158-178-0, e-ISBN: 978-2-35158-179-7; *Eds. Inge Rörig-Dalgaard and Ioannis Ioannou*
PRO 111: MSSCE 2016—Electrochemistry in Civil Engineering (2016), ISBN: 978-2-35158-176-6, e-ISBN: 978-2-35158-177-3; *Ed. Lisbeth M. Ottosen*
PRO 112: MSSCE 2016—Moisture in Materials and Structures (2016), ISBN: 978-2-35158-178-0, e-ISBN: 978-2-35158-179-7; *Eds. Kurt Kielsgaard Hansen, Carsten Rode and Lars-Olof Nilsson*
PRO 113: MSSCE 2016—Concrete with Supplementary Cementitious Materials (2016), ISBN: 978-2-35158-178-0, e-ISBN: 978-2-35158-179-7; *Eds. Ole Mejlhede Jensen, Konstantin Kovler and Nele De Belie*
PRO 114: MSSCE 2016—Frost Action in Concrete (2016), ISBN: 978-2-35158-182-7, e-ISBN: 978-2-35158-183-4; *Eds. Marianne Tange Hasholt, Katja Fridh and R. Doug Hooton*
PRO 115: MSSCE 2016—Fresh Concrete (2016), ISBN: 978-2-35158-184-1, e-ISBN: 978-2-35158-185-8; *Eds. Lars N. Thrane, Claus Pade, Oldrich Svec and Nicolas Roussel*
PRO 116: BEFIB 2016—9th RILEM International Symposium on Fiber Reinforced Concrete (2016), ISBN: 978-2-35158-187-2, e-ISBN: 978-2-35158-186-5
PRO 117: 3rd International RILEM Conference on Microstructure Related Durability of Cementitious Composites (2016), ISBN: 978-2-35158-188-9, e-ISBN: 978-2-35158-189-6; *Eds. Changwen Miao, Wei Sun, Jiaping Liu, Huisu Chen, Guang Ye and Klaas van Breugel*
PRO 118: International Conference on Advances in Construction Materials and Sytems (2017), ISBN: 978-2-35158-190-2, e-ISBN: 978-2-35158-191-9

RILEM Reports (REP)

Report 19: Considerations for Use in Managing the Aging of Nuclear Power Plant Concrete Structures (ISBN: 2-912143-07-1); *Ed. D. J. Naus*
Report 20: Engineering and Transport Properties of the Interfacial Transition Zone in Cementitious Composites (ISBN: 2-912143-08-X); *Eds. M. G. Alexander, G. Arliguie, G. Ballivy, A. Bentur and J. Marchand*
Report 21: Durability of Building Sealants (ISBN: 2-912143-12-8); *Ed. A. T. Wolf*
Report 22: Sustainable Raw Materials—Construction and Demolition Waste (ISBN: 2-912143-17-9); *Eds. C. F. Hendriks and H. S. Pietersen*
Report 23: Self-Compacting Concrete state-of-the-art report (ISBN: 2-912143-23-3); *Eds. Å. Skarendahl and Ö. Petersson*
Report 24: Workability and Rheology of Fresh Concrete: Compendium of Tests (ISBN: 2-912143-32-2); *Eds. P. J. M. Bartos, M. Sonebi and A. K. Tamimi*
Report 25: Early Age Cracking in Cementitious Systems (ISBN: 2-912143-33-0); *Ed. A. Bentur*
Report 26: Towards Sustainable Roofing (Joint Committee CIB/RILEM) (CD 07) (e-ISBN 978-2-912143-65-5); *Eds. Thomas W. Hutchinson and Keith Roberts*
Report 27: Condition Assessment of Roofs (Joint Committee CIB/RILEM) (CD 08) (e-ISBN 978-2-912143-66-2); *Ed. CIB W 83/RILEM TC166-RMS*

Report 28: Final report of RILEM TC 167-COM 'Characterisation of Old Mortars with Respect to Their Repair (ISBN: 978-2-912143-56-3); *Eds. C. Groot, G. Ashall and J. Hughes*

Report 29: Pavement Performance Prediction and Evaluation (PPPE): Interlaboratory Tests (e-ISBN: 2-912143-68-3); *Eds. M. Partl and H. Piber*

Report 30: Final Report of RILEM TC 198-URM 'Use of Recycled Materials' (ISBN: 2-912143-82-9; e-ISBN: 2-912143-69-1); *Eds. Ch. F. Hendriks, G. M. T. Janssen and E. Vázquez*

Report 31: Final Report of RILEM TC 185-ATC 'Advanced testing of cement-based materials during setting and hardening' (ISBN: 2-912143-81-0; e-ISBN: 2-912143-70-5); *Eds. H. W. Reinhardt and C. U. Grosse*

Report 32: Probabilistic Assessment of Existing Structures. A JCSS publication (ISBN 2-912143-24-1); *Ed. D. Diamantidis*

Report 33: State-of-the-Art Report of RILEM Technical Committee TC 184-IFE 'Industrial Floors' (ISBN 2-35158-006-0); *Ed. P. Seidler*

Report 34: Report of RILEM Technical Committee TC 147-FMB 'Fracture mechanics applications to anchorage and bond' Tension of Reinforced Concrete Prisms—Round Robin Analysis and Tests on Bond (e-ISBN 2-912143-91-8); *Eds. L. Elfgren and K. Noghabai*

Report 35: Final Report of RILEM Technical Committee TC 188-CSC 'Casting of Self Compacting Concrete' (ISBN 2-35158-001-X; e-ISBN: 2-912143-98-5); *Eds. Å. Skarendahl and P. Billberg*

Report 36: State-of-the-Art Report of RILEM Technical Committee TC 201-TRC 'Textile Reinforced Concrete' (ISBN 2-912143-99-3); *Ed. W. Brameshuber*

Report 37: State-of-the-Art Report of RILEM Technical Committee TC 192-ECM 'Environment-conscious construction materials and systems' (ISBN: 978-2-35158-053-0); *Eds. N. Kashino, D. Van Gemert and K. Imamoto*

Report 38: State-of-the-Art Report of RILEM Technical Committee TC 205-DSC 'Durability of Self-Compacting Concrete' (ISBN: 978-2-35158-048-6); *Eds. G. De Schutter and K. Audenaert*

Report 39: Final Report of RILEM Technical Committee TC 187-SOC 'Experimental determination of the stress-crack opening curve for concrete in tension' (ISBN 978-2-35158-049-3); *Ed. J. Planas*

Report 40: State-of-the-Art Report of RILEM Technical Committee TC 189-NEC 'Non-Destructive Evaluation of the Penetrability and Thickness of the Concrete Cover' (ISBN 978-2-35158-054-7); *Eds. R. Torrent and L. Fernández Luco*

Report 41: State-of-the-Art Report of RILEM Technical Committee TC 196-ICC 'Internal Curing of Concrete' (ISBN 978-2-35158-009-7); *Eds. K. Kovler and O. M. Jensen*

Report 42: 'Acoustic Emission and Related Non-destructive Evaluation Techniques for Crack Detection and Damage Evaluation in Concrete'—Final Report of RILEM Technical Committee 212-ACD (e-ISBN: 978-2-35158-100-1); *Ed. M. Ohtsu*

Report 45: Repair Mortars for Historic Masonry—State-of-the-Art Report of RILEM Technical Committee TC 203-RHM (e-ISBN: 978-2-35158-163-6); *Eds. Paul Maurenbrecher and Caspar Groot*

Contributors

Sofiane Amziane Institut Pascal, Clermont Université, Clermont-Ferrand, France

Florence Collet Laboratoire de Génie Civil et Génie Mécanique de Rennes, équipe Matériaux-Thermo-Rhéologie, Université de Rennes 1, Rennes, France; IUT Génie Civil Construction Durable, Rennes Cedex, France

Guillaume Delannoy Ifsttar, Mast/Cpdm, Université Paris-Est, Marne-La-Vallee Cedex 2, France

Gilles Escadeillas Laboratoire Matériaux et Durabilité Des Constructions, Toulouse, France; INSA/UPS Dép. Génie Civil, Toulouse, France

Yunhong Jiang BRE Centre for Innovative Construction Materials, University of Bath, Bath, UK

Christophe Lanos Laboratoire de Génie Civil et Génie Mécanique de Rennes équipe Matériaux-Thermo-Rhéologie, Université de Rennes 1, Rennes, France; IUT Génie Civil Construction Durable, Rennes Cedex, France

Mike Lawrence BRE Centre for Innovative Construction Materials, University of Bath, Bath, UK

Camille Magniont Laboratoire Matériaux et Durabilité Des Constructions, Toulouse, France; INSA/UPS Dép. Génie Civil, Toulouse, France

Sandrine Marceau Ifsttar, Mast/Cpdm, Université Paris-Est, Marne-La-Vallee Cedex 2, France

Vincent Nozahic Institut Pascal, Clermont Université, Clermont-Ferrand, France

Sara Pavía Department of Civil Engineering, Dublin 2, Ireland

Vincent Picandet Centre de Recherche, IRDL—Université de Bretagne Sud, Lorient, France

Mohammed Sonebi School of Planning, Architecture and Civil Engineering, Queen's University Belfast, Belfast, UK

List of Figures

List of Tables

Chapter 1
Chemical Composition of Bio-aggregates and Their Interactions with Mineral Binders

Camille Magniont and Gilles Escadeillas

Abstract This chapter gives the state of the art on the chemical composition of bio-aggregates and their interactions with mineral binders from the standpoint of bio-aggregate based building materials. The chemical composition of various bioresources included in bio-aggregate based building materials has been reviewed by comparing the results of 24 published references. This comparison highlights a large dispersion among the results of different references for the same bioresource, reflecting not only the effective variability of chemical composition due to agronomic, environmental and processing parameters but also the strong disparities in the results of common indirect gravimetric methods of biomass compositional analysis. However, the chemical composition of lignocellulosic particles can strongly impact their properties as bio-aggregates included in a mineral matrix. At early age, they can disturb the setting and hardening mechanisms of mineral binders; in the hardened state, they can modify the properties of the composite; and, finally, in the long term, they can influence durability.

Keywords Lignocellulosic particles · Chemical composition · Extractives · Mineral binder · Deleterious interactions · Setting delay · Durability

1.1 Introduction

The chemical composition of bio-aggregates can strongly influence their properties (compressibility, durability, chemical and/or physical interactions with mineral species etc.) and consequently those of the composite containing them (setting time, hydration mechanisms, mechanical properties, durability, etc.). Nevertheless, few

---

C. Magniont (✉) · G. Escadeillas
Laboratoire Matériaux et Durabilité Des Constructions, Toulouse, France
e-mail: camille.magniont@insa-toulouse.fr

C. Magniont · G. Escadeillas
INSA/UPS Dép. Génie Civil, Toulouse, France

S. Amziane and F. Collet (eds.), *Bio-aggregates Based Building Materials*, RILEM State-of-the-Art Reports 23, DOI 10.1007/978-94-024-1031-0_1

studies have explored the chemical composition of bio-aggregates from the standpoint of bio-aggregate based building materials.

Parts 1 and 2 of this chapter briefly recall the composition and microstructure of hemp stem. These two parts mainly focus on hemp because hemp shiv is the bio-aggregate that is the most widely used in building materials and the most studied in the literature. Part 3 aims to describe the chemical components of plant matter, and part 4 proposes a review of chemical composition analyses of different types of biomass that can be used as bio-aggregates. These data come from articles from materials, agricultural and biotechnology science, dealing with these plant particles as raw ingredients of building materials and also for paper and biodiesel production. These analyses consequently result from a large variety of laboratory analytical procedures. Other direct techniques can be used for chemical analyses and some are described here. Finally, parts 5 and 6 present respectively recent studies of surface characteristics of sunflower aggregates and the chemical interactions between plant matter and mineral binders.

1.2 Composition of Hemp Stem

From the centre to the outside of its cross section, hemp stem is composed of (Picandet 2013; British Columbia, Ministry of Agriculture and Food 1999):

- a hollow core,
- the pith or medulla layer, composed of soft, spongy parenchyma cells with intercellular spaces,
- the xylem layer, composed of thick woody tissues that support the plant and the vessels that ensure the transport of xylem sap (mainly water and minerals); this is the part that gives hemp shives,
- the cambium layer, the growth area where cambial activity produces secondary xylem on the inside and secondary phloem (tissue that channels elaborated sap) on the outside. This layer also constitutes the zone where fibres and shives separate during the defibration process,
- the phloem layer, which contains primary bast fibres (long fibres with low lignin content) and secondary fibres (generated from the cambium, they are shorter and have high lignin content). These fibres are encircled by cortical parenchyma cells,
- the epidermis layer, which provides protection for plant cells and is a zone of exchanges with the surrounding environment.

The different parts of hemp stem have been identified on the optical micrograph obtained after staining with 'carmino vert' of Mirande by Vignon et al. (1995) (Fig. 1.1). Cellulose-rich tissues from epidermis, phloem and pith are coloured pink whereas the lignin-rich cells of xylem appear in green.

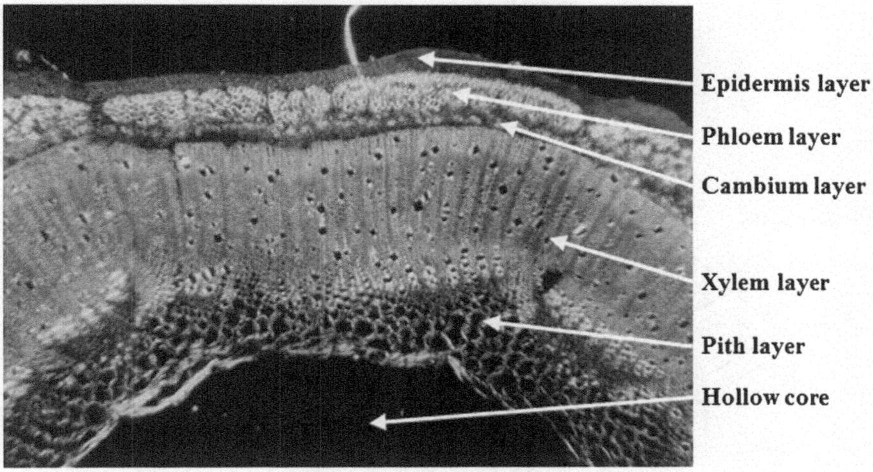

Fig. 1.1 Optical micrograph of a thin cross section of a hemp stem (×90) (Vignon et al. 1995)

1.3 Processing of Hemp Stem and Microstructure of Hemp Shiv

After harvesting, the typical industrial transformation process of hemp stem includes retting and mechanical defibration. Figure 1.2 is a schematic cross section of hemp stem.

Bast fibres are used as raw materials in the textile, insulation, paper and biopolymer industries.

Hurds or shives, the woody core particles, are used for animal bedding, garden mulch or lightweight aggregates in concrete.

In the next section, we focus on the chemical composition of these lignocellulosic aggregates that represent 60–80% of the total mass (Balciunas et al. 2013).

Lignocellulosic aggregates are mainly composed of xylem cells which are longitudinal, dead and empty of content. These tubes, serving for water transport, are responsible for the strongly anisotropic and porous microstructure of plant

Fig. 1.2 Schematic representation of hemp stem cross section

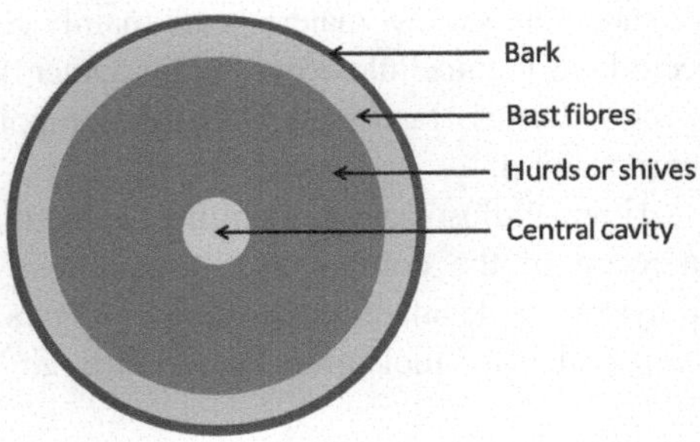

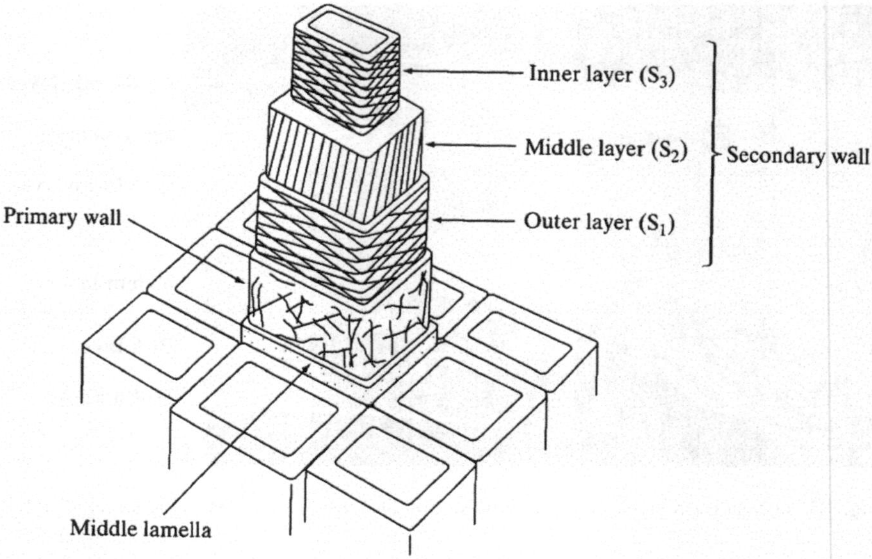

Fig. 1.3 Structure of plant cell walls (Dinwoodie 1989)

aggregates. Xylem cell walls themselves are complex composites; their structure is represented on Fig. 1.3.

The middle lamella is mainly composed of pectin that ensures the bonding between the cells.

Primary and secondary walls are made of cellulose microfibrils, which are bundles of cellulose molecules, embedded in a matrix of hemicellulose and pectin. Inside the secondary wall, three sublayers can be distinguished by the different orientation of the cellulose microfibrils. The presence of lignin in these layers brings rigidity and water-impermeability to the cell.

The chemical nature of these different components is described in Sect. 1.3.

1.4 Cell Wall Components

Lignocellulosic raw materials are mainly composed of three structural biopolymers: cellulose, hemicellulose and lignin. Other organic and inorganic substances, such as pectin, waxes, fats, water-soluble components and ash complete their chemical composition.

Holocellulose can be identified as the major constituent of plant aggregates; it is a result of the combination of cellulose and hemicellulose. Holocellulose is a polysaccharide made up of simple sugars. It is rich in hydroxyl groups that are responsible for moisture sorption through hydrogen bonding (Rowell et al. 2005).

1.4.1 Cellulose

Cellulose is a hydrophilic glucan polymer consisting of a linear chain of 1,4-β anhydroglucose units, each containing three alcoholic hydroxyl groups (Fig. 1.4) (Mohanty et al. 2005). The number of glucose units in a cellulose molecule is referred to as the degree of polymerization (DP) and, the average DP of cellulose in natural fibres is around 10,000. This characteristic can significantly affect the mechanical performance of the fibres.

Cellulose contains regions that are highly crystalline or amorphous depending on their packing density. Most wood-derived cellulose contains more than 65% of crystalline regions (Rowell et al. 2005). Accessible and non-accessible cellulose can also be distinguished (Rowell et al. 2005). Concerning crystalline cellulose, only the surface part is accessible whereas the majority of amorphous cellulose is accessible. This concept is very important as it can have a significant impact on the durability of cellulose, affecting its moisture sorption capacity, its chemical interaction with aggressive agents and its sensitivity to microorganisms. Cellulose is resistant to most solvents, including strong alkali, but can be hydrolysed by strong acids.

1.4.2 Hemicellulose

Hemicellulose combines many different polysaccharides composed of 5 and 6 carbon ring sugars plus glucuronic and galacturonic acids (Vassilev et al. 2012). Hemicelluloses have lower DP than cellulose and present a heterogeneous, amorphous, branched structure.

They form the supporting matrix for cellulose microfibrils (Mohanty et al. 2005).

Hemicelluloses are highly hydrophilic, easily hydrolysed by acids and soluble in diluted alkali solutions. This property could notably affect the durability of plant aggregates in an alkaline mineral matrix (cement, lime etc.) and the bonding mechanisms at the interface in these composite materials.

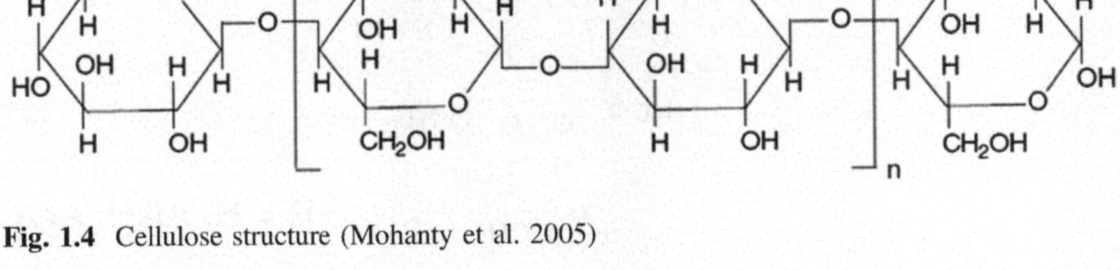

Fig. 1.4 Cellulose structure (Mohanty et al. 2005)

1.4.3 Lignin

Although the exact chemical nature of lignin remains obscure, most of the functional groups and units have been identified. Lignins can be described as amorphous, highly complex, three-dimensional, mainly aromatic polymers of phenylpropane units (Rowell et al. 2005; Davidson 2013). A structure of soft wood lignin proposed in (Mohanty et al. 2005) is represented in Fig. 1.5.

During the synthesis of plant cell walls, the production of cellulose and hemicelluloses is accompanied by the filling of the spaces between these two polysaccharides by lignins, binding them together. This is called the lignification process and leads to the stiffening of cell walls. The main roles of lignin in plants is thus to support them and to protect them from chemical and physical degradation (Mohanty et al. 2005).

Moreover, lignin is generally resistant to microbial attack. The nature and the amount of lignin thus affect the durability and the biodegradability of the different plant materials.

Lignin also plays an important role in water transport, providing the hydrophobic surface that allows water circulation within plants. Lignin is mainly found in the middle lamella of the fibre bundle, and also in the woody core and the epidermal and cortical cells of the plant stem (Lewin 2006). The lignin content in

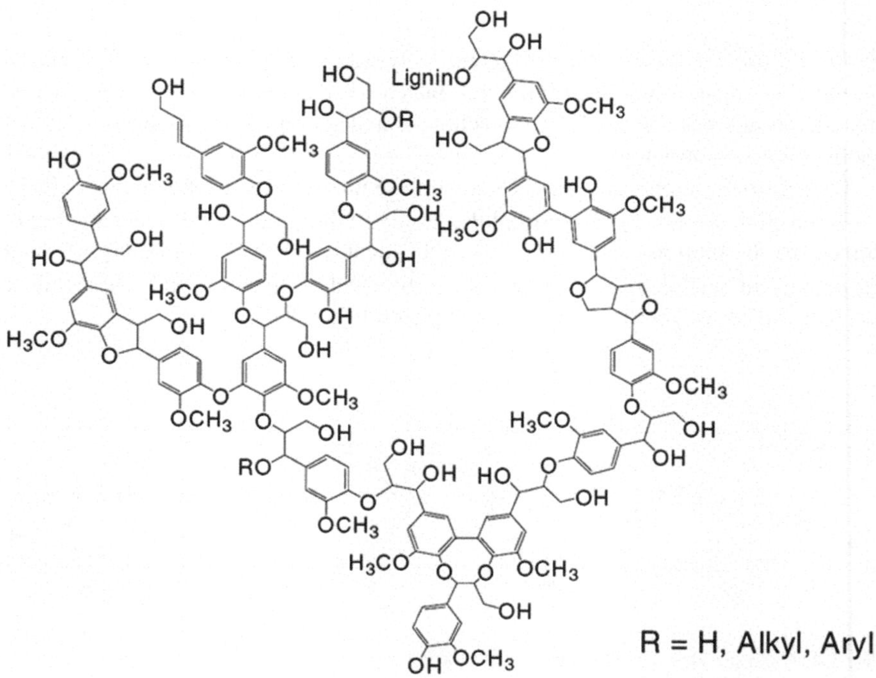

Fig. 1.5 Proposed model structure for soft wood lignin (Mohanty et al. 2005)

plant aggregates mostly extracted from the woody core (xylem) of the stem is thus high in comparison with plant fibres.

1.4.4 Pectin

Pectin is a collective name for different forms of polysaccharides made up of repeating units of α-1,4-linked galacturonic acid. Pectins are contained in the primary cell wall of most plants but they are also a major component of the middle lamella and constitute up to 5% of walls in woody tissues. The pectins ensure a variety of functions including mechanical properties, cell-cell adhesion, wall porosity and binding of ions (Mohnen 2008).

They can be easily hydrolysed at high temperatures and are degraded by bacteria and mould during the retting process that is applied to fibrous plants (including hemp and flax) to separate the fibres from the non-fibrous part (Picandet 2013).

Pectins can create gels when divalent cations, like calcium or magnesium, form cross-linkages between adjacent polymers. This affinity with cations could interact with setting mechanisms of mineral binders such as Portland cement or lime used in bio-aggregate based composites.

1.4.5 Extractives

This category assembles different non-structural chemical components that can be easily extracted using polar or non-polar solvents (usually water, ethanol, toluene and ether). Extractives include some carbohydrates (monosaccharides, starches etc.), lipids (fats, oils, waxes etc.), proteins, hydrocarbons (terpenes etc.), phenolic compounds and inorganic materials. Their concentration in plants is relatively low (<10%) but can be locally higher. For example, wax accumulates on the cuticle synthesized by the epidermal cells to provide a protective barrier against drying and the entry of microorganisms into the plant (Akin 2010).

Others can be responsible for different properties of the plant material: colour, smell and durability (Rowell et al. 2005). In the case of bio-aggregates incorporated into a mineral matrix, these extractives can interact with the mineral species of the binder and degrade its properties (see Sect. 1.6).

1.4.6 Ash

Ash is defined as the inorganic matter of a biomass. These components can be structural, i.e. bound in the physical structure of the biomass, or extractable, if they can be removed by washing or extraction (Sluiter et al. 2010).

According to Vassilev et al. (2010), the inorganic compounds can be divided into three categories: crystalline (mineral species from phosphates, carbonates, silicates, chlorides, sulfates, oxides and hydroxides, nitrates, etc.), semi crystalline (poorly crystallized forms of some silicates, phosphates, hydroxides, etc.) and amorphous (amorphous phases such as various glasses, silicates, etc.).

The ash content is usually low, although this parameter can vary considerably among plants as it depends on genetic and environmental factors, and also on physiological and morphological characteristics of the crops (Vassilev et al. 2010).

According to Picandet (2013), this mineral content can be significantly higher in plants with high silica (SiO_2) content, especially rice or wheat straw. On the contrary, whatever the time of harvesting (from 30 to 120 days) the amount of silica is very low in hemp stems (<1.5%) (Kamat 2000).

For most plant fibres, the structural mineral content remains below 2% according to Akin (2010) but can be much higher locally in the plant. Akin cites the case of calcium accumulated in flax epidermis up to an amount 6 times that in the fibre tissue. Some plants (flax, hemp and cotton) can also accumulate heavy metals such as lead (Pb), copper (Cu), zinc (Zn) and cadmium (Cd) (Angelova et al. 2004). This last point is also highlighted by Linger et al. (2005), who shows the good possibility of cadmium phytoextraction from hemp. This point could nevertheless induce some health concerns for further use as bio-aggregates in building applications.

1.4.7 Water

In plant materials, the moisture content consists of water with dissolved free ions. Its content normally varies in the range of 10–60% but it can reach even 80–90% in some raw biomass species (Vassilev et al. 2010).

Water is largely present in the cell walls and can influence some of their properties. According to Khanna (2008), for example, hydrated walls are more flexible and extensible than non-hydrated walls. The moisture content of plant aggregates could significantly affect their macroscopic properties, especially their compressibility. This characteristic can largely modify the fresh and hardened performances of composites including these bio-aggregates.

1.5 Chemical Composition of Bio-aggregates

1.5.1 Chemical Composition of Bio-aggregates Measured by Indirect Methods

This part brings together data on the chemical composition of different bioresources that have been studied as bio-aggregates incorporated in a mineral matrix in previous works.

The values reported in are related to different plant aggregates: hemp and flax shives; corn cobs and stalks; sunflower stalks, pith and marrow-less stalks; wheat straw; and lavender stems. Typical compositions of hemp and flax fibres used for mineral matrix reinforcement are also reported for comparison.

The data collected come from the literature on bio-aggregate based building materials and also from published studies of lignocellulosic feedstock characterization mainly for biofuel applications. E.g., a recent study by Vassilev et al. (2010) gave an extended review of cellulose, hemicellulose, lignin and bulk extractives contents of 93 varieties of biomass.

The analytical methods applied in these studies can differ from one lab to another. They are mainly indirect gravimetric methods based on successive fractionation steps with different solvents, associated with washings, filtrations, evaporations and weighings or assays. Previous studies have given detailed reviews of the different existing wet chemical analytical methods that have been published (Krasznai 2012; Hayes 2004; Milne et al. 1990).

Historically, most of these experimental procedures come from two main sectors: the wood and food industries (Carrier et al. 2011). The general structures of these two groups of experimental procedures differ slightly as shown in Fig. 1.6.

Numerous indirect wet-chemical methods have been developed and published for more than 100 years. Among the most commonly used are the methods proposed by Klason (1922), Saeman (Saeman and Bubl 1945; Saeman et al. 1954) or Van Soest and Wine (1968). Some of them constitute the basis of current standards of ASTM (American Society for Testing Materials), TAPPI (Technical Association of the Pulp and Paper Industry), or of the Laboratory Analytical Procedures (LAPs) of NREL (National Renewable Energy Laboratory). But some procedures have also been adapted and modified locally and so details concerning the complete procedure are often missing.

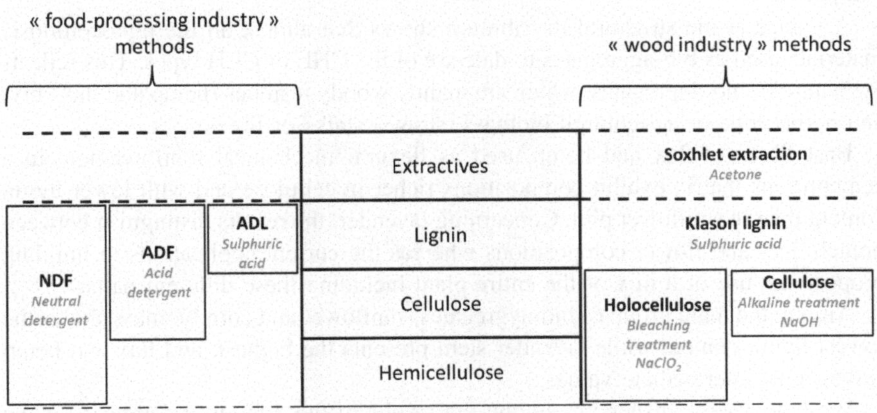

Fig. 1.6 Synthetic sketch of the two main analytical procedures to access the biomass composition adapted from (Carrier et al. 2011)

Sluiter et al. (2010) identified numerous differences between existing methods of biomass compositional analysis based on a sulfuric acid hydrolysis:

- drying method (air dried, dried at 100 or 105 °C),
- nature of the extraction solvents (alcohol/benzene, hot and cold water, acetone/water etc.),
- extraction time,
- sample amount,
- temperature, time, biomass to acid ratio, and H_2SO_4 concentration of primary and secondary hydrolysis.

The variations of these parameters among the procedures applied can induce significant disparities in the results and make it difficult to compare compositional data from interdisciplinary research groups (Krasznai 2012). Furthermore, by analysing 154 replicate samples of the same feedstock, with the same procedure by 7 analysts in 2 different laboratories, Templeton et al. (2010) demonstrated how small method changes (dysfunctioning solvent extractor and leaking gasket) could strongly affect the uncertainties of these empirical methods.

A detailed comparison of the different wet chemical procedures applied to biomass in the literature is given in Krasznai (2012), where the author highlights the limitations of each method. Due to incomplete fractionation steps or, conversely, solubilization of unexpected materials, these procedures can lead to overestimation or underestimation of some lignocellulosic compounds.

Finally, the accuracy of the results depends largely on the choice of hydrolysis conditions, which should be adapted for lignocellulosic feedstocks of different natures (soft wood, hard wood, herbaceous species, etc.).

The structural composition of bio-aggregates, i.e. the cellulose, hemicellulose and lignin contents (normalized to 100%) is represented in Fig. 1.7 as proposed in Vassilev et al. (2010). This presentation allows 6 subgroups to be distinguished, characterized by consecutively decreasing quantities of the three structural components.

A glance at the structural distribution shows that almost all the lignocellulosic materials used as bio-aggregates to date are of the CHL or CLH types. This reflects the nature of bio-aggregates, which are mainly woody residues (hemp and flax shiv) and herbaceous or agricultural biomass (straws, stalks or fibres).

Bast fibres of flax and hemp used as flexural mechanical reinforcement in a cementitious matrix exhibit compositions richer in cellulose and with lower lignin content than in sunflower pith. Concerning lavender, the results distinguish between branch, leaf and flower compositions whereas the current applications in building propose the use of a mix of the entire plant including these different parts.

Among the main group of bio-aggregates, sunflower and corn biomass shows the lowest lignin content while lavender stem presents the highest, and flax and hemp shives have intermediate values.

We note a large dispersion among the results of the different references for the same bioresource. For example, for hemp shiv (12 references), the cellulose,

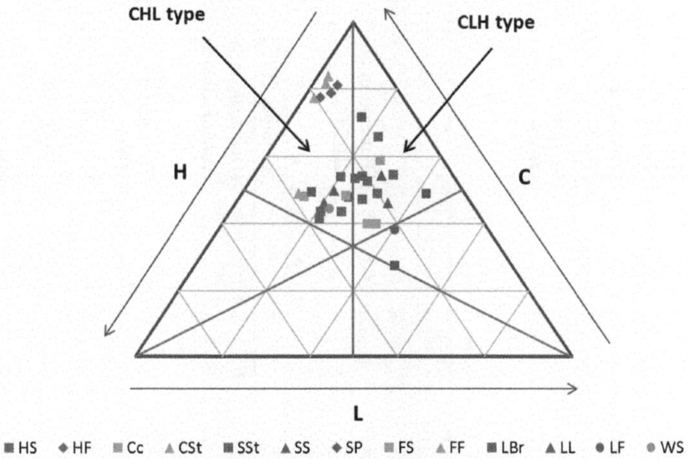

Fig. 1.7 Structural composition of bio-aggregates (C: cellulose; H: Hemicellulose; L: Lignin)

hemicellulose and lignin contents range respectively from 34.5 to 52%, from 9 to 34.5% and from 18 to 28%. Several parameters can explain these variations. The first one, as already mentioned, is the disparity in the analytical procedures applied to determine the chemical composition but effective differences in the proportions of the chemical components can be induced by other factors:

- the geographical location,
- the age of the plant,
- the climate,
- the soil conditions,
- the variety,
- the spatial location within the plant,
- the processing method applied to obtain the bio-aggregates, etc.

The study carried out on lavender (Kaloustian et al. 2000) followed the evolution of its chemical composition with the seasons and revealed large variations in cellulose and hemicellulose contents of lavender branches and leaves between winter and summer.

Concerning the extractives and ash contents of the bio-aggregates, the collected results are very disparate (see Table 1.1). First, depending on the wet chemical method applied, some studies only determined a bulk extractives content, whereas others detailed the pectin, wax/fat, protein and ash contents or at least Soxhlet extractives, ash and others. However, these categories are not always distinct. For example, ash content can be partly included in Soxhlet extractives or in "others". As a result, the comparison between the different studies becomes quite complicated.

Nevertheless, by summing pectin, wax, fat, and protein contents or considering Soxhlet extractives, the bulk extractives contents of bio-aggregates can be compared. They range from 4 to 12.4%. No correlation can be revealed with the nature

Table 1.1 Chemical composition of various types of lignocellulosic biomass used as bio-aggregates in building materials

Aggregate nature		Cellulose	Hemicellulose	Lignin	Pectin	Waxes, fats	Soxhlet extraction	Ash	Protein	Others	Reference
Hemp hurds	HS	36–41	31–37	19–21							Balčiūnasa et al. (2013)
Hemp hurds/shiv	HS	44	25	23	0.6	1.8		1.2	1.6		Gandolfi et al. (2013)
Corn cob	CC	48.1	37.2	14.7			7				Vassilev et al. (2010)
Corn stalks	CSt	49	37.9	13.1			10.5				
Sunflower stalks	SSt	71.6	12.3	16.1							
Flax shives	FS	39.9	26.8	33.3			7.7				
Wheat straw	WS	44.5	33.2	22.3			12.4				
Hemp shiv	HS	47.3	18.3	21.8			6	3.7		6.6	Diquelou et al. (2015)
Hemp shiv	HS	45.6	17.8	23.3			5.1	2.6		8.1	
Hemp shiv	HS	49.2	21	21.9			6.2	3.5		1.6	
Hemp shiv	HS	52	9	18				2			(Kanabat 2015)
Hemp shiv	HS	48	12	28	6	1		2	3		(Hustache and Arnaud 2008)
Hemp shiv	HS	48	12	28	6	6					Sedan (2007)
Hemp shiv	HS	48	21–25	17–19							Thomsen et al. (2005)
Hemp shiv	HS	37.7	26.8	22.1			8.87				De Groot (1994)
Hemp core	HS	38	31	18							Bedetti and Ciaralli (1976) cited in Van der Werf (1994)

(continued)

Table 1.1 (continued)

Aggregate nature		Cellulose	Hemicellulose	Lignin	Pectin	Waxes, fats	Soxhlet extraction	Ash	Protein	Others	Reference
Hemp core	HS	34.5	17.8	20.8							Van der Werf (1994)
Hemp shiv	HS	48	12	28	6	4		2			Garcia-Jaldon et al. (1998)
Hemp shiv	HS	44	18	28	4	1		2			Vignon et al. (1996)
Flax shiv	FS	34.2	21.3	30.2				1.2			Buranov and Mazza (2007) cited in Picandet (2013)
Flax shiv	FS	53	13	24				>2			Fengel and Wegener (1989)
Sunflower stalk	SSt	42.1	29.7	13.4	5.9	1		7.9			Jimenez and Lopez (1993)
Marrow-less sunflower	SS	38.6	22.8	16.2				12.2			Jimenez and Lopez (1993)
Marrow-less sunflower	SS	41.4	30	18.3				8.9			Khristova et al. (1996)
Sunflower marrow	SP	47.4	9.4	3.5	6			20.4			Yin et al. (2007)
Hemp fibre	HF	76.12	12.28	5.65	1.55		3.29				Kokkala (1987)
Hemp fibre	HF	74.4	17.9	3.7	0.9		0.8				Dhakal et al. (2007)
Flax fibre	FF	80.1	14.9	3.1							Khalil et al. (2003)

(continued)

Table 1.1 (continued)

Aggregate nature		Cellulose	Hemicellulose	Lignin	Pectin	Waxes, fats	Soxhlet extraction	Ash	Protein	Others	Reference
Flax fibre	FF	67	11	2							Lilholt et al. (1999) cites in
Flax fibre	FF	71.2	18.6	2.2	2		6.0				Batra (1998), Baley (Baley 2002)
Lavender branches	LB	33.6	13.9	25							Kaloustian et al. (2000) cited in Lesage-Meessen et al. (2015)
Lavender leaves	LL	13.1	12.7	21.7							
Lavender branches	LB	42.7	13	23.1							
Lavender leaves	LL	19.9	3.6	17.1							
Lavender flowers	LF	22.4	12.6	23.6							

of the aggregate as, within the eight results concerning hemp shives, the extractives content ranges from 4 to 12%.

Concerning ash content, for hemp and flax shives, it is less than 5%; but, for sunflower aggregates, it is significantly higher and reaches 20% in the pith. Yin et al. (2007) attribute this result to the high proportion of soluble inorganic elements resulting from the substances taken up by the plant.

The differences between the references can be largely attributed to variations in experimental protocols. Moreover, the measurement uncertainties will be relatively higher for components present in small quantities, such as extractives and ash.

This idea is confirmed by Diquelou's study (Diquelou et al. 2015), which determined, with the same chemical procedure, the composition of three different hemp shives, from different geographic zones and with different times of harvesting and different types of retting and processing. Cellulose, hemicellulose, lignin, extractives and ash contents were quite similar among the three hemp shives. The only major difference appeared in the non-identified fraction which ranged from 1.6 to 8.1%.

This part has shown the difficulty of obtaining accurate and comparable data for the chemical composition of bio-aggregates. A proven chemical procedure that is accepted worldwide is required. Nevertheless, any indirect quantification method based on various fractionation steps would result in a time consuming and expensive method. Moreover some of these procedures involve the consumption of large quantities of solvents, with economic, environmental and safety concerns.

For these reasons, other methods of characterization have been developed to allow fast, low-cost analysis. Some of them have been or could be applied to the characterization of bio-aggregates; they are the subject of the next part.

1.5.2 Other Methods to Characterize Biomass Chemical Composition

With the development of the biofuel industry and, more generally, of biorefinery applications, the analysis of lignocellulosic biomass feedstock has become a major concern. As previously explained, the traditional indirect empirical procedures need qualified technicians, time and money to obtain accurate results and, in any case, remain very sensitive to slight modifications of the experimental protocol.

Scientists are involved in the development of innovative methods and are looking for rapid, inexpensive techniques that would allow real-time chemical analysis of biomass.

Some additional analytical methods can also complete the traditional chemical analysis and identify the hydrolysates produced during the different fractionation steps. For example, chromatography methods are used for the quantification and analysis of carbohydrates. Nevertheless, few studies have been performed on bio-aggregates for building applications. Gandolfi et al. (2013) used High Performance Liquid Chromatography (HPLC) to identify the nature of extracted free amino acids and monosaccharide mixture resulting from the hydrolysed fraction of hemp shives. In the same article, he also explored the nature of lipid extractives using Gas Chromatography with Mass Spectroscopy (GC/MS) and detected mainly fatty acids, alkanes, aldehydes and sterols. Diquelou et al. (2015), analysed the carbohydrate fraction of water extracts of hemp shiv with High Performance Anion Exchange (HPAE) Chromatography and evidenced different proportions of each monomer in the three hemp shives studied. Diquelou also applied the Bradford colorimetric assay to determine the quantity of proteins in the shiv water extracts. Finally, the amount of phenolic and lignin-like compounds in the water extracts was directly determined with UV spectrometry. This technique is adequate to accurately estimate lignin content as lignin absorbs more strongly in the UV and visible regions than cellulose or hemicellulose, even though the identification of an appropriate calibration standard can present some difficulties (Kline et al. 2010). In any case, for biomass quantification purposes, a preliminary fractionation step would be necessary and would then present the same disadvantages as traditional compositional analysis based on biomass hydrolysis.

However, other non-invasive methods have been proposed recently, such as Near Infrared (NIR) spectroscopy or Fourier Transform Infrared (FTIR) spectroscopy.

FTIR has been used for qualitative comparison of hemp shiv before and after immersion in an aggressive solution of diluted cement (Diquelou et al. 2015). The authors highlighted the disappearance of different peaks attributed to the degradation of pectin or hemicellulose, and to lignin. Nozahic and Amziane (2012) used Attenuated Total Reflectance (ATR) in conjunction with FTIR to directly characterize the chemical composition of the surface of hemp shiv. The results of this study will be presented in the next section. In this work, too, the spectroscopic data were only qualitatively exploited.

Nevertheless, these methods can permit a quantitative chemical analysis of biomass, although, to the best of our knowledge, they have never been applied to bio-aggregates for building materials. As presented in Jin and Chen (2007; Krasznai et al. 2012) NIRS and FTIRS spectral analysis can be applied for rapid, accurate, non-destructive quantification of biomass components (ash, moisture, cellulose, hemicellulose and klason lignin) if coupled with efficient calibration models. The construction of these calibration models needs data on chemical composition obtained by traditional wet chemical techniques and chemical absorption spectra obtained by spectroscopy, and uses a statistical approach, such as Partial Least Squares (PLS). As emphasized in Sluiter et al. (2010) the condition for obtaining good data quality from NIR is then the good data quality of the wet chemical procedures used for calibration.

A detailed study was conducted by Sanderson et al. (1996) on 121 samples covering a large range of biomass materials. For the different components, calibration models were built from NIRS spectra (see Fig. 1.8) and wet chemistry data.

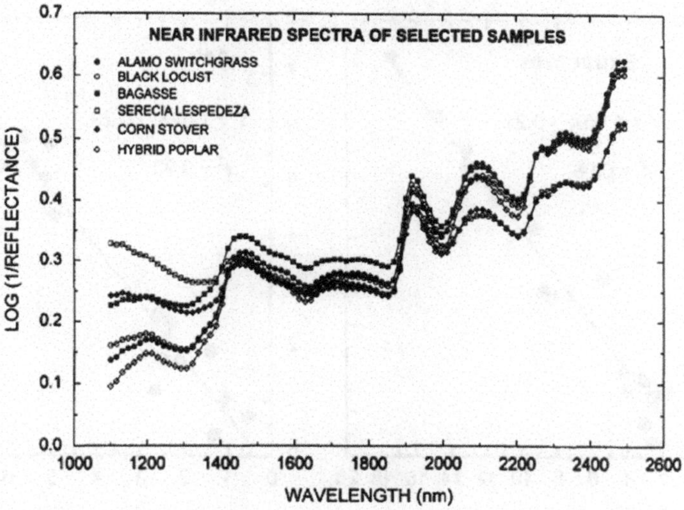

Fig. 1.8 Representative near infrared spectra of various biomass feedstocks (Sanderson et al. 1996)

These calibration curves (see Fig. 1.9) were then used to predict the chemical composition of a set of 20 samples. The authors concluded on the good ability of NIRS to predict the chemical composition of a broad range of biomass feedstocks, but they also highlighted the need to test a much larger quantity of samples to establish robust calibrations.

Thermogravimetric analysis can also be applied to quantify cellulose and hemicellulose contents as shown in Carrier et al. (2011). The authors used this alternative method to characterize two different biomasses: fern and wood. The degradation of hemicellulose, α-cellulose and lignin occur respectively in the temperature ranges of 200–300, 250–350 and 200–500 °C. Applying a deconvolution computation of the DTG curves (see Fig. 1.10), Carrier et al. (2011) identified the contents of the three structural components and compared them to the results obtained by classical wet methods. The calibration curves obtained were then used for the chemical analysis of cellulose pulps by ATG. They concluded that the accuracy of the thermogravimetric method was good for the determination of hemicellulose and α-cellulose contents in cellulose pulps but they noted the incapacity of the method to obtain the lignin content because of large deviations in the correlation curves.

Finally, numerous studies in the literature have used X-ray diffraction (XRD) to characterize lignocellulosic materials. This method is mainly applied to assess the cristallinity of cellulose, which influences its tensile mechanical performance (Thygesen et al. 2007). Although most of the components of biomass present a

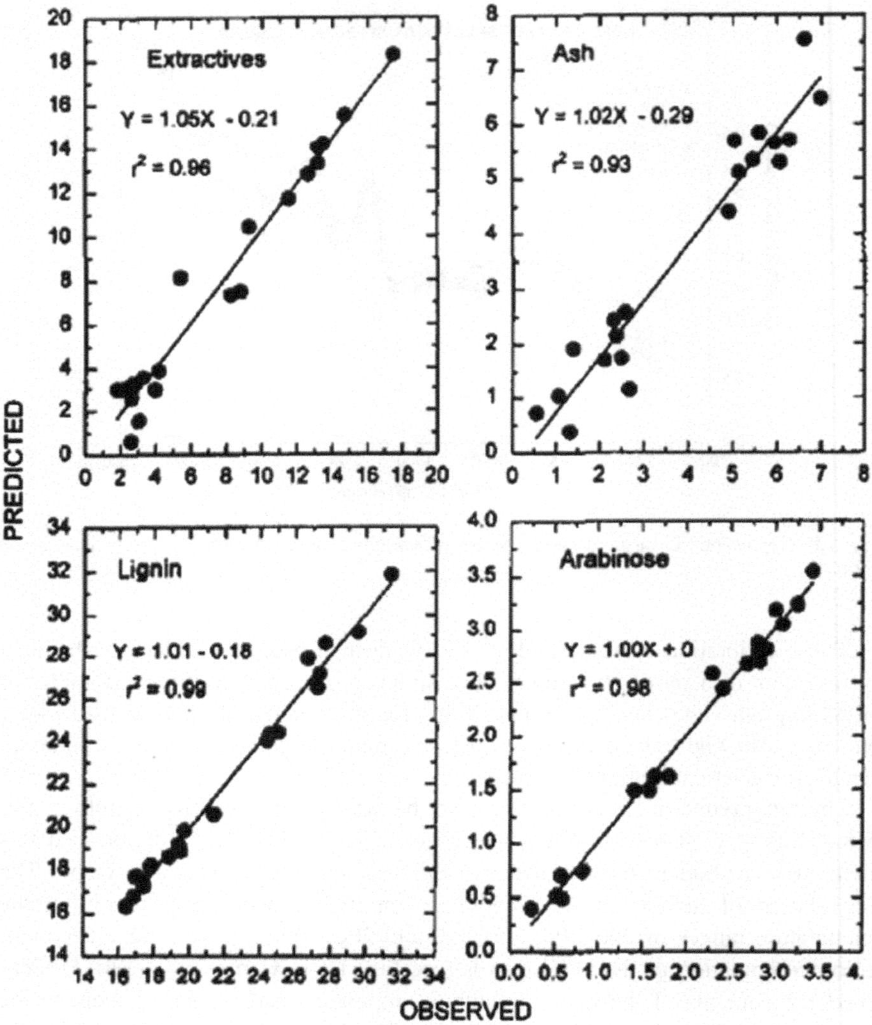

Fig. 1.9 Regression of NIRS predicted values for extractives, ash, lignin and arabinose on observed laboratory values (Sanderson et al. 1996)

non-crystalline character, some mineral constituents can, nevertheless, be easily identified by XRD. Vassilev et al. (2010) detected opal, calcium oxalates, calcite and halite in plum pits, walnut shells, corn cobs, beech wood chips and rice husks. The authors also emphasized that XRD could reveal useful information concerning the amorphous parts; the positions and shapes of the halos can serve as an additional method to help describe the chemical composition of the lignocellulosic compounds.

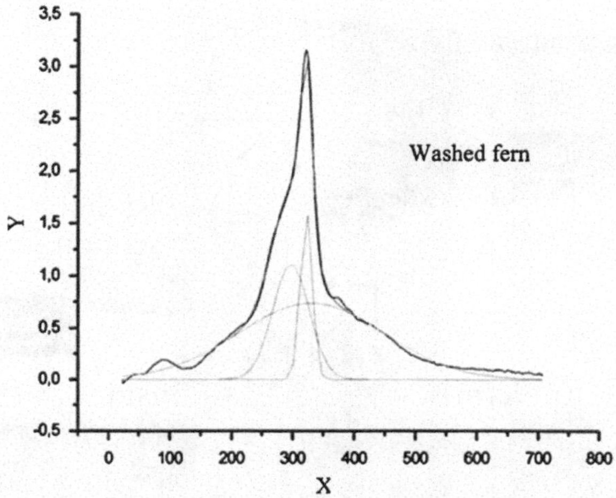

Fig. 1.10 Deconvolution computation of DTG curves from Origin software with X: Temperature (°C) and Y: DTG-d(m/m0)/dt (%/°C). *Black* Experimental DTG; *Green* Calculated peaks from the deconvolution computation with Origin software; *Red* Simulated DTG *curve* (Carrier et al. 2011)

1.6 Surface Characterization of Bio-aggregates—Adhesion Between Lignocellulosic Aggregates and a Mineral Binder

The methods presented in part 1.4 give the chemical composition of milled lignocellulosic compound, but, as stressed by Nozahic and Amziane (2012), sunflower aggregate, for example, is largely heterogeneous (see Fig. 1.11).

It has been shown elsewhere that this is also the case for hemp shiv (Nozahic 2012). Depending on their location inside the stem of the plant, these lignocellulosic aggregates can be of three distinct types: one is part of the epidermis, another can contain pith and the last is mainly composed of xylem cells.

This anisotropic character at macroscopic scale induces strong heterogeneity in the chemical composition of the aggregate. The chemical components of bio-aggregates can cause some undesired interactions with the matrix (as will be developed in Sect. 1.6) but can also enhance the adhesion with the matrix. As shown by Coutts and Kightly (1984), hydrogen bonding and/or hydroxyl bridges between the hydroxyl groups of mineral paste and the covalent hydroxyl, phenolic or alcoholic groups of cellulose and lignin of wood can play a major role in the bonding of wood-fibre-reinforced cement (WFRC). The same phenomena may occur in bio-aggregate based building materials.

Attenuated Total Reflectance (ATR) in conjunction with FTIR has been used to analyse the surface chemistry of sunflower aggregates (inside and epidermis) (Nozahic and Amziane 2012). A qualitative comparison of the spectra (Fig. 1.12) allowed these authors to conclude that there are significant differences between

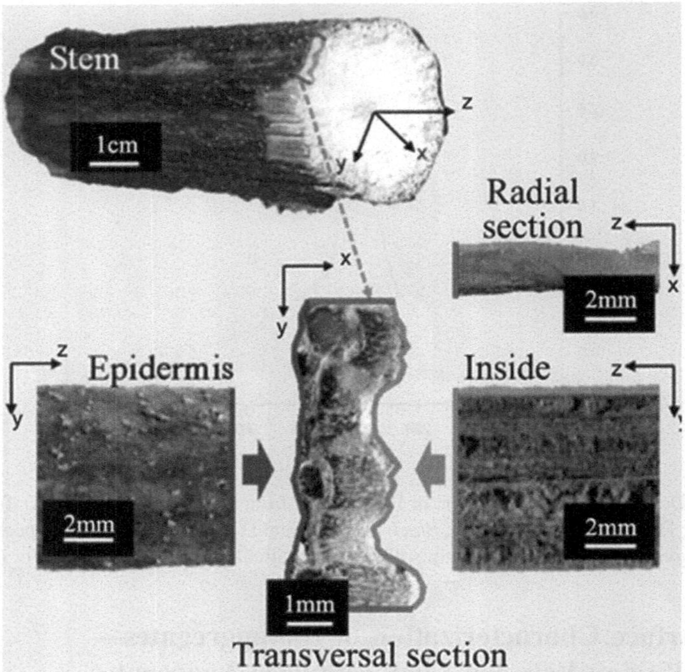

Fig. 1.11 Macroscopic aspect of sunflower stem and three views (inside, epidermis and transversal slice) of raw aggregate (Nozahic and Amziane 2012)

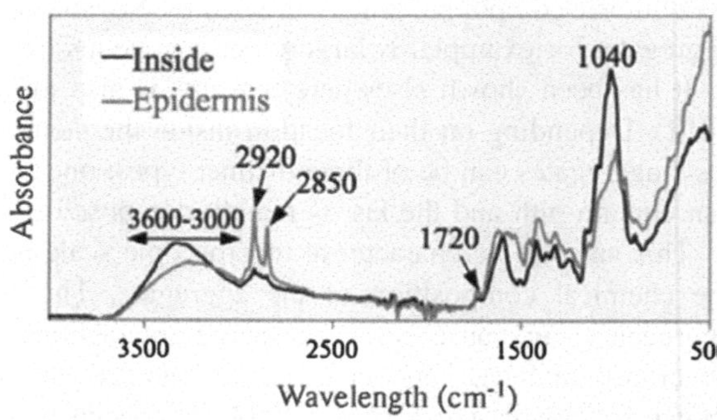

Fig. 1.12 ATR-FTIR analysis of sunflower particle inside and epidermis (Nozahic and Amziane 2012)

interior and epidermis. The first is richer in polysaccharides (the large peak between 3125 and 3600 cm^{-1} results from −OH hydroxyl group stretching) whereas the epidermis presents higher wax and fat contents (stretching vibration of CH_2, at 2920 and 2850 cm^{-1}).

In the same study, the local chemistry analysis by FTIR was also used to explore the impact of three different treatments of the sunflower aggregates: $Ca(OH)_2$, linseed oil and paraffin wax. With $Ca(OH)_2$ treatment, Nozahic observed:

- lowered peaks and, in particular, a strong decrease in hydroxyl –OH stretching bands, attributed to the hydrogen bonding with Ca^{2+},
- precipitation of calcite,
- the disappearance of C=O stretching of carboxylic acid in pectin, caused by the removal of pectin in the alkaline medium or by Ca^{2+} ion trapping by these components.

The other two treatments only resulted in a large increase of the CH_2 methylene group stretching peak. The author explains this result by the deposit of waxy or oily components on the surface of the bio-aggregates.

1.7 Chemical Interactions Between Bio-aggregates and Mineral Binders

The literature about bio-aggregate based building materials highlights the chemical interactions that can occur between lignocellulosic particles and mineral binders. These interactions can take place at different times:

- at early age, they can disturb the setting and hardening mechanisms of mineral binders,
- in the hardened state, they can modify the properties of the composite,
- in the long term, they can influence the durability of the material.

This part aims to present a synthesis of the studies performed on these interactions, focusing on the consequences on composite performances, the mechanisms of interaction (when described) and the influential parameters. The last paragraph will describe the corrective actions proposed to overcome these interactions.

Few studies have been conducted specifically on bio-aggregate based building materials, but similar concerns are largely studied in the sector of wood cement and natural-fibre-reinforced cement/concrete. Results from these domains are thus discussed in this section as the mechanisms involved and their consequences can be common.

1.7.1 Short-Term Interactions Between Lignocellulosic Particles and Mineral Binders

At the moment of mixing and during the early hardening of mineral binders, the interactions between mineral species and the lignocellulosic compounds can induce marked disorders: setting delays, changes in the quality and the quantity of hydrates or complete inhibition of the binder setting.

A significant study by Diquelou et al. (2015) explored the chemical interactions between hemp or flax shives and Portland cement CEM I 42.5 R matrix. After three

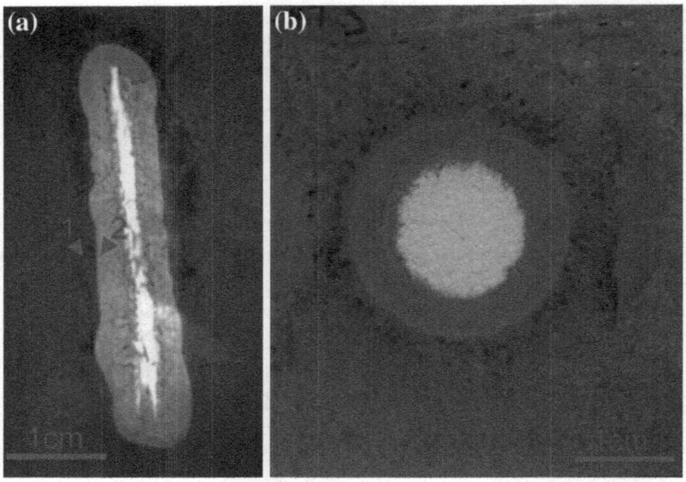

Fig. 1.13 Photograph taken after 3 days of hydration of cement paste with two kinds of plant aggre-gate inclusion: one particle of shiv (arrow 1 and 2 indicate sampling points of FT-IR analy-sis) (**a**); pellet made of shiv powder (**b**) (Diquelou et al. 2015)

days of hydration, the authors carried out FT-IR analysis. This showed the formation of a halo surrounding the hemp particle and corresponding to some non-hydrated cement (see Fig. 1.13). To complete the analysis, they proposed a new method based on the use of a labelled flax pellet enriched with carbon-13 isotope (^{13}C). This technique allowed them to monitor the migration of extractives into the paste and to correlate it with the hydration rate of the paste determined through TGA analysis. They distinguished three distinct zones around the pellet (see Fig. 1.13):

- in the first zone (0–3 mm from the pellet) the inhibition of the cement setting was almost total,
- in the second zone (3–8 mm from the pellet) the setting of the paste was only delayed,
- in the third zone (more than 8 mm from the pellet), the cement setting was not affected.

In the same study (Diquelou et al. 2015) the authors pointed out a decrease in C–S–H (calcium silicate hydrate) formation after 7 days and evaluated the deficit in portlandite ($Ca(OH)_2$) amount to be around 40% after 28 days in comparison with neat cement. The inhibition of the formation of some hydrates is also demonstrated in (Dinh 2014), where a pozzolanic binder was associated with hemp shives. XRD and TGA analyses of the composites after 28 days revealed considerable modifications in the hardening process of the pozzolanic binder in presence of hemp shiv (residual portlandite, undetected calcium aluminate phases and a decrease in the amount of C–S–H). A higher calcium carbonate content was also detected, which is in accordance with the results presented in Govin et al. (2006) on cement mixed with wood particles. The authors attributed this phenomenon to the release of carbon dioxide induced by the alkaline degradation of wood particles, which was responsible for the carbonation of portlandite.

Many other studies confirm the deleterious effects of different lignocellulosic materials on the setting and hardening kinetics and mechanisms of mineral binders. Among others, hemp fibres (Sedan et al. 2008), wood particles (Jorge et al. 2004; Na et al. 2014; Govin et al. 2006; Vaickelionis and Vaickelioniene 2006), cereal straw (Irle and Simpson 1996; Eusebio and Suzuki 1990, cited in Soroushian et al. 2004), arhar stalks (Aggarwal et al. 2008), sugar cane bagasse (Bilba et al. 2003) and coir particles (Menezes Brasileiro and Roche Vieira 2013) were shown to negatively impact the setting and early hardening of cement paste. Wood particles can also be associated with a plaster matrix (calcium sulphate hemihydrate). Boustingorry et al. (2005) emphasized that poplar and forest pine extracts clearly delayed the hydration of hemihydrate. Finally, the setting of a pozzolanic binder (mix of lime and metakaolin) was also affected by the presence of lavender stalks (Ratiarisora et al. 2015).

In the wood concrete sector, the compatibility between wood species and cement is generally assessed indirectly by calorimetric measurements. The decrease of heat release during cement hydration in presence of wood particles in comparison with neat cement is used to calculate various indicators, supposedly correlated with the extent of inhibition associated with the different wood particles.

A review of these indicators can be found in Jorge et al. (2004) and (Wei et al. 2000). Direct indicators such as the maximum temperature of hydration (T_{max}) or the time needed to reach this temperature (t_{max}) were used. More complex indicators were also proposed, such as C_A factor (ratio of the amounts of heat released by cement hydration with or without wood) (Semple and Evans 1998), ratio of setting time of wood-cement composite to that of neat cement (Olorunnisola 2008), initial rate of temperature rise (Sandermann et al. 1960), or other inhibitory indexes calculated from the relative differences in T_{max}, t_{max} and rate of temperature rise in presence or absence of wood (Weatherwax and Tarkow 1964; Hofstrand et al. 1984). Although they have been called into question by some authors (Hachmi et al. 1990; Lee and Hong 1986), these methods have been largely applied in wood-cement composites science to classify wood species in regard to their compatibility with cement and, recently, to cement mortar composites reinforced with cellulosic fibres (Ardanuy et al. 2011). Nevertheless, to the best of our knowledge, at present, no comparable study has assessed the chemical compatibility of biobased aggregates with different mineral binders.

Chemical interactions between lignocellulosic particles and mineral binders at early age involve complex mechanisms, which are not yet fully and precisely understood. Several hypotheses have been proposed in the literature.

The deleterious species responsible for the interactions are of two types: water extractives and products of the alkaline attack generated by the cement or lime-based matrix on the plant aggregates or fibres. Different studies have highlighted the distinct impact of these two components by separating their effects. Diquelou et al. (2015), for example, compared the effects of hemp shiv and water washed hemp shiv on the hydration of Portland cement CEM I 42.5 R. Govin et al. (2005) monitored the hydration of Portland cement CEM I 52.5 R mixed with leaching solutions of wood in water, lime and cement. Although the relative proportion of the interaction attributed to one and to the other differed between the two studies, both underlined the strong delaying effect of both water extractives and alkaline degradation products.

Among the numerous studies focusing on this subject (Diquelou et al. 2015; Sedan et al. 2008; Jorge et al. 2004; Na et al. 2014; Vaickelionis and Vaickelioniene 2006; Boustingorry et al. 2005; Weatherwax and Tarkow 1964; Yasuda et al. 2002; Miller and Moslemi 1991; Semple and Evans 2002; Le Troëdec et al. 2011; Pehanich et al. 2004), different families of extractives were identified as potential retarding agents for mineral binder hydration:

- monosaccharides such as glucose, galactose, mannose, rhamnose, arabinose, xylose,
- polysaccharides such as sucrose, arabinogalactan, xylan, starch, hemicellulose, pectin,
- phenolic compounds such as tannins,
- terpenes,
- organic acids such as acetic acid or fatty acid.

Nevertheless, observations differed among the various publications. Species with a strong inhibiting effect in one case were shown to have negligible impact in another [acetic acid and phenolic compounds in Boustingorry et al. (2005), terpenes in Miller and Moslemi (1991)]. Khazma et al. (2008) even demonstrated the benefits of incorporating large quantities of sucrose into a flax shaves and Portland cement CEM I 52.5 composite (increased mechanical performance and decreased setting time). These differences could result from the large number of influencing parameters that will be detailed at the end of this part.

Moreover, as pointed out in Diquelou et al. (2015) the exhaustive identification of water extracts of three distinct hemp shives is not easy to achieve through common wet chemical analyses. A fraction of 15–30% in mass remains unidentified and the authors conclude that these unknown molecules may have a potentially major effect on cement setting.

As mentioned above, the degradation of lignocellulosic aggregates in the alkaline medium of the mineral binders can produce additional molecules with a negative effect on binder hydration. These reactions have been extensively studied in the pulping industry, where they are used to produce cellulose pulp from wood or fibre crops. In the area of wood-cement composites, hemicellulose is identified as the most sensitive component of wood particles. Hemicellulose is hydrolysed under soft alkaline conditions and this reaction leads to the formation of carboxylic acids (such as glycolic, pyruvic, malic or *o*-salicylic acids) that can act as strong retarding agents for cement hydration (Govin et al. 2005). Hemicellulose degradation is also observed on hemp shives (Diquelou et al. 2015) and cellulosic fibres (Ardanuy et al. 2011) immersed in a Portland cement solution and on hemp fibres subjected to alkaline treatment (NaOH) (Sedan et al. 2008; Le Troëdec et al. 2011). Concerning the degradation of lignin, contradictory results can be found in the literature: the alkaline medium of cement partially dissolves lignin in hemp shiv (Diquelou et al. 2015) but does not affect its content in cellulosic fibres (Ardanuy et al. 2011). This point might be important as Bilba et al. (2003) noted the significant retarding effect of lignin on the hydration of Portland cement. Cellulose is less impacted by the alkaline environment, even though amorphous cellulose can be hydrolysed (Govin et al. 2005). Bilba et al. confirm its low impact on cement hydration.

Beyond the nature of the deleterious molecules responsible for the early age interactions between plant material and mineral binders, several studies have focused on their mechanisms of interaction (Sedan et al. 2008; Jorge et al. 2004; Na et al. 2014; Govin et al. 2006; Vaickelionis and Vaickelioniene 2006; Sedan et al. 2007; Frybort et al. 2008). In addition, these chemical interactions can be compared with those studied in the context of organic admixtures for cement, i.e. polysaccharides, sugar acids or lignosulfonates used as water retention agents, water reducing agents or set-retarding agents (Garci Juenger and Jennings 2002; Peschard et al. 2004; Thomas and Birchall 1983; Bishop and Barron 2006). The findings of these studies highlight the existence of distinct ways of action:

- trapping of calcium ions present in the binder solution by plant components. The consequent decrease in the concentration of Ca^{2+} ions induces a delay in the nucleation process of $Ca(OH)_2$ and C–S–H. This phenomenon has been observed in cement and calcium hydroxide solutions mixed with wood (Jorge et al. 2004) and hemp fibres (Sedan et al. 2007, 2008). In these references, the authors concluded on the specific role of pectin, which can react with Ca^{2+} ions in an alkaline environment to form a very stable "egg box" structure,
- formation of a thin adsorption layer on the surface of cement grains (Na et al. 2014; Vaickelionis and Vaickelioniene 2006; Frybort et al. 2008; Peschard et al. 2004). This less permeable coating slows down the hydration process by preventing the migration of water to the anhydrous part of the grain,
- poisoning of the nucleation sites by adsorption on the first $Ca(OH)_2$ and C–S–H hydrates (Garci Juenger and Jennings 2002; Thomas and Birchall 1983). In the first step, sugars would increase the dissolution rate of anhydrous cement but, in the second step, their adsorption onto the first hydrates would inhibit their growth,
- formation of calcium carbonate (Govin et al. 2006). Govin et al. evidenced the early carbonation of cement paste in the presence of wood. They attributed this reaction to the release of carbon dioxide due to alkaline degradation of wood. Nevertheless, this conversion of portlandite to calcium carbonate is not necessarily negative.

In the cited articles, numerous parameters thus appear to have a significant influence on the short-term chemical interactions between lignocellulosic compounds and mineral matrixes:

- the nature of the inhibitory molecules,
- the concentration of plant extracts. The degree of inhibition exerted by wood extracts and polysaccharides on cement hydration has been found to be correlated with their concentration (Vaickelionis and Vaickelioniene 2006; Boustingorry et al. 2005; Wei et al. 2000; Garci Juenger and Jennings 2002; Peschard et al. 2004),
- the nature of the binder (Govin et al. 2006; Vaickelionis and Vaickelioniene 2006; Bilba et al. 2003; Boustingorry et al. 2005; Frybort et al. 2008). Different binders (cement and plaster) can be influenced differently by the same molecules. Most of the results reported in the literature concern Portland cement hydration and hardening, but the mechanisms involved in the case of lime or pozzolanic binders might be significantly different. The alkalinity and the nature of the main ions of a given binder might strongly modify the mechanisms of degradation of lignocellulosic compounds and their interactions with mineral species (Bishop and Barron 2006). In fact, as shown by Peschard et al. (2004) and Thomas and Birchall (1983), the chemical interactions between mineral binders and plant products depend on the mineralogical nature of the binder phases. Concerning cementitious binders for example, C_3A content appears as a key parameter (Peschard et al. 2004; Bishop and Barron 2006). pH also influences the interactions as reported by Sedan et al. (2007) and Jorge et al. (2004). Finally, the addition of a pozzolanic mineral material to cement can reduce the deleterious effect of wood extractives on cement setting (Vaickelionis and Vaickelioniene 2006),
- the production process of plant aggregates. Diquelou et al. (2015) showed that extracts of ground-retted hemp shiv contain more sugars and unidentified products than non-retted hemp shiv. Their impacts on cement hydration and hardening are consequently more severe. The authors put forward the hypothesis that the microbial activity during the retting process led to the extraction of additional products, either by microorganisms' metabolism or by cell wall degradation. In contrast, the removal of pectin due to the retting process (Thygesen et al. 2013) could reduce the capacity of hemp material to trap calcium ions. The retting process would then influence short-term interactions between hemp and cement in opposite ways, reducing the pectin content on the one hand but increasing the solubility of plant components on the other,
- the wood species involved. This influences the extractives content and their chemical compositions (Jorge et al. 2004; Wei et al. 2000). Generally, hardwoods have much greater impact on the hydration of cement than softwoods do, because of their higher hemicellulose content (Weatherwax and Tarkow 1964). Other authors have stated that more extractives are found in deciduous trees than in firs (Vaickelionis and Vaickelioniene 2006),
- the part of the wood involved. Heartwood is reputed to impact cement hydration more severely than sapwood because of its higher solubility, which leads to a higher proportion of potentially inhibiting molecules (Jorge et al. 2004; Na et al. 2014),

– the storage conditions of the wood. Several studies listed by Na et al. (2014) show that storing logs in outdoor conditions for weeks or months tends to reduce their sugar and starch contents and, consequently, to decrease the impact of the wood on cement hydration.

1.7.2 Medium and Long-Term Chemical Interactions Between Plant Particles and Mineral Binder

A few references deal with medium-term interactions between plant particles and mineral binders and their consequences on the mechanical performance of the composites.

The most severe disorder observed on hardened composite including lignocellulosic compounds is poor cohesion between the plant compounds and the mineral matrix. This phenomenon is usually associated with a powdering of the binder. Diquelou et al. (2015) explored this point by comparing the development of the compressive strength of cement paste mixed with a solution containing hemp shive extractives with that of neat cement paste. They reported a decrease in compressive strength of 25% linked to the reduced amount of hydrates (C–S–H and portlandite).

Comparable results are reported in Ratiarisora et al. (2015) on a composite made of lavender stalks incorporated in a pozzolanic matrix. Its very weak compressive strength is attributed to the deleterious chemical interactions between lavender aggregates and the mineral binder. This hypothesis is confirmed by the comparison of the compressive performance of neat pozzolanic paste with that of paste containing lavender extractives. The latter exhibits a deficit of 20% in compressive strength after 49 days.

Dinh (2014), working on the same binder but in association with hemp shives, highlights strong modifications of the hardening mechanisms in presence of plant aggregates. After 28 days of curing, the amounts of residual portlandite observed in the composite were larger than in the pure paste and quantities of hydrated phases (straetlingite, ettringite and carboluminates) were smaller. The author concludes that there was partial inhibition of the pozzolanic reaction due to hemp shiv components. These interactions are strongly linked with the nature of the binder, as, in the same study, the results obtained with a commercial binder did not show that kind of deleterious interactions.

These medium-term defects could be the direct consequences of early age interactions, the mechanisms of which were described in the previous part. But other long-term interactions between plant particles and mineral species can occur, affecting the durability of the composite. To the best of our knowledge, no specific study has been made of these phenomena in bio-aggregate based materials. Nevertheless, the literature on the durability of plant-fibre-reinforced cement-based composites is very abundant. A recent review by Ardanuy et al. (2015) synthesizes the latest research on this subject, including the ageing mechanisms. Although these articles deal with short or pulp fibres, which are much smaller than plant aggregates, common chemical mechanisms could be evidenced. The main phenomenon explaining the loss of durability of cellulose-fibre-reinforced cement-based composite is the mineralization of the reinforcement fibres by cement hydration products. Several references listed by Ardanuy et al. (2015) report the diffusion and reprecipitation of hydrated calcium compounds in the porosity of the fibres. This induces a loss of flexibility and/or a weakening of the fibres, resulting in the embrittlement of the composites. Wei and Meyer (2015) complete this description with an additional degradation of the amorphous components of the fibres (lignin and hemicellulose) followed by the alkali degradation of cellulose. This proceeding degradation allows cement hydration products to infiltrate into the cell wall, resulting in the mineralization of the natural fibre.

Similar reactions may occur between plant aggregates and binders but their consequences might not be as negative. The aim of reinforcing cement-based mortar with plant fibres is to improve their flexural behaviour. The loss of ductility induced by fibre mineralization is then considered to lead to poor durability. However, in the case of bio-aggregate based building material, such a chemical interaction could even be beneficial, considering, for example, the compressive performance of the composite. In a review (Frybort et al. 2008), Frybort reports two studies on cement-bonded particle boards that evidenced the diffusion of cement ions into the wood particles and the partial filling of the porosity by cement hydrates. The author concludes that the adhesive forces between the matrix and the wood inclusions are improved. In another study on hemp concrete (Magniont et al. 2012), the mineralization of hemp aggregates after one year of ageing is evidenced through SEM observations. This phenomenon could be responsible for the continuous enhancement of compressive strength observed on the composite between 28 days and 1 year. The progressive mineralization of hemp shives, which are initially very compressible, might stiffen them by filling the porosity with calcium mineral hydrates. Nevertheless, to the best of our knowledge, there is no existing study exploring these phenomena over longer periods or using accelerating ageing test methods. References assessing the impact of such chemical interactions on other properties of the composites (thermal, hygric, acoustical etc.) are not found in the literature either.

1.7.3 Corrective Treatments to Enhance the Compatibility Between Lignocellulosic Materials and Mineral Binders

Numerous studies have been conducted in order to propose corrective methods limiting the negative effects of chemical interactions between lignocellulosic particles and mineral binders. These methods aim to pretreat the plant particles or to modify the mineral binder, and differ depending on the time the deleterious chemical interaction lasts: short, medium or long term.

At the short term, most of the procedures found in the literature concern wood concrete and aim to overcome the delay in setting and hardening of the binder. To achieve this objective, two main types of methods are applied:

- the first one consists of eliminating the deleterious components of the wood particles. This removal can be performed using various aqueous extraction methods: with cold or hot water (Diquelou et al. 2015; Eusebio et al. 2000; Kavvouras 1987) and with alkaline solutions (Alberto et al. 2000; Soroushian and Hassan 2012). The fermentation of sawdust (Simatupang and Handayani 2001) induces a decrease in its total sugar content and significantly improves its compatibility with cement. The treatment of the wood material by fungi (Thygesen et al. 2005, cited in Frybort et al. 2008) and the long-term storage of wood in outdoor conditions can effectively reduce the interaction (Frybort et al. 2008). Finally, coating the particles is also described in (Na et al. 2014) as a method for preventing the contact between wood constituents and cement,
- the second one aims to accelerate the setting and hardening mechanisms of cement in order to offset the inhibitory effects of lignocellulosic components. Numerous cement curing accelerators have been studied in the literature, mainly calcium chloride, ferric chloride, magnesium chloride, aluminium sulfate and sodium silicate (Matoski et al. 2013). Semple et al. (2004) compare the efficiency of 137 inorganic compounds for the manufacture of *Acacia mangium* cement composites. They conclude that the compounds bring specific benefits, with the dual ability to accelerate cement hydration and to form complexes with the inhibitory extractives. More recently, Matoski et al. (2013) have explored the possible use of *Pinus* wood to compose wood-cement panels. When 4 accelerating admixtures ($CaCl_2$, $MgCl_2$, Na_2SiO_3 and Al_2SO_4) were considered, the best results were obtained with calcium chloride. This low-cost admixture (Na et al. 2014) appeared as one of the most efficient accelerators (Eusebio et al. 2000; Okino et al. 2005; Wei et al. 2000). Beyond the use of curing accelerators, the partial substitution of cement by pozzolans [silica fume (Okino et al. 2005), opoca (Vaickelionis and Vaickelioniene 2006)] has proved to successfully limit the setting delay in presence of wood. Finally, a common technique largely described in the literature on wood- and lignocellulosic-waste cement composites (Na et al. 2014; Frybort et al. 2008; Karade 2010) is the injection of CO_2 as a curing process. The early carbonation of Portland cement reduces the pressing time of the components and improves the mechanical performance of the cured panels. Nevertheless, a homogeneous injection of CO_2 into the whole element can be a technical challenge, especially for thick elements (blocks). The incorporation of water-soluble carbonates into the mix could be an alternative source of carbon dioxide.

Authors often propose combining various corrective treatments to manufacture wood-cement composites with adequate performances. The corrective methods presented above and applied to the wood particles could be easily transposed to the treatment of bio-aggregates (Diquelou et al. 2015). However, accelerating methods that work well for cement will not necessarily be as effective on lime or pozzolanic binders. Further investigation is required to conclude on this point.

In addition to those already presented, various treatments can be applied to process wood or to improve some of its properties. These treatments can have a significant impact on the chemical interactions with the mineral binder. Thermal treatment (Govin et al. 2006; Boustingorry et al. 2005) and the water-vapour explosion process (Wei et al. 2003) induce a clearly negative effect on the setting of cement. The application of high temperature and/or pressure conditions to the wood modifies the chemical composition of its extractives by a partial degradation of its main components. In consequence, the low-molecular weight, water-soluble polysaccharide content, responsible for the inhibitory effect, increases.

Another group of pretreatment methods aims at improving the bonding between plant particles and mineral binder and consequently the mechanical performance of the composite in the medium term. Some of the methods already cited, alkaline solution treatments (Sedan et al. 2008; Le Troëdec et al. 2011) or hydrophobic coating (Monreal et al. 2011), are used for that purpose. Other authors propose the use of mineral binder treatment (Dinh 2014; Monreal et al. 2011). Nozahic and Amziane (2012) assess the ability of three distinct treatments (soaking in $Ca(OH)_2$ solution, application of linseed oil, and paraffin wax coating) to improve the adhesion process of sunflower aggregate with a pumice-lime binder. The three treatments induce modifications in the hygroscopic behaviour of the particle but only $Ca(OH)_2$ and paraffin wax promote better adhesion with the binder. A very simple, low-cost mineral pretreatment suitable for industrial production was proposed in (Dinh 2014). Prior to their incorporation into the binder, hemp shives were treated with a part of the same binder. After 28 days, hemp concrete mixed with pretreated hemp shives exhibited a 48% increase in compressive strength in comparison with that containing untreated hemp shives.

In order to prevent the durability problems of plant-fibre-reinforced cement, preventive methods have been proposed. In a similar manner to that applied to treat short-term interactions, two kinds of procedure can be carried out: modifying the matrix or modifying the fibres (Ardanuy et al. 2015).

The aggressiveness of the matrix can be reduced by a partial substitution of cement by pozzolanic additions such as silica fume (Gram 1988; Soroushian et al. 1994) or metakaolin (Toledo Filho et al. 2009; Wei and Meyer 2015). Accelerated carbonation is another alternative. Almeida et al. (Almeida et al. 2013) significantly improved the durability of eucalyptus-pulp-reinforced cementitious composites through a 3 day curing phase until the samples were completely carbonated. According to the authors, the higher mechanical strength after the ageing process was attributable to the reduction of $Ca(OH)_2$ content, the lower porosity and the good fibre-matrix adhesion.

Finally, appropriate physical or chemical treatments of the fibres prior to their incorporation into the matrix can successfully prevent long-term deleterious interactions between fibres and matrix. The following are worth noting:

- hornification of the fibres, i.e. subjecting them to drying and rewetting cycles (Ardanuy et al. 2015),
- hydrophobic coating (Gram 1988; Juarez et al. 2007; Asprone et al. 2011),
- soaking in a silica fume slurry (Toledo Filho et al. 2003),
- surface treatment with silanes (Bilba and Arsène 2008; Tonoli et al. 2009).

1.8 Conclusion

The chemical composition of lignocellulosic particles can strongly impact their properties as bio-aggregates included in a mineral matrix. These bioresources are mainly composed of three structural biopolymers: cellulose (C), hemicellulose (H) and lignin (L). Other secondary components such as pectin, various extractives and ash complete their chemical composition.

The chemical composition of various bioresources included in bio-aggregate based building materials have been reviewed by comparing the results of 24 published references. This review shows that they are mainly of the CHL or CLH types because of their nature, i.e. mainly wood residues (hemp and flax shives) and herbaceous or agricultural biomass (straws, stalks or fibres). This comparison highlights a large dispersion among the results of different references for the same bioresource, reflecting not only the effective variability of chemical composition due to agronomic, environmental and processing parameters but also the strong disparities in the results of common indirect gravimetric methods of biomass compositional analysis. Similar variations are observed in the extractives contents, which are, in fact, largely responsible for the negative chemical interactions between lignocellulosic compounds and mineral binders. In order to overcome this difficulty, scientists are currently involved in the development of alternative techniques mainly based on infrared spectroscopy, thermogravimetric analyses and X-ray diffraction. It is to be hoped that these works will lead to a fast, low-cost method for the chemical analysis of bio-aggregates in the near future.

One of the main reasons why the chemical characterization of bioparticles is so important concerns the deleterious interactions that can take place with mineral binders.

At early age, they can disturb the setting and hardening mechanisms of mineral binders; in the hardened state, they can modify the properties of the composite; and, finally, in the long term, they can influence durability.

Even though these chemical interactions have been largely studied in the area of wood-concrete and natural-fibre-reinforced mortars and although the interaction mechanisms with a matrix of Portland cement are fairly well known in that context, very few studies have dealt with these concerns in bio-aggregate based building materials such as hemp concretes. In particular, there are hardly any studies focusing on the chemical interactions between lime-based or pozzolanic binders and lignocellulosic particles.

Forthcoming studies should be based on the existing results in order to propose quick, easy methods for assessing the compatibility of potential bio-aggregates with binders of different natures. When deleterious interactions are observed, corrective treatments based on the pretreatment of plant particles or on binder design optimization will have to be developed. This constitutes an essential step towards the large scale development of adequate manufacturing processes for bio-based building materials having valuable properties in use and guaranteed durability.

References

Aggarwal, L.K., Agrawal, S.P., Thapliyal, P.C., Karade, S.R.: Cement-bonded composite boards with arhar stalks. Cement Concr. Compos. **30**, 44–51 (2008)

Akin, D.E.: Chemistry of plant fibres. In: Müssig, J. (ed.). Industrial Applications of Natural Fibres, Structure, Properties and Technical Application, pp. 13–22. Wiley (2010)

Alberto, M.M., Mougel, E., Zoulalian, A.: Compatibility of some tropical hardwoods species with Portland cement using isothermal calorimetry. Forest Prod. J. **50**(9), 83–88 (2000)

Almeida, A.E.F.S., Tonoli, G.H.D., Santos, S.F., Savastano, H.: Improved durability of vegetable fiber reinforced cement composite subject to accelerated carbonation at early age. Cement Concr. Compos. **42**, 49–58 (2013)

Angelova, G., Ivanova, V., Delibaltovab, V., Ivanova, K.: Bio-accumulation and distribution of heavy metals in fibre crops (flax, cotton and hemp). Ind. Crops Prod. **19**(3), 197–205 (2004)

Ardanuy, M., Claramunt, J., Toledo Filho, R.D.: Cellulosic fiber reinforce cement-based composites: a review of recent research. Constr. Build. Mater. **79**, 115–128 (2015)

Ardanuy, M., Claramunt, J., García-Hortal, J.A., Barra, M.: Fiber-matrix interactions in cement mortar composites reinforced with cellulosic fibers. Cellulose **18**, 281–289 (2011)

Asprone, D., Durante, M., Prota, A., Manfredi, G.: Potential of structural pozzolanic matrix-hemp fiber grid composites. Constr. Build. Mater. **27**, 2867–2874 (2011)

Balciunas, G., Vejelis, S., Vaitkus, S., Kairyte, A.: Physical properties and structure of composite made by using hemp hurds and different binding materials. Proc. Eng. **57**, 159–166 (2013)

Balčiūnasa, G., Vėjelisb, S., Vaitkusc, S., Kairytėd, A.: Physical properties and structure of composite made by using hemp hurds and different binding materials. Proc. Eng. **57**, 159–166 (2013)

Baley, C.: Analysis of the flax fibres tensile behaviour and analysis of the tensile stiffness increase. Compos. A **33**, 939–948 (2002)

Bedetti, R., Ciaralli, N.: Variazione del contenuto délia cellulosa durante il periodo vegetativo délia canapa. [Variation of the cellulose content during the vegetative period of hemp.] Cellulosa e Carta **26**, 27–30 (1976)

Bilba, K., Arsène, M.A.: Silane treatment of bagasse fiber for reinforcement of cementitious composites. Compos. A **39**, 1488–1495 (2008)

Bilba, K., Arsene, M.A., Ouensanga, A.: Sugar cane bagasse fibre reinforced cement composites. Part I. Influence of the botanical components of bagasse on the setting of bagasse/cement composite. Cement Concr. Compos. **25**, 91–96 (2003)

Bishop, M., Barron, A.R.: Cement hydration inhibition with sucrose, tartatic acid, and lignosulfonate: analytical and spectroscopic study. Ind. Eng. Chem. Res. **45**, 7042–7049 (2006)

Boustingorry, P., Grosseau, P., Guyonnet, R., Guilhot, B.: The influence of wood aqueous extractives on the hydration kinetics of plaster. Cem. Concr. Res. **35**, 2081–2086 (2005)

British Columbia, Ministry of Agriculture and Food: Industrial Hemp (Cannabis sativa L.). Speciality Crops Factsheet (1999)

Buranov, A.U., Mazza, G.: Fractionation of flax shives by water and aqueous ammonia treatment in a pressurized low polarity water (PLPW) extractor. J. Agric. Food Chem. **55**, 8548–8555 (2007)

Carrier, M., Loppinet-Serani, A., Denux, D., Lasnier, J.M., Ham-Pichavant, F., Cansell, F., Aymonier, C.: Thermogravimetric analysis as a new method to determine the lignocellulosic composition of biomass. Biomass Bioenergy **35**, 298–307 (2011)

Coutts, R.S.P., Kightly, P.: Bonding in wood fibre-cement composites. J. Mater. Sci. **19**, 3355–3359 (1984)

Davidson, B.H.: Plant cell walls: basics of structure, chemistry, accessibility and the influence on conversion. In: Wyman, C.E. (ed.). Aqueous Pretreatment of Plant Biomass for Biological and Chemical Conversion to Fuels and Chemicals, 1st edn. (2013)

De Groot, B.: Alkaline hemp woody core pulping—impregnation characteristics, kinetic modelling and papermaking qualities (1994)

Dhakal, H.N., Zhang, Z.Y., Richardson, M.O.W.: Effect of water absorption on the mechanical properties of hemp fibre reinforced unsaturated polyester composites. Compos. Sci. Technol. **67**, 1674–1683 (2007)

Dinh, T.M.: Contribution to the development of precast hempcrete using innovative pozzolanic binder. PhD thesis, Université de Toulouse, France (2014)

Dinwoodie, J.M.: Wood, nature's cellular, polymeric, fibre-composite. Institute of Metals (1989)

Diquelou, Y., Gourlay, E., Arnaud, L., Kurek, B.: Impact of hemp shiv on cement setting and hardening: influence of the extracted components from the aggregates and study of the interfaces with the inorganic matrix. Cement Concr. Compos. **55**, 112–121 (2015)

Eusebio, E.A., Suzuki, M.: Production and properties of plant materials cement bonded composites. Bulletin of the Experimental Forest Laboratory, Tokyo University of Agriculture and Technology, p. 27. (1990)

Eusebio, D.A., Soriano, F.P., Cabangon, R.J., Evans, P.D.: Manufacture of low-cost wood-cement composites in the Philippines using plantation-grown Australian species: I, pp. 105–114. Eucalyptus. Wood-Cement Composites in the Asia-Pacific Region, Canberra, Australia (2000)

Fengel, D., Wegener, G.: Wood Chemistry, Ultrastructures, pp. 56–58. Reactions. Walter de Gruyter Press, New York (1989)

Frybort, S., Mauritz, R., Teicbhinger, A., Müller, U.: Cement bonded composites—a mechanical review. Bioresources **3**(2), 602–626 (2008)

Gandolfi, S., Ottolina, G., Riva, S., Fantoni, G.P., Patel, I.: Complete chemical analysis of Carmagnola hemp hurds and structural features of its components. BioResources **8**(2), 2641–2656 (2013)

Garci Juenger, M.C., Jennings, H.M.: New insights into the effects of sugar on the hydration and microstructure of cement pastes. Cem. Concr. Res. **32**, 393–399 (2002)

Garcia-Jaldon, C., Dupeyre, D., Vignon, M.R.: Fibres from semi-retted hemp bundles by steam explosion treatment. Biomass Bioenergy **14**, 251–260 (1998)

Govin, A., Peschard, A., Fredon, E., Guyonnet, R.: New insights into wood and cement interaction. Holzforschung **59**, 330–335 (2005)

Govin, A., Peschard, A., Guyonnet, R.: Modification of cement hydration at early ages by natural and heated wood. Cement Concr. Compos. **28**, 12–20 (2006)

Gram, H.E.: Durability of natural fibres in concrete, in Concrete technology and design, Natural fibre reinforced cement and concrete, N. Swamy, Blackie and Son Ltd, UK (1988). 1988

Hachmi, M., Moslemi, A.A., Campbell, A.G.: A new technique to classify the compatibility of wood with cement. Wood Sci. Technol. **24**(4), 345–354 (1990)

Hayes, D.J.: An examination of Irish Feedstocks For Biorefineries. PhD Transfer Thesis, Dept. of Chemical and Environmental Sciences, University of Limerick (2004)

Hofstrand, A.D., Moslemi, A.A., Garcia, J.F.: Curing characteristics of wood particles from nine northern Rocky Mountain species mixed with Portland cement. Forest Prod. J. **34**(2), 57–61 (1984)

Hustache, Y., Arnaud, L.: Synthèse des connaissances sur les bétons et mortiers de chanvre. Fibres Recherche Développement, Lhoist, Construire en Chanvre (2008)

Irle, M., Simpson, H.: Agricultural residues for cement-bonded composites. In: Moslemi, A. (ed.). Inorganic-Bonded Wood and Fiber Composite Materials Conference Proceedings. vol. 5, pp. 54–58. (1996)

Jimenez, L., Lopez, F.: Characterization of paper sheets from agricultural residues. Wood Sci. Technol. **27**, 468–474 (1993)

Jin, S., Chen, H.: Near-infrared analysis of the chemical composition of rice straw. Ind. Crops Prod. **26**, 207–211 (2007)

Jorge, F.C., Pereira, C., Ferrera, J.M.F.: Wood-cement composites: a review. Holz Roh Werkst. **62**, 370–377 (2004)

Juarez, C., Duran, A., Valdez, P., Fajardo, G.: Performance of "Agave Lechuguilla" natural fiber in portland cement composites exposed to severe environment conditions. Build. Environ. **42**(3), 1151–1157 (2007)

Kaloustian, J., Pauli, A.M., Pastor, J.: Chemical and thermal analysis of the biopolymers in the lavandin. J. Appl. Polym. Sci. **77**, 1629–1641 (2000)

Kamat, J.: Effect of harvesting time on the physical, chemical and pulping properties of hemp Cannabis. In: Sativa, L. Thesis for the degree of Master of Science in Forestry, Faculty of Forestry, University of Toronto (2000).

Kanabat, technical document. La Chanvrière de l'Aube. http://www.chanvre.oxatis.com (2015). Accessed 20 April 2015

Karade, S.R.: Cement-bonded composites from lignocellulosic wastes. Constr. Build. Mater. **24**, 1323–1330 (2010)

Kavvouras, P.K.: Suitability of Quercus conferta wood for the manufacture of cement-bonded flakeboards. Holzforschung **41**, 159–163 (1987)

Khalil, A., Rozman, H.D., Ahmad, N.N., Ismail, H.: Acetylated plant-fiber-reinforced polyester composite: a study of mechanical, hydrothermal, and aging characteristics. Polym.-Plast. Technol. Eng. **39**(4), 757–781 (2003)

Khanna, P.: Cell and Molecular Biology. I.K. International (2008)

Khazma, M., El Hajj, N., Goullieux, A., Dheilly, R.M., Queneudec, M.: Influence of sucrose addition on the performance of a lignocellulosic composite with a cementitious matrix. Composite: Part A **39**, 1901–1908 (2008)

Khristova, P., Yossifov, N., Gabir, S.: Particle board from sunflower stalks: preliminary trials. Bioresour. Technol. **58**, 319–321 (1996)

Klason, P.: Contribution to a more exact knowledge of the chemical composition of spruce wood, part I. Pap. Trade J. **74**(18), 45–51 (1922)

Kline, L.M., Hayes, D.G., Womac, A.R., Labbé, N.: Simplified determination of lignin content in hard and soft woods via UV-spectrophotometric analysis of biomass dissolved in ionic liquids. BioResources **5**(3), 1366–1383 (2010)

Kokkala, M.: Fire tests on loose-fill insulation materials. Technical Research Centre of Finland, Research reports 485 (1987)

Krasznai, D.J.: Multivate characterization of lignocellulosic biomass and graft modification of natural polymers. Master of Applied Science degree, Queen's University, Kingston, Ontario, Canada (2012)

Krasznai, D.J., Champagne, P., Cunningham, M.F.: Quantitative characterization of lignocellulosic biomass using surrogate mixtures and multivariate techniques. Bioresour. Technol. **110**, 652–661 (2012)

Le Troëdec, M., Rachini, A., Peyratout, C., Rossignol, S., Max, E., Kaftan, O., Fery, A., Smith, A.: Influence of chemical treatments on adhesion properties of hemp fibres. J. Colloid Interf. Sci. **356**, 303–310 (2011a)

Le Troëdec, M., Rachini, A., Peyratout, C., Rossignol, S., Max, E., Kaftan, O., Fery, A., Smith, A.: Influence of chemical treatments on adhesion properties of hemp fibres. J. Colloid Interf. Sci. **356**, 303–310 (2011b)

Lee, A.W.C., Hong, Z.: Compressive strength of cylindrical samples as an indicator of wood-cement compatibility. Forest Prod. J. **36**(11/12), 87–90 (1986)

Lesage-Meessen, L., Bou, M., Sigoillot, J.C., Faulds, C.B., Lomascolo, A.: Essential oils and distilled straws of lavender and lavandin: a review of current use and potential application in white biotechnology. Appl. Microbiol. Biotechnol. (2015). doi:10.1007/s00253-015-6511-7
Lewin, M. (ed.): Handbook of Fiber Chemistry. CRC Press, (2006)
Lilholt, H., Toftegaard, H., Thomsen, A.B., Schmidt, A.S., Natural composites based on cellulosic fibres and polypropylene matrix. Their processing and characterization. Proceedings of ICCM 12, Paris; July 1999. p. 9
Linger, P., Ostwald, A., Haensler, J.: Cannabis sativa L. growing on heavy metal contaminated soil: growth, cadmium uptake and photosynthesis. Biol. Plant. **49**(4), 567–576 (2005)
Magniont, C., Escadeillas, G., Coutand, M., Oms-Multon, C.: Use of plant aggregates in building ecomaterials. Eur. J. Environ. Civil Eng. **16**(SUP1), 17–33 (2012)
Matoski, A., Hara, M.M., Iwakiri, S., Machado Casali, J.: Influence of accelerating admixtures in wood-cement panels: characteristics and properties. Maringa **35**(4), 655–660 (2013)
Menezes Brasileiro, G.A., Roche Vieira, J.A.: Use of coir pith particles in composites with Portland cement. J. Environ. Manage. **131**, 228–238 (2013)
Miller, D.P., Moslemi, A.A.: Wood-cement composites: effect of model compounds on hydration characteristics and tensile strength. Wood Fibre Sci. **23**(4), 472–482 (1991)
Milne, T.A., Brennan, A.H., Glenn, B.H.: Sourcebook of methods of analysis for biomass and biomass-conversion process. Solar Technical Information Program, U.S. Department of Energy (1990)
Mohanty, A.K., Misra, M., Drzal, L.T.: Natural Fibers, Biopolymers, and Biocomposites. CRC Press (2005)
Mohnen, D.: Pectin structure and biosynthesis. Curr. Opin. Plant Biol. **11**(3), 266–277 (2008)
Monreal, P., Mboumba-Mamboundou, L.B., Dheilly, R.M., Quéneudec, M.: Effects of aggregate coating on the hygral properties of lignocellulosic composites. Cement Concr. Compos. **33**, 301–308 (2011)
Na, B., Wang, Z., Wang, H., Lu, X.: Wood-cement compatibility review. Wood Research **59**(5), 813–826 (2014)
Nozahic, V., Amziane, S.: Influence of sunflower aggregates surface treatments on physical properties and adhesion with a mineral binder. Composite: Part A **43**, 1837–1849 (2012)
Nozahic, V.: Vers une nouvelle démarche de conception des bétons de végétaux lignocellulosiques basée sur la compréhension et l'amélioration de l'interface liant/végétal. PhD thesis, Université Blaise Pascal, Clermont-Ferrand, France (2012)
Okino, E.Y.A., Souza, M.R., Santan, M.A.E., Alves, M.V., Sousa, M.E., Texeira, D.E.: Physico-mechanical properties and decay resistance of Cupressus spp. cement-bonded particleboards. Cement Concr. Compos. **27**, 333–338 (2005)
Olorunnisola, A.O.: Effects of pre-treatment of rattan (*Laccosperma secundiflorum*) on the hydration of Portland cement and the development of a new compatibility index. Cement Concr. Compos. **30**(1), 37–43 (2008)
Pehanich, J.L., Blankenhorn, P.R., Silsbee, M.R.: Wood fiber surface treatment level effects on selected mechanical properties of wood fiber-cement composites. Cem. Concr. Res. **34**(1), 59–65 (2004)
Peschard, A., Govin, A., Grosseau, P., Guilhot, B., Guyonnet, R.: Effect of polysaccharides on the hydration of cement paste at early ages. Cem. Concr. Res. **34**, 2153–2158 (2004)
Picandet, V.: Characterization of plant-based aggregates. In: Amziane, S., Arnaud, L. (eds.) Bio-aggregate-based Building Materials—Application to Hemp Concretes. Civil Engineering and Geomechanics series, pp. 75–116. ISTE and WILEY (2013). ISBN: 978-1-84821-404-0
Ratiarisora, V., Magniont, C., Ginestet, S., Oms, C., Escadeillas, G.: Impact of lavender bio-aggregates on the setting and hardening of a pozzolanic binder. 8th International Symposium on Cement Based Materials for a Sustainable Agriculture, Iasi, Romania, October 22–25th (2015)
Rowell, R.M., Pettersen, R., Han, J.S., Rowell, J.S., Tshabalala, M.A.: Chapter 3. Cell wall chemistry. In: Rowel, R.M. (ed.). Handbook of Wood Chemistry and Wood Composites. (2005). ISBN 0-8493-1588-3

Saeman, J.F., Bubl, J.L., Harris, E.E.: Quantitative saccharification of wood and cellulose. Ind. Eng. Chem. Anal. Ed. **17**(1), 35–37 (1945)

Saeman, J.F., Moore, W.E., Mitchell, R.L., Millet, M.A.: Techniques for the determination of pulp constituents by quantitative paper chromatography. Tappi J. **37**(8), 336–343 (1954)

Sandermann, W., Preusser, H.J., Schwiens, W.: The effects of wood extractives on the setting of cement-bonded wood materials. Holzforschung **14**(3), 70–77 (1960)

Sanderson, M.A., Foster, A., Collins, M., Johns, D.K.: Compositional analysis of biomass feedstocks by near infrared reflectance spectroscopy. Biomass Bioenergy **11**(5), 361–370 (1996)

Sedan, D.: Etude des interactions physico-chimiques aux interfaces fibres de chanvre/ciment. Influence sur les propriétés mécaniques du composite. PhD thesis, Université de Limoges. Groupe d'étude des matériaux hétérogènes (2007)

Sedan, D., Pagnoux, C., Chotard, T., Smith, A., Lejolly, D., Gloaguen, V., Krausz, P.: Effect of calcium rich and alkaline solutions on the chemical behavior of hemp fibre. J. Mater. Sci. **42**, 9336–9342 (2007b)

Sedan, D., Pagnoux, C., Smith, A., Chotard, T.: Mechanical properties of hemp fibre reinforced cement: Influence of the fibre/matrix interaction. J. Eur. Ceram. Soc. **28**, 183–192 (2008)

Semple, K.E., Evans, P.D.: Compatibility of some Australian acacias with Portland cement. Holz Roh Werkstoff **56**(1), 24 (1998)

Semple, K.E., Evans, P.D.: Screening inorganic additives for ameliorating the inhibition of hydration of Portland cement by the Heartwood of *Acacia mangium.* In: Wood-cement Composites in the Asia-Pacific Region. Proceedings of a workshop held at Rydges Hotel, Canberra, Australia, 10 December 2000, pp. 23–39 (2002)

Semple, K.E., Cunningham, R.B., Evans, P.D.: Manufacture of wood-cement composites from Acacia mangium: mechanistic study of compounds improving the compatibility of Acacia mangium heartwood with Portland cement. Wood Fiber Sci. **39**(1), 120–131 (2004)

Simatupang, M.H., Handayani, S.A.: Fermentation of saw dust from freshly cut rubber wood to improve its cement compatibility. Holz als Roh- und Werkstoff **59**, 27–28 (2001)

Batra, S.K.: Other long vegetable fibers. In: Lewin, M., Pearce, E.M. (ed.). Handbook of fibre chemistry. Handbook of fibre Science and Technology. Marcel Dekker, Volume IV, Fibre Chemistry pp. 505–575. New York (1998)

Sluiter, J.B., Ruiz, R.O., Scarlata, C.J., Sluiter, A.D., Templeton, D.W.: Compositional analysis of lignocellulosic feedstocks. 1. Review and description of methods. J. Agric. Food Chem. **58**, 9043–9053 (2010)

Soroushian, P., Hassan, M.: Evaluation of cement-bonded strawboard against alternative cement-based siding products. Constr. Build. Mater. **34**, 77–82 (2012)

Soroushian, P., Shah, Z., Won, J.P., Hsu, J.W.: Durability and moisture sensitivity of recycled wastepaper-fiber-cement composites. Cement Concr. Compos. **16**, 115–128 (1994)

Soroushian, P., Aouadi, F., Chowdhury, H., Nossoni, A., Sarwar, G.: Cement-bonded straw board subjected to accelerated processing. Cement Concr. Compos. **26**, 797–802 (2004)

Templeton, D.W., Scarlata, C.J., Sluiter, J.B., Wolfrum, E.J.: Compositional analysis of lignocellulosic feedstocks. 2. Method uncertainties. J. Agric. Food Chem. **58**, 9054–9062 (2010)

Thomas, N.L., Birchall, J.D.: The retarding action of sugars on cement hydration. Cem. Concr. Res. **13**, 830–842 (1983)

Thomsen, A.B., Rasmussen, S., Bohn, V., Vad Nielsen, K., Thygesen, A.: Hemp raw materials: the effect of cultivar, growth conditions and pretreatment on the chemical composition of the fibres. Department: Biosystems Department, Risø-R-1507 (2005). ISSN 0106-2840 ISBN 87-550-3419-5

Thygesen, A., Liu, M., Meyer, A.S., Daniel, G.: Hemp fibres: enzymatic effect of microbial processing on fibre bundle structure. In: Madsen, B., Lilhlt, H., Kusano, Y., Faester, S., Ralph, B. (eds.). Proceedings of the 34th Riso International Symposium on Material Science. Department of Wind Energy, Riso Campus, Technical University of Denmark (2013)

Thygesen, A., Daniel, G., Lilholt, H., Thomsen, A.B.: Hemp fiber microstructure and use of fungal defibration to obtain fibers for composite materials. J. Nat. Fibers **2**, 19–37 (2005)

Thygesen, A., Thomsen, A.B., Daniel, G., Lilholt, H.: Comparison of composites made from fungal defibrated hemp with composites of traditional hemp yarn. Ind. Crops Prod. **25**, 147–159 (2007)

Toledo Filho, R.D., de Andrada Silva, F., Fairbain, E.M.R., De Almeida Melo Filho, J.: Durability of compression molded sisal fiber reinforced mortar laminates, Constr. Build. Mater. **23**, 2409–2420 (2009)

Toledo Filho, R.D., Ghavami, K., England, G.L., Scrivener, K.: Development of vegetable fibre-mortar composites of improved durability. Cement Concr. Compos. **25**, 185–196 (2003)

Tonoli, G.H.D., Rodrigues Filho, U.P., Savastano Jr., H., Bras, J., Belgacem, M.N., Rocco Lahr, F.A.: Cellulose modified fibres in cement based composites. Compos. Part A **40**, 2046–2053 (2009)

Vaickelionis, G., Vaickelioniene, R.: Cement hydration in presence of wood extractives and pozzolan mineral additives. Ceramics Silikaty **50**(2), 115–122 (2006)

Van der Werf, H.: Crop physiology of fibre hemp (Cannabis sativa L.), 1994 Netherlands. CEP—Gegevens Koninklijke Bibliotheek, Den Haag (1994)

Van Soest, P.J., Wine, R.H.: Determination of lignin and cellulose in acid-detergent fiber with permanganate. J. Assoc. Agric. Chem. **51**(4), 780–785 (1968)

Vassilev, S., Baxter, D., Andersen, L., Vassileva, C.: An overview of the chemical composition of biomass. Fuel **89**, 913–933 (2010)

Vassilev, S.V., Baxter, D., Andersen, L.K., Vassileva, C.G., Morgan, T.J.: An overview of the organic and inorganic phase composition of biomass. Fuel **94**, 1–33 (2012)

Vignon, M.R., Garcia-Jaldon, C., Dupeyre, D.: Steam explosion of woody hemp chènevotte. Int. J. Biomacromolecules **17**, 395–404 (1995)

Vignon, M.R., Dupeyre, D., Garcia-Jaldon, C.: Morphological characterization of steam exploded hemp fibers and their utilization in polypropylene-based composites. Bioresour. Technol. **58**, 203–215 (1996)

Weatherwax, R.C., Tarkow, H.: Effect of wood on setting of Portland cement. Forest Prod. J. **14** (12), 567–570 (1964)

Wei, J., Meyer, C.: Degradation mechanisms of natural fiber in the matrix of cement composites. Cem. Concr. Res. **73**, 1–16 (2015a)

Wei, J., Meyer, C.: Degradation mechanisms of natural fiber in the matrix of cement composites. Cem. Concr. Res. **73**, 1–16 (2015b)

Wei, Y.M., Zhou, Y.G., Tomita, B.: Hydration behavior of wood cement-based composite I: evaluation of wood species effect on compatibility and strength with ordinary Portland cement. J. Wood Sci. **46**, 296–302 (2000a)

Wei, Y.M., Zhou, Y.G., Tomita, B.: Study of hydration behaviour of wood cement-based composite II: effect of chemical additives on the hydration characteristics and strengths of wood-cement composites. J. Wood Sci. **46**, 444–451 (2000b)

Wei, Y.M., Tomita, B., Hiramatsu, Y., Miyatake, A., Fujii, T., Fujii, T., Yoshinaga, S.: Hydration behaviour and compressive strength of cement mixed with exploded wood fiber strand obtained by the water-vapor process. J. Wood Sci. **49**, 317–326 (2003)

Yasuda, S., Ima, K., Matsushita, Y.: Manufacture of wood-cement boards. VII: Cement-hardening inhibitory compounds of hannoki (Japanese alder, *Alnus japonica* Steud.). J. Wood Sci. **48**(3), 242–244 (2002)

Yin, Z., Pan, Z., Wang, C., Dong, Y., Ou, Y.: Composition, structure and mechanical properties of several natural cellular materials. Chin. Sci. Bull. **52**, 2903–2908 (2007)

Chapter 2
Porosity, Pore Size Distribution, Micro-structure

Mike Lawrence and Yunhong Jiang

Abstract The high porosity and microstructure of bio-aggregates are fundamental to their physical properties. Typically they have a low density and a complex pore structure. This has two principal effects. In the first instance, low density is associated with low strength, but also with low thermal conductivity. For this reason most bio-aggregates are not suitable for use as structural materials, but are eminently suited to act as a low density filler in composite materials conferring low thermal conductivity on the resulting bio-composite. The complex nature of their porosity results in a material that is able to readily adsorb moisture and humidity. This results in a material that has an exceptionally high moisture buffering capacity, a characteristic that is of great interest in building materials, because it tends to stabilise the internal environment of a building, thereby resulting in a much more healthy indoor environment. This chapter considers the range of methods that can be used to measure porosity and to characterise the microstructure of materials in general, and discusses how some of these techniques have been used on bio-aggregates. It also identifies opportunities to use novel techniques on bio-aggregates in order to improve our understanding of their porosity, pore size distribution, pore connectivity and microstructure, all of which are characteristics that are essential to the optimisation of the performance of bio-aggregates within the construction industry.

Keywords Porosity · Microstructure · Physisorption · Mercury Intrusion Porosimetry · Scanning Electron Microscopy · X-Ray Computed Tomography

---

M. Lawrence (✉) · Y. Jiang
BRE Centre for Innovative Construction Materials, University of Bath,
Bath BA2 7AY, UK
e-mail: m.lawrence@bath.ac.uk

S. Amziane and F. Collet (eds.), *Bio-aggregates Based Building Materials*,
RILEM State-of-the-Art Reports 23, DOI 10.1007/978-94-024-1031-0_2

2.1 Introduction

Porosity is a measure of the void fraction in a material. Voids can either be 'closed', and inaccessible or 'open' and connected to other voids and thence to the exterior of the material. The total porosity (ϕ) is defined by the ratio of the volume of void space (V_V) to the total, or bulk volume of the material (V_T):

$$\phi = \frac{V_V}{V_T}$$

Porosity can also be expressed as a percentage of the bulk volume of the material. In bio-aggregates the pores are formed during the growth phase of the plant and in the living plant they have the function of nutrient storage and transmission of water and nutrients from the roots to the rest of the plant. Plant stems have a low density of between 110 and 120 $kg.m^{-3}$ and a high porosity of the order of 70–80%. The specific properties of bio-aggregates, such as low density, high porosity and the complexity of pore size and pore structure, result in a material that has a low thermal conductivity, and which is highly absorbent, capable of absorbing up to four times its own weight in water. Once the shiv has been dried out it has a high moisture buffer value, conferring on it the ability to passively manage internal humidity levels when used as a building material (Latif et al. 2015).

Porosity is a rather easy parameter to define, but certainly not so easy to quantify. The reason is that the void/space in bio-aggregates can span from few nanometres to centimetres or larger. There is really no one method that can adequately cover this enormous range in scale. In addition, the porosity can be modified or changed by a variety of processes during the test such as deformation, hydrothermal alteration and producing secondary or fracture porosity. Finally, the pore shape and connection structure (open and closed) have a significant effect on the porosity results depending on the testing approach (Bismarck 2002; Brewer 2014; Chundawat 2011; Collet 2008; Donato 2012; Hamdi 2015; Dougal et al. 2006). The aims of this chapter is to summarize many of the available techniques that can be used to analyse the porosity of bio-aggregate materials. Manger et al. (1963) concluded that most of the total porosity measurements are variations on bulk volume/grain volume or bulk density/grain density approaches, and the apparent porosity measurements are made by variations of absorption methods for different fluids or gases. Anovitz et al. (2015) summarized 10 methods for measuring the porosity and pore size distribution (PSD) used on core or crushed rock materials (Fig. 2.1). This shows the range of pore sizes that each method is capable of measuring. It should be kept in mind that different techniques are based on different principles and have different capability for measurement. Depending on the natural properties of bio-aggregates, there is no best approach to determine their

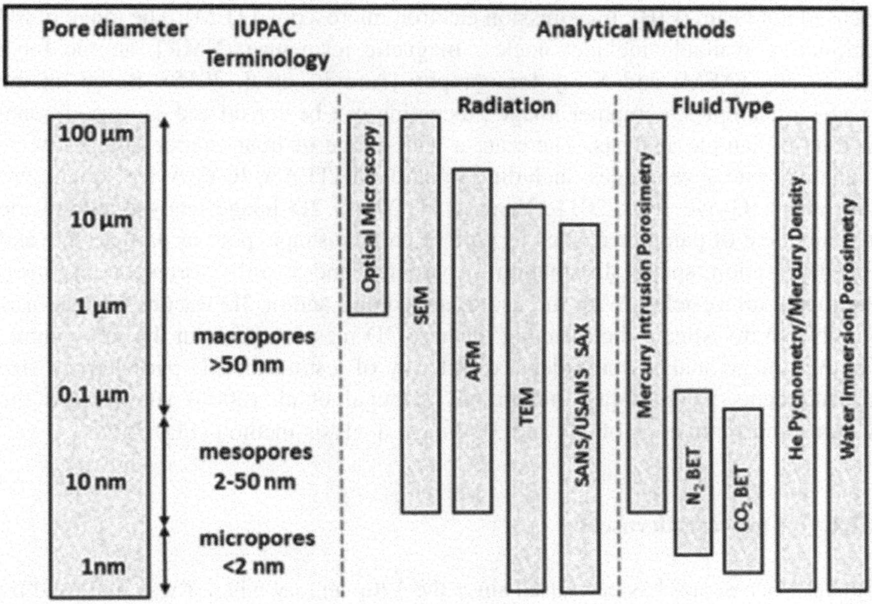

Fig. 2.1 Methods used to determine porosity and pore size distribution (PSD) (Anovitz and Cole 2015)

porosity. The combination of several techniques and comparing the results of pore structure from different methods may gain an insight into the complex pore system of bio-aggregates.

2.2 Techniques Used to Measure Porosity

2.2.1 Imaging Methods

With the development of digital images and computer software, image processing is a new and convenient method which is able to determine the pore size, pore size distribution, porosity and microstructure of bio-aggregates. There has been much progress in materials science, biology, and geology regarding the application of image analyse. It has become an important technique for the investigation of the porosity and particle size of bio-aggregates and bio-composites (Mermut et al. 2009; Ziel et al. 2008; Nimmo et al. 2004; Aydilek et al. 2002; Shen 2015). A broad range of imaging methods are available to describe the nature of porosity in bio-aggregates materials. 2D techniques that can be used include: optical light microscopy (OM), scanning electron microscopy (SEM) with energy dispersive X-ray spectroscopy (EDX), field emission scanning electron microscopy (FESEM), focused ion beam (FIB), transmission electron microscopy (TEM). The range of 3D techniques available include: nuclear magnetic resonance (NMR), atomic force microscopy (AFM) and X-ray tomography (Anovitz et al. 2015). Based on the images of samples, computer image processing can be considered as an additional method of sample analysis. There are a wide range of image processing software available to analyse images, including ImageJ, MATLAB, ICY, Avizo, Image Pro and others (Grove et al. 2011; Yang et al. 2014). 2D image analysis allows the measurement of parameters such as pore or particle shape, pore or particle size and size distribution, spatial distribution of particles, and also the corresponding measurements for vesicles. With the use of stereology and/or 3D texture models, it is possible to investigate the meaning of these 2D measurements in the 3D volume. The 3D image analysis provides a direct way of testing the 3D particle/pore size measurements using X-ray tomography. Dougal et al. (2006) summarized the features and limitations of 2D and 3D image analysis method (Fig. 2.2).

2.2.1.1 Optical Microscopy

Optical Microscopy has been used since the 17th century when it was first used by Robert Hooke to describe '…minute bodies made by magnifying glasses with observations and inquiries thereupon.' Since that time the optical microscope has been refined to produce the range of modern research microscopes used in laboratories today.

Transmitted Light Microscopy uses light that is transmitted from a source on the opposite side of the specimen to the objective lens. Normally the light is passed through a condenser to focus it on the specimen in order to maximise the amount of light available. The optimum set-up for specimen illumination and image generation is known as Köhler illumination after the man who invented it. It is used for most of the optical configurations listed below. The microscope techniques requiring a transmitted light path include bright field, dark field, phase contrast, polarisation and differential interference contrast optics. Transmitted light microscopy relies on preparing samples that are sufficiently thin to allow the passage of light. Some materials remain opaque even when ground to a thickness of 30 μm, and for this reason reflected light microscopes have been developed. The sample is often polished to a high degree in order to allow all the features to be seen in the same plane, and therefore remain in focus. Samples are illuminated from above through the objective.

Bright Field Microscopy is the most frequently used technique where no optical contrast methods are used. It uses transmitted light to view a specimen that contains inherent contrast or where the specimen has been stained to improve contrast. Figure 2.3 shows a transverse section of Hemp shiv that has been set in a blue stained resin under a ×10 magnification and Fig. 2.4 shows the central section of the same specimen under ×40 magnification.

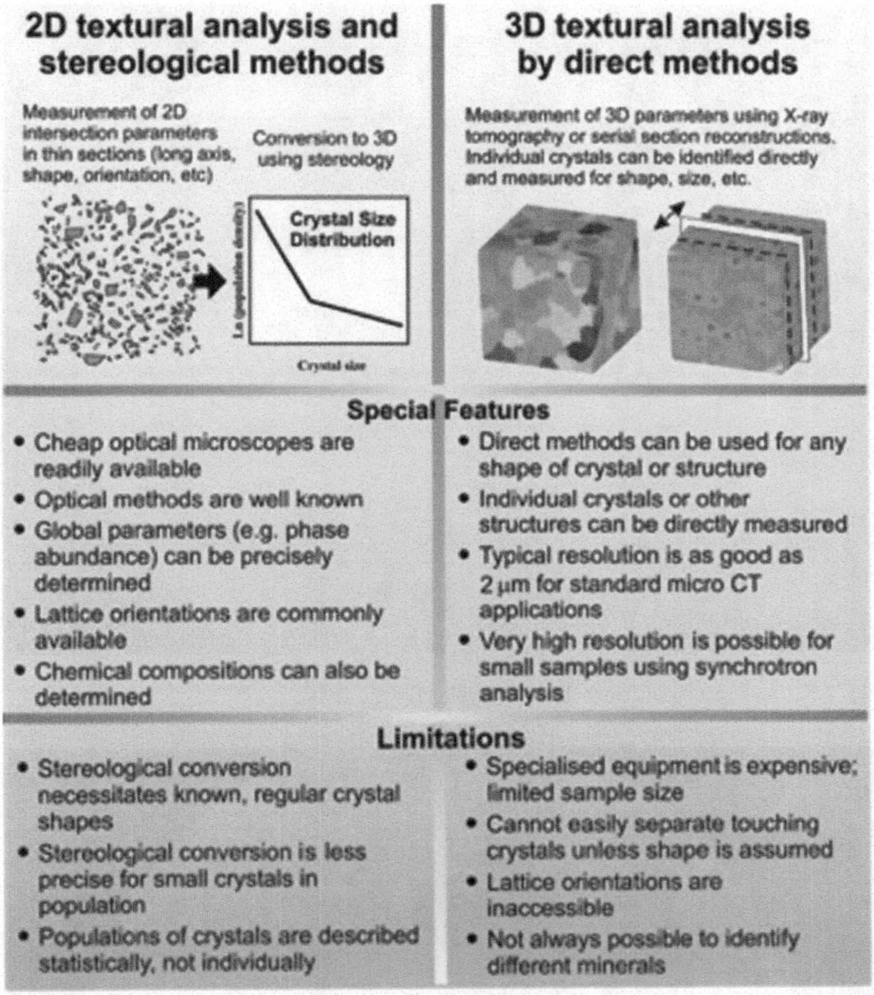

Fig. 2.2 Comparison of the imaging analysis approaches in 2D and 3D analysis (Dougal et al. 2006)

Dark Field Microscopy uses oblique illumination, and is used for the detection of micro-organisms in unstained smear preparations and diatom studies. Phase contrast exploits the phenomenon that light slows slightly when passing through biological specimens. When used in conjunction with phase contrast objective lenses which contain a corresponding phase plate, degrees of constructive and destructive interference occur which produce the characteristic light and dark features in the image.

Polarised Light Microscopy uses plane polarised light to analyse structures that are birefringent, which is to say structures that have two different refractive indices

Fig. 2.3 Cross-section of hemp shiv ×10 (author)

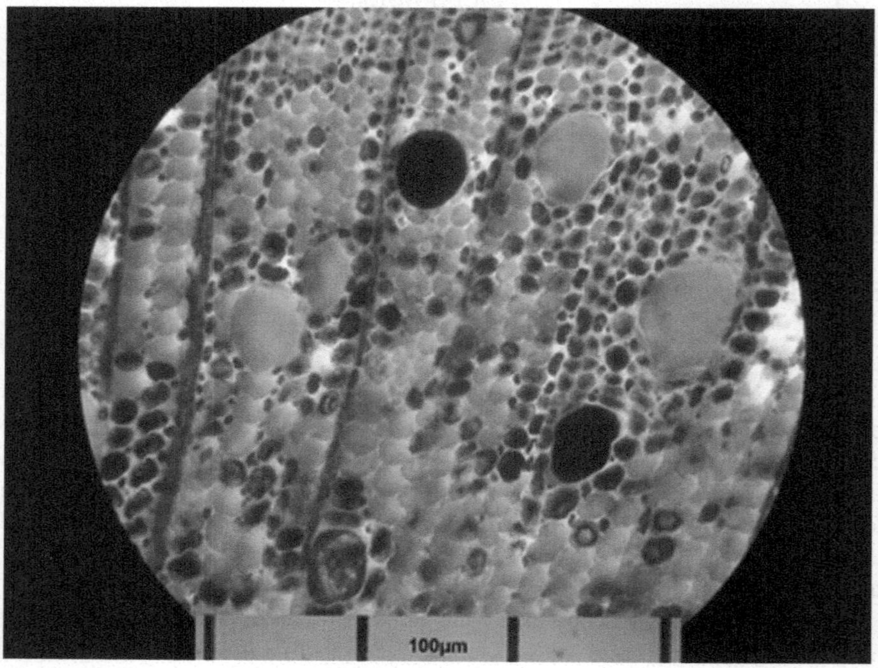

Fig. 2.4 Cross-section of hemp shiv ×40 (author)

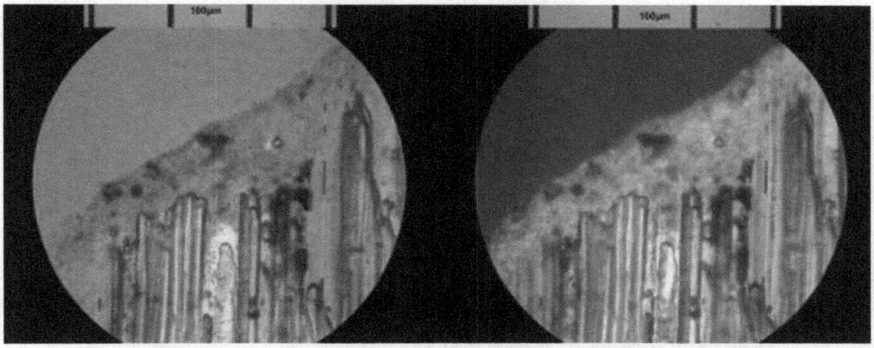

Fig. 2.5 Transverse section of hemp-lime in light polarised at 0° (*left*) and 90° (*right*) (author)

at right angles to each other (e.g. calcite). Figure 2.5 shows a transverse section of hemp-lime in light polarised at 0° (left) and 90° (right). This clearly shows the coating of lime binder around the bio-aggregate and demonstrates that the binder has only limited penetration into the capillaries of the material.

Differential Interference Contrast Microscopy is a more complex form of polarised light microscopy involving two slightly separate plane polarised beams of light to create a 3D-like image with shade of grey.

Stevulova et al. (2014) used reflected light microscopy to examine the surface morphology of hemp shiv and the impact of different surface treatments (Fig. 2.6).

2.2.1.2 Scanning Electron Microscopy

Scanning Electron Microscopy (SEM) is one of the most popular imaging techniques. Porosity can be measured by image analysis based on scanning electron microscopy incorporating digital image processing. In addition, the dimension, shape and the number of pores in bio-aggregates and bio-composites can be inspected by image processing analysis. The image analysis was developed using various mathematical morphology algorithms to provide a complete pore size distribution (PSD) curve for each sample. The main image processing tasks are sample preparation, specimen scanning process, image enhancement, pixel classification, and pixel clustering (Kaestner et al. 2008; Dougal et al. 2007). The key to be able to perform accurate digital porosity measurements is the ability to generate a porosity threshold image (one which separates the porosity voids from the rest of the objects in the image). The signals of the secondary electrons gives information on the surface topography. The backscattered electrons (BSE) gives complementary information of the chemical composition of the sample surface. High atomic number elements backscatter electrons more strongly than low atomic number elements, and thus appear brighter in an image. BSE are used to detect contrast

Fig. 2.6 Micrographs of original sample of hemp hurds (**a**) and chemically modified by NaOH (**b**), EDTA (**c**) and Ca(OH)2 (**d**) (40× magnification) (Stevulova et al. 2014)

between areas on the sample surface with different chemical compositions. In a backscattering electron image, the relative difference between signals, known as the contrast, C, can be obtained from the following equation:

$$C = \frac{\eta_1 - \eta_2}{\eta_1}$$

In which η_1 and η_2 are, respectively, the backscattering coefficients for high and low density materials. The minimum value of C in an image is defined by the threshold equation:

$$i_B > \frac{4\times10^{-12}}{\varepsilon C^2 t_f}$$

In which i_B is the minimum beam current required to provide contrast C, t_f is the time required to scan a 1000 × 1000 pixel frame (typical SEM pixel density for a photomicrograph), and ε is the detector "efficiency" (ratio of signal current to beam current). The ultimate resolution available with SEM is on the order of 0.5 μm (Zhao et al. 1992).

There are many factors affecting the calculated results, such as the thin section thickness, threshold value, and pore circularity (Anovitz et al. 2015; Marinello et al. 2008). Poor quality data can arise from the introduction of noise and inadequate or overzealous pre-processing methods, increasing user bias during thresholding. The kind of sample preparation required of the sample depends on whether it is electrically conducting or not. Non-conductive samples must first be sputter coated with an ultra-thin coating of an electrically-conducting material before imaging. Otherwise, samples will tend to charge when scanned by the electron beam leading to scanning faults and other image artifacts. The advantages of the SEM over optical petrography are greater depth of field and resolution, and a significantly higher magnification range. It can provide direct and detailed structural information including the shape and size of individual pore inside the bio-aggregates. The assumptions about pore shape are not made, but rather images capture this information directly. The disadvantage of SEM is that the obtained images are exactly two-dimensional grey scale image. The SEM images showed partly the inner structure of the samples and they cannot be considered completely as sections. It underestimates the pore radius due to only working part of the pore. However, Image-based analyses have been used to extract meaningful quantities which characterize pore structure and describe several spatial characteristics of porosity. Mathematical correction techniques are necessary to estimate unbiased pore body and opening sizes. Sizing techniques, a two-point correlation technique, and fractal analysis can be used to analysis the individual pore size, shape and distribution (Zhao et al. 1992).

During the past ten years, SEM image analysis has become an important tool for microscopic studies of bio-aggregates based materials. SEM images have been used to study aspects of bio-aggregates porosity by several researchers. Walker and Pavia (2014) studied the microstructure and pore size of hemp lime concrete using FESEM. The results showed that the morphology of hemp interface changed over time, from predominantly needle-shape at early ages to sponge and gel types at later ages. A wide distribution of pore size was evident in all binders. Most pores ranged from 200 to 2000 nm. Lubelli et al. (2013) studied porosity and pore size distribution of a wet poultice by using a FIB-SEM with a cold stage and MIP. The incremental and cumulative pore diameter distribution obtained using cryo-SEM image method showed the prevalence of pore radii in the range of 100 nm. The total pore fraction varies between 26 and 44% and the mean pore radius between 110 and 160 nm. The total pore area obtained by image analysis was 44%, which is significantly smaller than the total pore volume measured by MIP (57.93%). This is might be due to the absence of pores diameter larger than 1.2 μm in the studied cryo-SEM images (Lubelli et al. 2013). Chundawat et al. (2011) studied the shape, size (10–1000 nm), and spatial distribution of the pores on their location within the cell wall and the cell wall volume, ranged between 0.005 and 0.05 nm^2 per nm^3 by using TEM-image analysis. Figure 2.7 showed how water was likely incorporated in additional hydrates resulting in weight gain (7 wt%) of the hemp-lime concrete and abundant needle shaped hydrates are growing into the pores suggesting a reduction in pore size (Walker 2014).

There are a large number of papers which discuss SEM image analysis approaches to the analysis of the pore size distribution and pore shape of soil and membrane materials. However, there only few papers that use the SEM image analysis method to study porosity and pore distribution on bio-aggregates. Sassoni et al. (2014) have used SEM to visualise the microstructure of bio-aggregate composites (Fig. 2.8), but did not use the information to analyse the pore structure or pore size distribution.

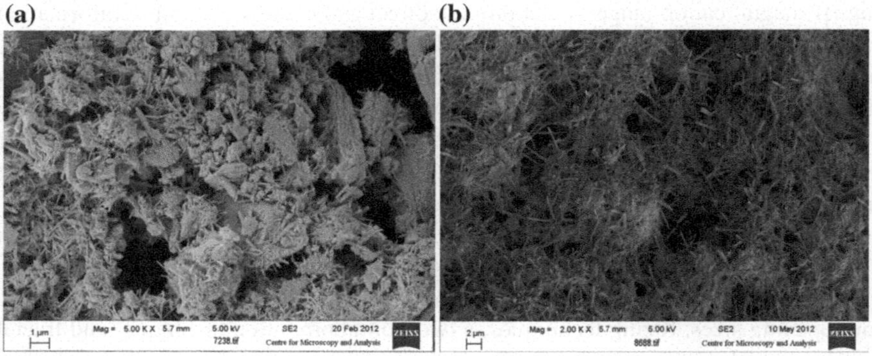

Fig. 2.7 Hydrates in hemp lime concrete made with commercial binder (**a**) and in-creased quantity of hydrates in the binder of hemp concrete saturated with water for 2 weeks (Walker et al. 2014)

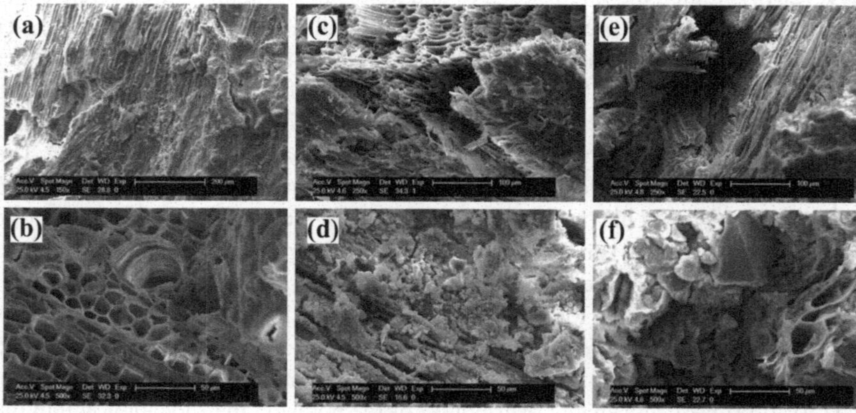

Fig. 2.8 Visualisation of the microstructure of low density (**a** and **b**), medium density (**c** and **d**) and high density (**e** and **f**) hemp shiv-magnesium oxide composites (Sassoni et al. 2014)

The SEM image analysis methods demonstrate the ability to quantify the nature of porosity in complex heterogeneous matrices. In the bio-aggregates, there is a hierarchy of pore types that range from micro-pore to organic meso-pore to fracture macro-pores. For future experimental work, SEM image analysis approach will be an important technique for analysing the porosity of bio-aggregate materials.

2.2.1.3 X-Ray Computed Tomography

X-ray computed tomography (XRCT) has been used to characterize the cellular microstructure and porosity of bio-aggregates materials. This method can produce the non-destructive and three-dimensional images to quantify the micro-structure such as pore size distribution, porosity and tortuosity of the porous network. X-ray radiography physics is based on the Beer-Lambert law (Maire et al. 2001; Dougal et al. 2007; Roche et al. 2010). Pore are distinguished in X-XRCT on the basis of their linear attenuation coefficient, μ. This parameter depends on the electron density of the sample, the effective atomic number of the sample, and the energy of the incoming X-ray beam. XRCT comprises an X-ray source, a rotation state on which the object is fixed, an X-ray detector and a reconstruction software (Dougal et al. 2007). Construction of a 3D map of a specimen, several X-ray radiographs of the sample are recorded at different viewing angles. This imaging information is then used in reconstruction software to recalculate the 3D map of the attenuation from the combination of the obtained radiographs. The Fig. 2.9 shows the stages from image capture through to processing and finally to reconstructed 3D virtual texture ready for quantification. The crucial point in applying tomography to materials science is the achievable spatial resolution. For a limit resolution of the order of 8 μm, a cone-beam system can be used with a classical micro-focus X-ray tube as the source. Most recent lab systems routinely achieve resolution as low as 5 μm. Larger samples have a lower resolution, limited by the number of pixels. The best quality images in terms of signal-to-noise ratio and spatial resolution allowing high-resolution micro-tomography are obtained on synchrotron radiation (Maire et al. 2012).

There are several disadvantages about XRCT porosity analysis. These include: 3D methods are not always applicable if samples are too large or small; it is very expensive and cannot distinguish touching grains of the same mineral; nor can it clearly separate different minerals with very similar properties.

Mostefai et al. (2015) investigated microstructural effects of hemp fibre and shiv addition on modified mortars by using X-ray tomography. Results showed that porosity level in hemp shiv modified mortar with regard to the weight content of 2 wt%, is 5.08%; whereas, porosity in the case of hemp fibre modified mortar is 3.54%. Both of hemp filler play a significant role in triggering the porosity content. The porosity of wood bark insulation board was studied by using XRCT

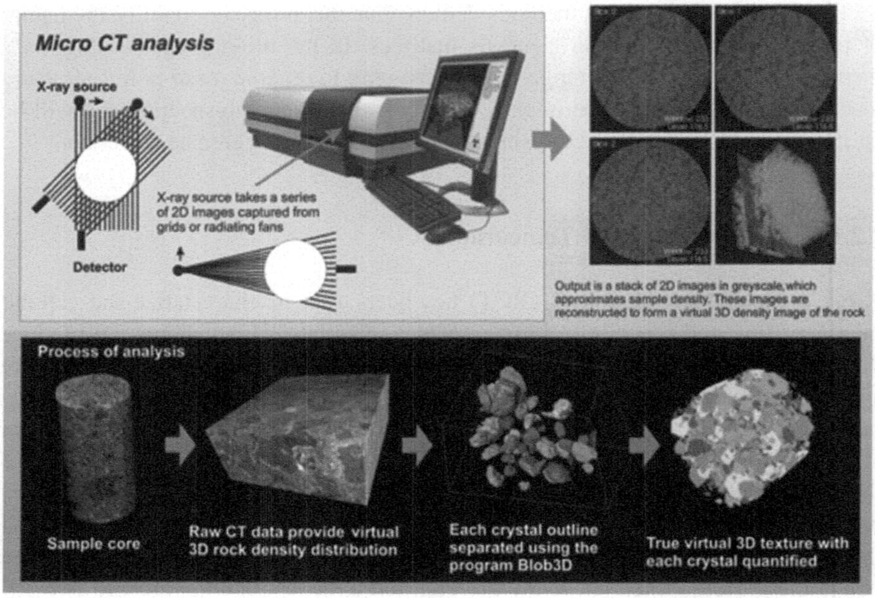

Fig. 2.9 Measuring 3D textures using XRCT techniques. X-rays are passed through the sample to procude a series of 2D images that are spaced close to each other (typically at resolutions of 7–30 μm). Sequential images are captured by rotating the sample or source. Using software such as Blob3D, the individual elements of the texture, such as crystals, can be sampled and quantified (Dougal et al. 2007)

(Fig. 2.10). The results showed that pore size distribution is clearly influenced by the panel density and the small pores (<1 mm^2) are predominantly pores within the bark materials itself. In addition, larger pores (>1 mm^2) are void due to imperfect stacking during the press process.

Pores smaller than 1 mm^2 account for between 8 and 30% of the pore area (Kain et al. 2015). William et al. (2015) studied the volumetric ratios change of visible air voids related to consolidation process in the level of macro scale by using XRCT. The results showed that the ratio of macro scale air voids to micro scale air voids changed significantly during the consolidating process. Hamdi et al. (2015) studied and compared the X-ray computed tomography and 2D image analysis on lignocellulosic fibres raw materials. The strengths and drawbacks of the applied imaging methods on lignocellulosic materials are listed in Table 2.1.

Tran et al. (2015) studied the porosity of coir fibre by using SEM image analysis and XRCT analysis. The results from SEM methods with software Leica QWin showed that the fibre porosity is in the range from 22 to 30% (Fig. 2.11). In reality, the lumen of each elementary fibre is not a cylinder. In this case, the volume of lumens can be underestimated when a smaller cross section is analysed. The porosity of the coir fibres using XRCT methods, ranges from 27 to 40%. The fibre

Fig. 2.10 **a** X-ray μ-tomography images of hemp fibre (MHFA) and shiv (MS22) modi-fied mortars. Representative dimensions are 26 Å ~ 22 Å ~ 21 mm^3 and 37 Å ~ 38 Å ~ 22 mm^3 for MHFA and MS22, respectively. **b** In-depth views showing porosity and hemp phase arrangement. Crop volumes are 24 Å ~ 21 Å ~ 16 mm^3 and 38 Å ~ 38 Å ~ 19 mm^3 for MHFA and MS22, respectively (Mostefai et al. 2015)

porosity analysed with this method is likely to be overestimated because the method is based on the densitometry principle, and there is only a small difference in density between coir fibres and air. In addition, coir fibres consist of various thin organic tissues, which may not be detected on the scanned images. The author concluded that Image analysis on SEM pictures will give a better estimation of the porosity of coir fibres.

In summary, XRCT offers a good tool to study the porosity and generally the internal structure of bio-aggregates. Based on the above discussion, the bio-aggregate porosity will be better estimated by SEM image analysis in a small level of scale (<1 μm). XRCT will give a better results on the porosity of bio-aggregate above 1–3 μm. Combining these two methods will improve accuracy of pore-size distributions and porosity for bio-aggregates.

Table 2.1 Characteristics of the X-ray CT and 2D scanning methods for lignocellulosic fibre size estimation

Imaging technique	Strengths	Drawbacks
X-ray CT	– Suitable for the detection of small fiber entities – High level of detail and ability to internally observe the structures – Reliable fiber size estimation for regular shapes with low rates of porosity and small amounts of large bundles – Does not reject fiber size estimation of hollow fibers (which can be filled)	– Fibers with a length greater than the maximum 3D Spatial resolution of X-ray microtomograph cannot be assessed – Fibers with a diameter smaller than the minimum 3D Spatial resolution of the X-ray microtograpgh cannot be assessed – Underestimates the fiber size distribution of a heterogeneous fiber elements
2D scanning	– Reliable fiber size estimation for regular fiber shapes with low rates of porosity and small amounts of large bundles – Relatively easy implementation and high reproductiblity	– Exponentially overestimates the fiber size distribution of heterogeneous fiber population with a high porosity level – Underestimates the fiber length distribution of heterogeneous fiber network with high rates of tortuosity – The fibers need to be manually spread over the scanner surface to prevent inter-fiber contacts

Hamdi et al. (2015)

Fig. 2.11 Image analysis to measure the porous area of the fibre cross-section using the software Leica QWin (Tran et al. 2015)

2.2.1.4 Nuclear Magnetic Resonance

NMR Spectroscopy has been used commercially in a variety of disciplines, ranging from oil exploration to food technology, to examine the moisture content of materials. The hydrogen atoms within the material, which is exposed to a large magnetic field, are excited by a pulse of radio waves, and subsequently relax back to their normal state releasing a characteristic signal. Measurement of the relaxation signal (T_1) allows the number of hydrogen atoms present and, with appropriate calibration for the material, the absolute water content to be evaluated. An additional feature of NMR is its ability to measure the amount of moisture in different physical states (i.e. chemically bound, physically bound and free liquid). Because the technique is tuned to excite the hydrogen nucleus, water molecules are particularly responsive to the technique, however, care is needed where polymer-based or organic materials are present as their hydrogen content may confuse results. The technique requires experience and sophisticated analysis to fulfil its potential performance (Phillipson et al. 2007).

Faure et al. (2012) investigated water transfer in hemp-lime using NMR. The relaxation time T_1 is known to be a probe of water mobility which depends on the water interaction with its environmental conditions: ions or paramagnetic components, media rigidity (for example due to polymer presence), pore size, etc. In complex media such as civil engineering materials, water present in the sample may experience different local environment: it may be confined in pores of different sizes, or may be physically or chemically trapped in specific microstructures. T_1 decreases with the pore size, and other more subtle relationships may be found between the porous structure and relaxation times of water inside. This work focused on the development of pore structure over time due to the hydration of the binder, and it is evident that the technique is effective as a method of measuring pore size distribution in bio-aggregate composites. Figure 2.12 shows the distribution of water in water-hemp mixtures with differing water to hemp mass ratios, and clearly shows the typical bi-modal pore size distribution that is known to occur in hemp shiv.

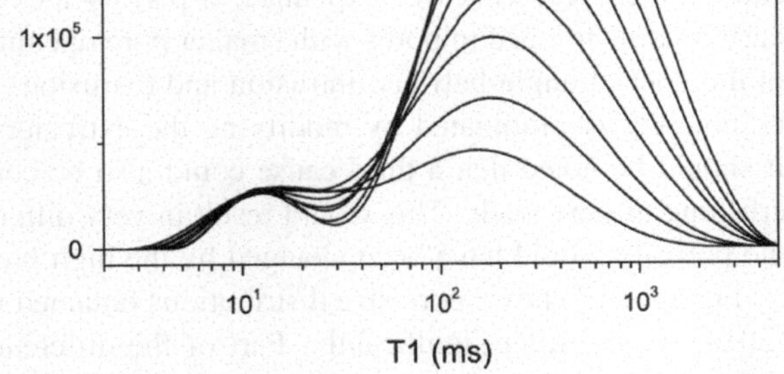

Fig. 2.12 T1 distribution for water-hemp mixtures after 18 h for different water to hemp mass ratios: (from *bottom* to *top*) 0.62, 0.86, 1.12, 1.37, 1.63, 1.89, 2.14, and 2.4 (Faure et al. 2012)

2.2.2 Other Methods

2.2.2.1 Mercury Intrusion Porosimetry

Mercury intrusion porosimetry (MIP) is a powerful technique which can be used to explore the structure of pores larger than about 3.5 nm. In MIP, the volume of liquid metal that penetrates a solid is measured as a function of applied pressure. Subsequent analysis is based on the capillary law governing liquid penetration into small pores. Since mercury is a non-wetting liquid for most materials (its contact angle is greater than 90°), an externally imposed pressure is required to force it into the pores of a porous solid.

The smaller the pore size, the greater the pressure required to force the mercury into the pore. In general, penetration data are analyzed using the Washburn equation (Washburn 1921). This relates the radius r of pores (assumed to be cylindrical) to the imposed pressure P as follows:

$$P = \frac{-2\gamma \cos \theta}{r}$$

where:

γ Interfacial energy (surface tension) of mercury
θ contact angle of mercury with the material

Common values of γ and θ (which assume interfaces involving a gas or vapour phase) are 485 mJ/m^2 and 140°. Whilst pores are rarely cylindrical, the Washburn equation is generally accepted as a practical method of analysing what are normally very complex pore systems (Lawrence 2007).

Boitryk and Pawluczuk (2014) have tested bio-composites (reed and sawdust bound with cement) using MIP in order to establish the impact of superplasticizer. Collet et al. (2008) tested hemp-lime renders and mortars using MIP. This study found that the pore size distributions of the composites were monomodal, with frequency peaks at about 0.9 μm. The vast majority of the pores seen were in the meso- and macro-pore range (respectively 96 and 94% of the intruded mercury volume). They also identified a significant hysteresis between the intrusion and extrusion curves. This was explained in part by the 'ink-bottle' effect where mercury became trapped in pores with smaller pore openings, and partly by a difference in the contact angle between intrusion and extrusion. In some cases hysteresis can be completely eliminated by modifying the extrusion contact angle (León 1998). It should be noted that a third cause could also be considered, associated with the crushing of pore walls. This would result in very different extrusion curves because the porosity would have been changed by the high pressures involved in intrusion.

Figure 2.13 shows pore size distributions obtained by Collet (2008) and Boitryk (2014) which differ significantly. Part of the difference is likely to be due to the different binders used, and potentially also due to differences in hemp shiv itself,

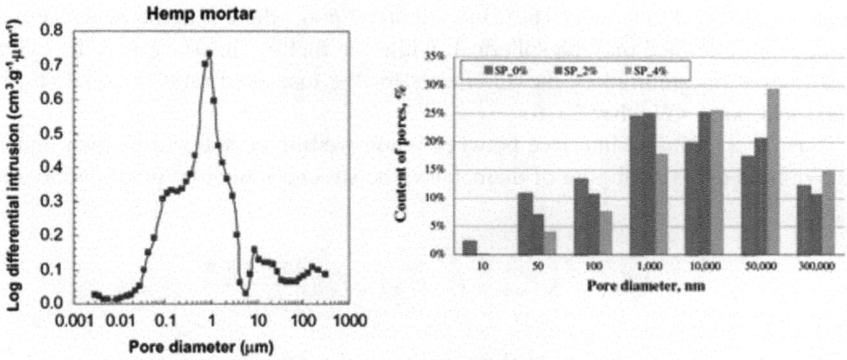

Fig. 2.13 Pore size distribution of hemp-lime composites using MIP. Collet (2008) *left*; Boitryk (2014) *right*

but it is evident that a more systematic study of the use of MIP to characterise the pore size distribution of bio-aggregates is required.

There do not appear to be any MIP studies of bio-aggregate on its own.

2.2.2.2 Thermoporometry

A small crystal of a liquid melts at a lower temperature than the bulk liquid, as given by the Gibbs-Thomson equation (Jackson and McKenna 1990). Thus if a liquid is imbibed into a porous material, and frozen, the melting temperature will provide information on the pore-size distribution. The detection of the melting can be done by sensing the transient heat flows during phase-changes using differential scanning calorimetry.

For an isolated spherical solid particle of diameter x in its own liquid, the Gibbs-Thomson Equation for the structural melting point depression can be written:

$$\Delta T_m(x) = T_{mB} - T_m(x) = T_{mB}\frac{4\sigma_{sl}}{H_f \rho_s x}$$

where:

T_{mB} Bulk Melting temperature
σ_{si} solid-liquid interface energy (per unit area)
H_f bulk enthalpy of fusion (per gram of material)
ρ_s density of solid

Very similar equations may be applied to the growth and melting of crystals in the confined geometry of porous systems. However the geometry term for the crystal-liquid interface may be different, and there may be additional surface energy terms to consider, which can be written as a wetting angle term cosφ. The angle is usually considered to be near 180°. In cylindrical pores there is some evidence that the freezing interface may be spherical, while the melting interface may be cylindrical, based on preliminary measurements for the measured ratio for $\Delta T_f/\Delta T_m$ in cylindrical pores (Webber 2010).

Thus for a spherical interface between a non-wetting crystal and its own liquid, in an infinite cylindrical pore of diameter x, the structural melting point depression is given by:

$$\Delta T_m(x) = T_{mB} - T_m(x) = -T_{mB}\frac{4\sigma_{sl}\cos\varphi}{H_f \rho_s x}$$

There has not been any published work which uses thermoporometry to characterize the pore structure of bio-based aggregates, although theoretically it should produce more representative results than MIP because it is conducted at atmospheric pressure which makes it less likely to crush pore walls during the characterization.

Landry (2005) has shown that Thermoporometry can be conducted using organic liquids as a probe, and cyclohexane is suggested to be a convenient organic liquid because its fusion temperature is 279.7 K (6.6 °C), very close to water. Cyclohexane is hydrophobic and non polar which would present a different interaction with the surface of the bio-aggregate that water. As a probe liquid, cyclohexane provides a complete baseline resolution at a 0.06 K/min scanning rate for all pore sizes. Another advantage to using hydrocarbons for thermoporometry characterisation of large pore materials is the larger temperature depression it offers compared with water.

2.2.2.3 Physisorption

The term "physical adsorption" or "physisorption" refers to the phenomenon of gas molecules adhering to a surface without the formation of a chemical bond at a pressure less than the vapor pressure. The attractions between the molecules being adsorbed and the surface are relatively weak and definitely not covalent or ionic. Some adsorption process is accompanied by absorption, which is the penetration of the fluid into the solid phase. It is sometimes difficult to distinguish between adsorption and absorption. This interaction is generally the result of a van der Waals interaction. On the other hand, chemical adsorption, called chemisorption, occurs when a molecule or atom is adsorbed to a surface by forming a chemical bond. Chemisorption is limited to monolayer coverage (Sing 1985; Forrest 2012; Condon 2006; Rouquerol 2014). Physical adsorption takes place on all surfaces and can forms multiple layers under proper conditions. Chemisorption, however, is highly selective and only proceeds as long as the adsorptive can make direct contact with the surface. It is therefore a single-layer process. The enthalpy of chemisorption is often much greater than that of physical adsorption. The physical sorption energy usually not exceed 80 kJ/mole, with typical energies being considerably less due to relatively weak Van der Wall's forces. Physically adsorbed molecules may diffuse along the surface of the adsorbent and typically are not bound to a specific location on the surface. Being only weakly bound, physical adsorption is easily reversed. A chemical bond involves sharing of electrons between the adsorbate and the adsorbent, with typical energies up to about 600 kJ/mole for C–N bonds and 800 kJ/mole for chemical bonds. Due to the bond strength, chemical adsorption is difficult to reverse (Webb 2003; Rouquerol 2014).

For physisorption, the principle measurement performed as an adsorption experiment is the measurement of the adsorption isotherm. The adsorption isotherm is the measurement of amount adsorbed versus adsorptive pressure at constant temperature. The slightest change in the shape of the plotted isotherm in indicative of a particular surface feature. Analyses of physical adsorption isotherm data reveal the total surface area, mesopore and micropore volume and area, total pore volume, the distribution of pore volume and area by pore size, and surface energy distribution. This is the easiest measurement to make. Another type of measurement is calorimetry. There are various forms of calorimetry but the most accurate methods are very difficult to perform and only a few examples are available in the literature. There are principal methods to measure the adsorption isotherm, volumetric and gravimetric. In both methods the adsorbent is held at a constant temperature, usually near or at the boiling point of the adsorptive. The amount adsorbed is measure in the case of the volumetric system by measuring the pressure change and comparing this to the expected pressure change if the adsorbent were absent. In the case of the gravimetric measurement the amount adsorbed is indicated by the mass gain. The most common measurement of the isotherm is volumetric method. This method has the advantage that it is simplest and relatively inexpensive. It has the disadvantage of a greater uncertainty in the results. A low cost alternative to the volumetric is the flow or carrier gas system. The disadvantage of this method is that the results are very uncertain and normally does not yield the isotherm. Generally, the gravimetric method is more accurate and precise, however such instrumentation is more expensive and requires a little more skill and patience to operation.

When a polar molecule is adsorbed on an ionic or polar surface various types of specific interactions may contribute to the adsorption energy. A useful general expression for the adsorption energy, E_0, at very low surface coverage was first proposed by Barrer (1966) in the form of the sum:

$$E_0 = E_D + E_R + E_P + E_{F\mu} + E_{FQ}$$

in which E_D and E_R represent the non-specific dispersion and repulsion contributions and the terms E_P, $E_{F\mu}$ and E_{FQ} represent, respectively, the three types of specific contributions: the polarization, field-dipole and field gradient-quadrupole energies. The adsorbent-adsorbate interactions must be studied at very low surface coverage. It is only under these conditions that we can eliminate, or at least minimize, the adsorbate-adsorbate interactions. Calorimetry measures the temperature change as

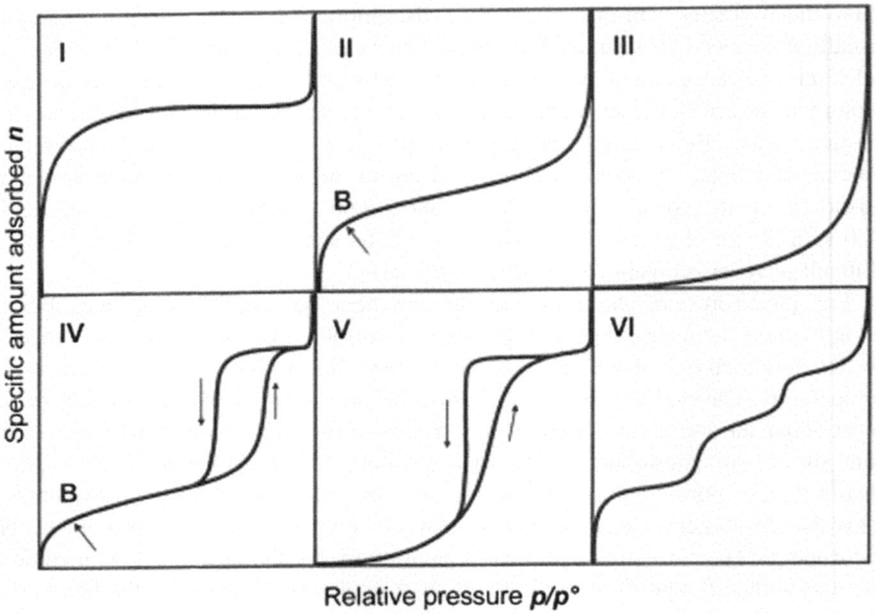

Fig. 2.14 The six main types of gas physisorption isotherms, according to the IUPAC classification (Sing et al. 1985)

the adsorption occurs. This along with a heat capacity measurements of the resultant adsorbate-adsorbent combination yields the heat of adsorption as a function of pressure. Calorimetry is not widely used since accurate calorimetry is extremely difficult to perform and requires a great amount of time and effort (Condon 2006).

Since physisorption is a complex process involving various interactions, the majority of these isotherms may conveniently be grouped into six classes in the IUPAC classification (Fig. 2.14).

- I: Microporous materials (e.g. Zeolite and Activated carbon)
- II: Non porous materials (e.g. Nonporous Alumina and Silica)
- III: Non porous materials and materials which have the weak interaction between the adsorbate and adsorbent (e.g. Graphite/water)
- IV: Mesoporous materials (e.g. Mesoporous Alumina and Silica)
- V: Porous materials and materials that have the weak interaction between the adsorbate and adsorbent (e.g. Activated carbon/water)
- VI: Homogeneous surface materials (e.g. Graphite/Kr and NaCl/Kr)

Several models have been developed to aid in the use of experimental data at the gas-solid interface. Each of these relies upon different assumptions that may affect the model's validity for a given surface. The Langmuir theory is used for ideal localized monolayer adsorption. The most widely used is the Brunauer, Emmett and Teller (BET) and its various modifications including the Brunauer, Deming,

Deming and Teller (BDDT). It is widely used for multilayer adsorption. The mechanisms involved in the BET model will be mainly dealt with next section. The calculation of the pore size distribution is performed by various methods based on the use of the Kelvin equation. Another widely used isotherm, especially for porous material, is the Dubinin-Radushkevich (DR) isotherm. Gibbs adsorption equation is to describe the adsorbed phase on available surface or in micropores. There are several other methods to describe the porosity, such as Barrett Joyner Halenda method (BJH), Density functional theory method (DFT) and Alpha S method (αs) and so on (Sing 1985; Brunauer 1938; Condon 2006).

2.2.2.4 Nitrogen Adsorption and BET Analysis

Of all the many gases and vapours, which are readily available and could be used as adsorptive, nitrogen has remained universally pre-eminent. Nitrogen gas is generally employed as the probe molecule and is exposed to a solid under investigation at liquid nitrogen conditions (i.e. 77 K). The surface area of the solid is evaluated from the measured monolayer capacity and knowledge of the cross-sectional area of the molecule being used as a probe. For the case of nitrogen, the cross-sectional area is taken as 16.2 Å^2/molecule. In the early 1930s, it was realized that multilayer adsorption of nitrogen can occur at liquid nitrogen temperature (77 K). Emmet and Brunauer came to the empirical conclusion that the beginning of the middle almost linear section of a Type II isotherm (Point B in Fig. 2.15) was the point most likely

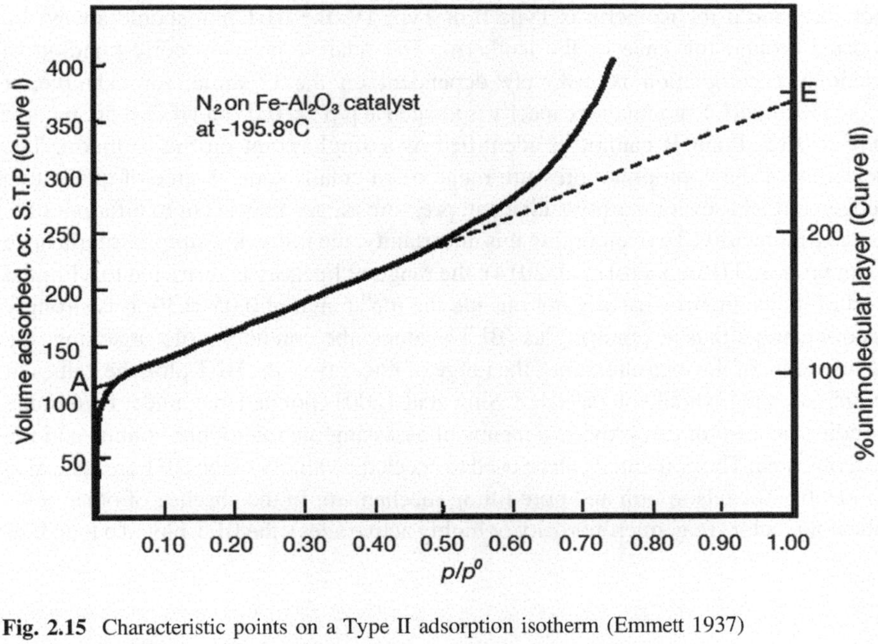

Fig. 2.15 Characteristic points on a Type II adsorption isotherm (Emmett 1937)

to correspond to monolayer completion. Their work prepared the way for the development of the BET theory in 1938. From the amount adsorbed at Point B, Emmett and Brunauer went on to calculate the surface area by assuming the monolayer to be molecularly close packed (Emmett 1937). To use of nitrogen adsorption for pore size analysis dates from the late 1940s. It is based on the application of the Kelvin equation, with a correction for the multilayer thickness on the pore walls. One of the computational method devised by Barrett, Joyner and Halenda (BJH) in 1951 remains the most popular way of deriving the pore size distribution from an appropriate nitrogen isotherm (Sing et al. 1998, 2001 and 2004).

Generally, two stages are involved in the evaluation of the surface area from physisorption isotherm data by the BET method. First, it is to construct the BET plot and to derive the value of the monolayer capacity, $\mathbf{n}_m$. The second stage is the calculation of the specific surface area. The BET equation is conveniently expressed in the linear form:

$$\frac{p/p^0}{n(1-p/p^0)} = \frac{1}{\boldsymbol{n}_m C} + \frac{C-1}{\boldsymbol{n}_m C}\left(\frac{p}{p^0}\right)$$

where $\mathbf{n}$ ($=n^a/m^s$) is the amount adsorbed at a relative pressure p/p^0 and $\mathbf{n}_m$ ($=n^a/m^s$) is the monolayer capacity. In the BET theory, the parameter C is exponentially related to E_1 (the first-layer adsorption energy). A reliable analysis of the BET plot requires a certain number of experimental points: 10 is, we consider, a minimum in the exploratory range of relative pressures from 0.01 to 0.30. The location and extent of the linear region of a BET plot are dependent on the system and operational temperature, and if the isotherm is Type II or Type IV, the BET plot should always be located around the knee of the isotherm. The relative pressure corresponding to monolayer completion is inversely dependent on the C value, for example, if $C > 350$, the BET monolayer capacity is located at $p/p^0 < 0.05$ and if $C < 50$, $\mathbf{n}_m$ is at $p/p^0 > 0.18$. Point B cannot be identified as a single point on the isotherm. The selection of the appropriate pressure range often entails some degree of qualitative judgement and several narrow, adjacent, pressure ranges may seem to offer possible ranges of linearity. To overcoming this uncertainty, the following simple criteria have been proposed (Rouquerol et al. 2014): the range of linearity is restricted to a limited part of the isotherm—usually not outside the p/p^0 range of 0.05–0.30. It is strongly recommended that in reporting as (BET) values, the conditions of outgassing, the temperature of the measurements, the range of linearity of the BET plot, the values of p^0, n_m^a, a_m and C should all be stated. Sing et al. (2001) pointed that, under favourable conditions, a t-plot can provide a means of assessing the micropore volume and the external area. The α_s method can be used to check the validity of the BET area and also to identify the adsorption and pore filling mechanism. In the absence of other complicating factors (e.g. microporosity or highly active sites), the BET plot of a type II or type IV isotherm does appear to provide a fairly reliable assessment. It is not appropriate to apply the BET method to type III, type V or type VI isotherms. For a porous materials, or one that has an unsmooth surface, the BET surface area is generally appreciably larger than its non-porous analog. BET experiments are typically conducted to a relative pressure, P/P^0, of approximately 0.3 at 77 K, where P^0 is the saturation pressure. At relative pressures above the point at which a N_2 monolayer has formed on the solid, capillary condensation occurs within the pore structure of the material such that the smaller pores are filled more easily and consecutively larger pores are filled as pressure is increased. When the saturation point is approached, i.e., P/P^0 is approximately 1.0, the internal pore structure of the material contains condensed (liquid) nitrogen. The total pore volume can be calculated by assuming that the density of liquid nitrogen in the pores is the same as that bulk liquid nitrogen. Nitrogen sorption is suitable to characterize materials with pores with the range of ~2 nm to below ~150 nm. For materials containing larger pores, mercury porosimetry is the preferred experimental technique and spans the pore range from 3.5 to 2000 nm. Sample preparation prior sorption analysis is a key aspect of material characterization. Caution must be used when heating some common samples because melting, dehydration, sintering, and decomposition are processes that can drastically alter the surface properties of the sample.

Collet et al. (2008) studied the porosity and pore structure of hemp wool, lime hemp render and hemp mortar by using BET methods with water vapour as probe agent. The results showed that the isotherms obtained are S-shaped and can be classified as type II according to IUPAC classification. The specific surface area for hemp mortar and hemp wool were 80.82 and 88.77 m^2/g, respectively. Collet et al. (2013) also showed that the isotherms of two kinds of hemp wool are classified as type II and the specific surface area is 111 m^2/g for the first hemp wool. Bismarck et al. (2002) showed all natural fibres have very small specific surface areas about 0.5 m^2/g, which is just slightly bigger than the calculated geometric surface area (As, geo 0.38 m^2/g at a fibre diameter $d_f \sim 14$ μm (compare SEM-micrographs) and a density of flax fibres of $\rho = 1.47$ g/cm^3). Yin et al. (2015) compared the changes in micropore and mesopores in the wood cell walls of sapwood and heartwood. The results showed that specific surface area of sapwood ranged from 1.255 to 2.08 m^2/g, but specific surface area of heartwood ranged from 0.078 to 1.058 m^2/g. Brewer et al. (2014) that BET surface area of the slow pyrolysis biochars increased exponentially with pyrolysis temperature, from <1 to 317 ± 16 m^2/g for the wood biochars and from <1 m^2/g to 387 ± 6 m^2/g for the grass biochars. BET surface area was low, <10 m^2/g, for all of the intermediate pyrolysis biochars. Rachini et al. (2012) showed the specific surface areas of natural, ethanol/water extracted and silane treated (at different concentrations) hemp fibres were about 0.7 m^2/g. Before BET characterization, the fibres (2 cm) were grounded onto very small pieces (500 nm) using a wood grinder.

2.2.2.5 Dynamic Vapour Sorption

Dynamic Vapour Sorption (DVS) is a gravimetric technique that measures the speed and amount of a solvent that is absorbed by a sample. It is an automated alternative to traditional sorption techniques. Within building materials the technique is primarily used to measure the sorption of water vapour, although using other solvents can provide some useful insights into pore structure and surface area.

Traditional water vapour sorption isotherms are conducted gravimetrically using saturated salt solutions which generate known relative humidity. For each humidity value, a sorption isotherm indicates the corresponding water content value at a given, constant temperature. If the composition or quality of the material changes, then its sorption behaviour also changes. Because of the complexity of sorption processes, the isotherms cannot be determined by calculation, but must be recorded experimentally for each product.

The main application of DVS is to measure water sorption isotherms. In general, a vapor sorption isotherm shows the equilibrium amount of vapor sorbed as a function of steady state relative vapor pressure at a constant temperature. For water sorption isotherms, water relative vapor pressure is more commonly expressed as relative humidity. In a DVS experiment this is accomplished by exposing a sample to a series of step changes in relative humidity and monitoring the mass change as a function of time. The sample mass must be allowed to reach gravimetric equilibrium at each step change in humidity before progressing to the next humidity level. Then, the equilibrium mass values at each relative humidity step are used to generate the isotherm. Isotherms are typically divided into two components: sorption for increasing humidity steps and desorption for decreasing humidity steps. Sorption can be further divided into adsorption (sorbate located on the surface) and absorption (sorbate penetrates the bulk). Figure 2.16 shows a typical DVS reference plot and isotherm for hemp-lime using water vapour. The chapter on hygric properties discusses sorption isotherms in detail and the extent to which they have been used with bio-aggregates

Figure 2.17 shows DVS data for hemp shiv using cyclohexane. This test allows an insight to be gained into the BET surface area and sorption kinetics as well as into porosity and diffusion. The cyclohexane sorption study (Fig. 2.17 up) on the sample indicates that there is a very small interaction between cyclohexane vapour and the hemp sample. The sorption isotherm (Fig. 2.17 down) shows the cyclohexane sorption capacity for hemp at 95% P/P^0 to be less than 1% by mass. The BET surface areas can be calculated from the BET model with cyclohexane as probe and give a specific surface area of 2.463 m^2/g with a regression (R2) of linear fit of 99.846%. Table 2.2 shows that for the hemp shiv sample, for steps from 60% P/P^0 to 90% P/P^0, there is a steady decrease in the diffusion constant almost linearly with increasing partial vapour pressures of cyclohexane. Although the hemp sample shows a straightforward sorption/desorption mechanism (Fig. 2.17 up), the isotherm (Fig. 2.17 down) shows a distinct Type IV mechanism with a characteristic hysteresis loop which can be associated with a mesoporous structure.

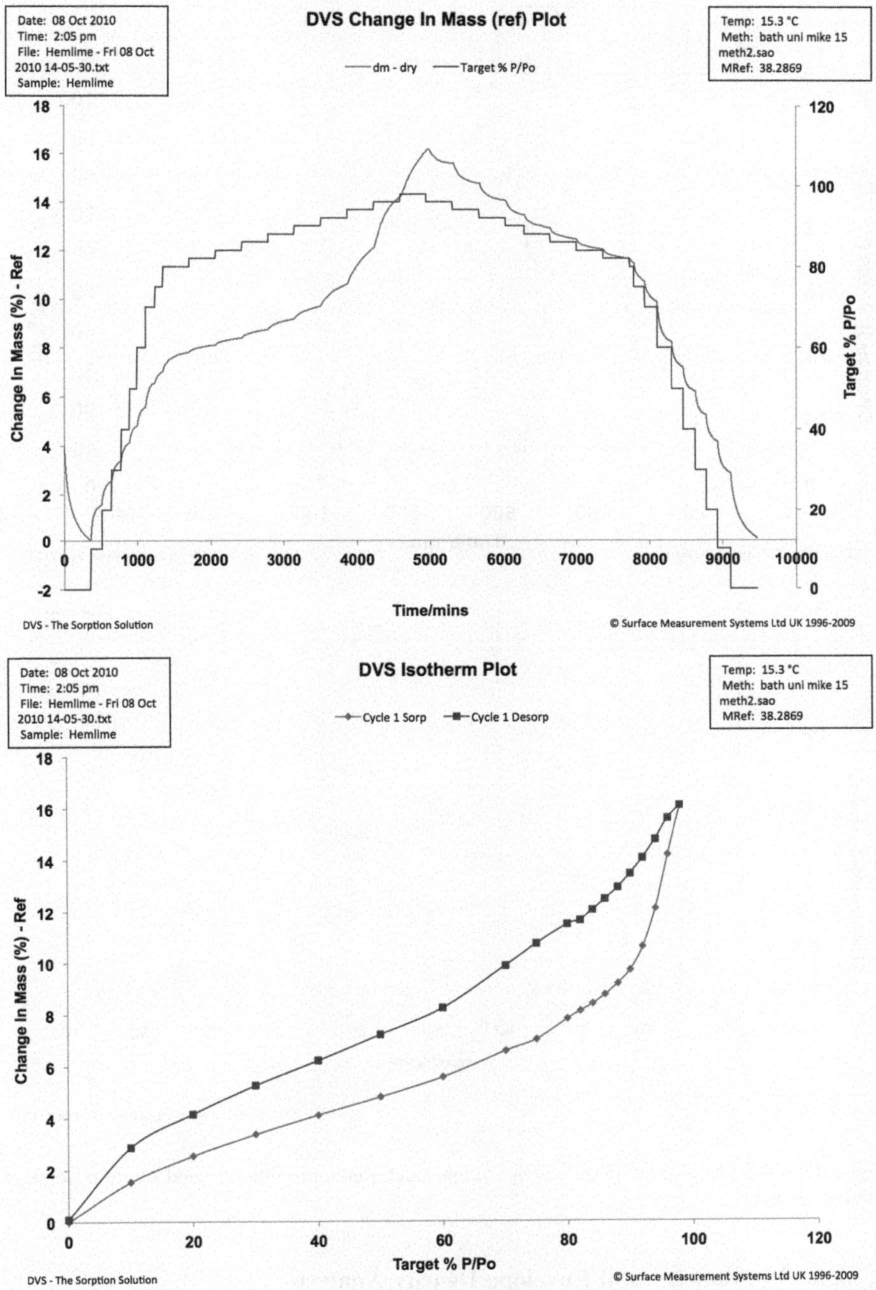

Fig. 2.16 Typical DVS reference plot (*up*) and sorption isotherm (*down*) for hemp-lime (author)

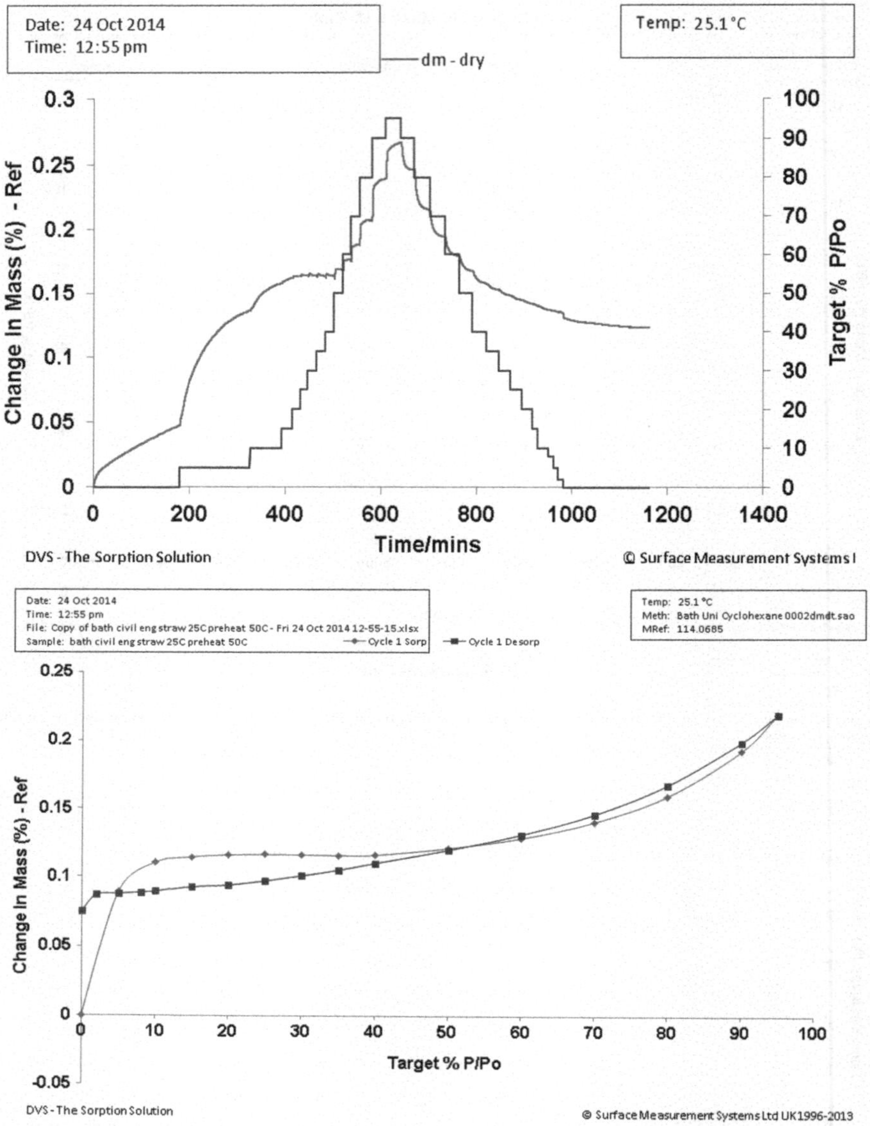

Fig. 2.17 DVS plots for hemp shiv using cyclohexane. Reference plot (*up*) and sorption isotherm (*down*) (author)

2.2.2.6 Pycnometry and Envelope Density Analysis

Pycnometry is used for measuring the density and the pore volume based on Boyle-Mariaotte's law of volume-pressure relationships. Details of the construction of simple gas pycnometers were published more than fifty years ago. There are

Table 2.2 Diffusion coefficients for hemp shiv sample

Hemp shiv			
Sample	%P/P^0	Temperature (°C)	Diffusion coefficient (cm^2/s)
Hemp	60–70	25.0	2.57E−04
Hemp	70–80	25.0	1.65E−04
Hemp	80–90	25.0	1.13E−04

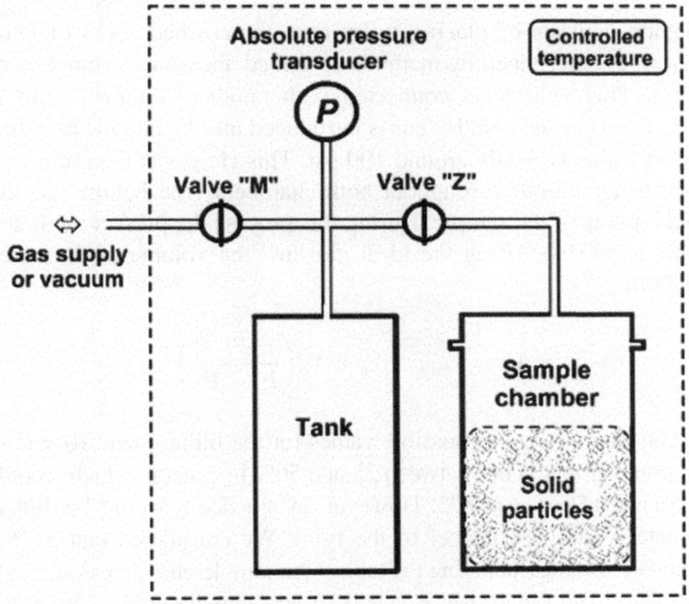

Fig. 2.18 Diagram of a constant volume gas pycnometer. The sample-chamber and the tank, initially filled with gas at two different pressures, are connected by opening vale 'z'. The final gas pressure indicated of how much of the sample-chamber volume is occupied by the solid particles (Tamari 2004)

three kinds of gas pycnometers reported in literature: 'constant-volume', 'variable volume' and 'comparative' (Kummer 1945; Tamari 2004). The 'constant-volume' gas pycnometer was considered because of its widespread use. This technique uses the gas displacement method to measure volume accurately. As shown in Fig. 2.18, inert gases, such as helium or nitrogen, are used as the displacement medium. The sample is sealed in the instrument compartment of known volume, the appropriate inert gas is admitted, and then expanded into another precision internal volume. The pressures observed upon filling the sample chamber and then discharging it into a second empty chamber allow computation of the sample solid phase volume. Helium molecules rapidly fill pores as small as one angstrom in diameter; only the solid phase of the sample displaces the gas. Dividing this volume into the sample weight gives the gas displacement density.

This method has been widely used to determine the volume and the density of bio-aggregates materials. Three fundamental hypotheses are made:

(1) the gas inside the pycnometer behaves ideally (i.e. its compressibility is negligible and it does not adsorb on solids),
(2) the sample and the pycnometer's components are rigid, and
(3) the pycnometer is gas-tight and the expanding gas quickly reaches a static equilibrium.

The method consists of placing a dry core (or crushed rock) of known bulk volume (V_{bulk}, as determined by methods described above) in a container of known volume (V_a). This volume is connected with another container with a known volume (V_b) that is evacuated. He gas is introduced into V_a and the pressure (P_1) set to an arbitrary value typically around 100 psi. This He gas is then released into V_b and allowed to equilibrate throughout both chambers. The helium gas then penetrates into the pores of the sample. During this process the pressure will decrease to a new stable level (P_2). Using the ideal gas law, the volume of the pores can be calculated from

$$V_v = V_{bulk} - V_a - V_b\left(\frac{P_2}{P_2 - P_1}\right)$$

It must consider a range of possible values for the filling factor ($0 < \varphi < 1$). The porosity of granular media lies between 25 and 50% in general, which would give the theoretical range $0.50 < \varphi < 0.75$. However, in practice it would be difficult to fill the pycnometer's sample-chamber to the brim. We considered that 5–25% of the sample-chamber volume (including the tube at the sample-chamber side) might be free of solid particles. The range $0.40 < \varphi < 0.70$ was thus thought to be realistic for a sample-chamber filled with as many solid particles as practical (Tamari 2004; Anovitz 2015). This method is recognized as one of the most reliable techniques for obtaining true, absolute, skeletal, and apparent volume and density. This technique is non-destructive as it uses the gas displacement method to measure volume. Inert gases, such as helium or nitrogen, are used as the displacement medium. Density calculations using the gas displacement method are much more accurate and reproducible than the traditional Archimedes water displacement method (Zauer et al. (2013).

Mwaikambo et al. (2001) studied the porosity of plant fibres by helium pycnometry method. The results showed that the porosity of hemp fibres is 2.46% and the porosity of Sisal and Jute are similar (10.85 and 11.36%, respectively). Zauer et al. (2013) studied porosity of wood by pycnometry method using helium and nitrogen as a displace gas. The results in Table 2.3 showed how this can lead to misinterpretation of the cell wall density or porosity of wood determined by gas pycnometry. The results clearly showed that the sample geometry and the dimensions

Table 2.3 Average cell wall densities and porosities of oven dry native spruce, maple and ash in dependence of sample dimension or geometry as well as displacement gas (helium and nitrogen) (Zauer et al. 2013)

Wood species	Thickness (mm)	Thickness direction	Helium		Nitrogen	
			Cell-wall denstiy (g cm^{-3})	Porosity (%)	Cell-wall denstiy (g cm^{-3})	Porosity (%)
Spruce	2	Long	1.47	71.4	1.40	70.3
		Tang	1.29	67.3		
	6	Long	1.38	69.9		
		Tang	1.31	67.4		
Maple	2	Long	1.51	65.8	1.46	64.7
		Tang	1.43	64.3		
	6	Long	1.43	64.0		
		Tang	1.40	63.1		
Ash	2	Long	1.36	52.2	1.35	51.3
		Tang	1.35	51.3		
	6	Long	1.32	50.4		
		Tang	1.33	51.1		

significantly affect the calculated values. This is primarily due to the inaccessibility of some uncut wood cell lumen. Thus, the determined cell wall volume is falsely too high, and consequently, the calculated cell wall density or porosity falsely too low.

Donato (2012) studied the porosity of waterlogged woods by using helium pycnometry. The porosity of waterlogged wood ranged between 83.4 and 90.6%. The author believed the porosity values calculated using the helium pycnometer are reliable because helium may easily penetrate even in the smallest cavities of the cell wall. Eventual differences between the porosity values obtained in wet and dry conditions might be ascribed to different structure assumed by the cell wall and different water and helium penetration, which is larger for the inert gas. Gershon et al. (2010) studied the porosity of light wood and dark using pycnometry method. This results in a volume fraction of porosity of 0.82 for the dark and 0.80 for the light.

2.3 Conclusion

There are a large number of methods for determining porosity. However, the porosity of bio-aggregates is difficult to quantify due to the complex pore size and shape. For example, the size of pore in the bio-aggregate are between a few nanometers to several millimeters. There is really no one method that can characterize pores in this enormous range in scale. In this chapter, the theory, advantage and disadvantage of several common methods using for determination of porosity have been briefly described and summarised. In essence, the different techniques, each based on different inherent assumptions, have their own capabilities and advantages. A combination of several methods is most likely to give a good understanding of the size, shape and structure of pore in the bio-aggregates materials.

The study of bio-aggregates is still in its infancy and their unique characteristics demand the development of novel methods or the adaptation of existing methods in order to satisfactorily characterise their microstructure. There is a need for robust and comprehensive studies to be made into these materials in order that their performance can be satisfactorily modelled using Building Physics models to confidently predict the performance of dwellings constructed form these materials.

References

Anovitz, L.M., Cole, D.R.: Characterization and analysis of porosity and pre structures. Rev. Mineral. Geochem. **80**, 61–164 (2015)

Aydilek, A.H., ASCE, A.M., Oguz, S.H., ASCE, M., Edil, T.B.: Digital image analysis to determine pore opening size distribution of nonwoven geotextiles. J. Comput. Civil Eng. **16**(4), 280–290 (2002)

Barrer, R.M.: Specificity in physical sorption. Journal of colloid interface science **21**, 415–434 (1966)

Bismarck, A., Askargorta, I., Agranberri, S., Jurgen, L., Thomas, W., Bernhard, S., Artemis, S.I., Limbach, H.H.: Surface characterization of flax, hemp and cellulose fibers; surface properties and the water uptake behaviour. Polym. Compos. **23**(5), 872–894 (2002)

Boitryk, M., Pawluczuk, E.: Properties of a lightweight cement composite with an ecological organic filler. Constr. Build. Mater. **51**, 97–105 (2014)

Brewer, C.E., Chuang, V.J., Masiello, C.A., Gonnermann, H., Gao, X., Dugan, B., Driver, L.E., Panzacchi, P., Zygourakis, K., Davies, C.A.: New approaches to measuring biochar density and porosity. Biomass Bioenergy **66**, 176–185 (2014)

Brunauer, S., Emmett, P.H., Teller, E.: Adsorption of gases in multimolecular layers. J. Am. Chem. Soc. **60**, 309–319 (1938)

Chundawat, S.P.S., Donohoe, B.S., Sousa, L., da Costa, E., Thomas, A., Umesh, P., Lu, F., Ralph, J., Himmel, M.E., Balan, V., Dale, B.E.: Multi-scale visualization and characterization of lignocellulosic plant cell wall deconstruction during thermochemical pretreatment. Energy Environ. Sci. **4**, 937 (2011)

Collet, F., Bart, M., Serres, L., Miriel, J.: Porous structure and water vapour sorption of hemp-based materials. Constr. Build. Mater. **22**, 1271–1280 (2008)

Collet, F., Chamoin, J., Pretot, S., Lanos, C.: Comparison of the hygric behaviour of three hemp concretes. Energy Build. **62**, 294–303 (2013)

Condon, J.B.: Surface area and porosity determinations by physisorption measurements and theory. Elsevier, Oxford (2006)

Donato, I.D., Lazzara, G.: Porosity determination with helium pycnometry as a method to characterize waterlogged woods and the efficacy of the conservation treatments. Archaeometry **54**(5), 906–915 (2012)

Dougal, A., Jerram, Kent, & Adam, J.R.: An overview of modern trends in petrography: textural and microanalysis of igneous rocks. J. Volcanol. Geoth. Res. **154**, 158 (2006)

Dougal, A., Jerram, Higgins, & Micahael, D.: 3D analysis of rock textures: quantifying igneous microstructures. Elements **3**, 239–245 (2007)

Emmett, P.H., Brunauer, S.: The use of low temperature van de Waals adsorption isotherms in determining the surface area of iron synthetic ammonia catalysts. J. Am. Chem. Soc. **59**(8), 1553–1564 (1937)

Faure, P., Peter, U., Lesueur, D., Cossot, P.: Water transfers within Hemp Lime concrete followed by NMR. Cem. Concr. Res. **42**, 1468–1474 (2012)

Forrest, S.C.: Physical adsorption of gases onto mesoporous silica material SBA-15. Chemistry publications and other works (2012)

Gershon, A.L., Bruck, H.A., Xu, S., Sutton, M.A., Tiwari, V.: Multiscale mechanical and structural characterizations of Palmetto wood for bio-inspired hierarchically structured polymer composites. Mater. Sci. Eng. C **30**, 235–244 (2010)

Grove, C., Jerram, D.A.: jPOR: an imageJ macro to quantify total optical porosity from blue-stained thin sections. Comput. Geosci. **37**, 1850–1859 (2011)

Kain, G., Johann, C.P., Marius-Catalin, B., Bernhard, P., Klaus, R., Alexander, P.: Analyzing wood bark insulation board structure using X-ray computed tomography and modelling its thermal conductivity by means of finite difference method. J. Compos. Mater. **0**(0), 1–12 (2015)

Hamdi, S.E., Delisee, C., Malvestio, J., Silva, N.D., Duc, A.L., Beaugrand, J.: X-ray computed microtomography and 2D image analysis for morphological characterization of short lignocellulosic fibres raw materials: a benchmark survey. Compos. A **76**, 1–9 (2015)

Jackson, C.L., McKenna, G.B.: The melting behavior of organic materials confined in porous solids. J. Chem. Phys. **93**(12), 9002–9011 (1990)

Kaestner, A., Lehmann, E., Stampanoni, M.: Imaging and image processing in porous media research. Adv. Water Resour. **31**(9), 1174–1187 (2008)

Kummer, F.A., Cooper, A.W.: Soil porosity determination with the air pycnometer as compared with the tension method. Agric. Eng. **26**, 21–23 (1945)

Landry, M.R.: Thermoporometry by differential scanning calorimetry: experimental considerations and applications. Thermochim. Acta **433**, 27–50 (2005)

Latif, E., Lawrence, R., Shea, A., Walker, P.: Moisture buffer potential of experimental wall assemblies incorporating formulated hemp-lime. Build. Environ. **93**(2), 199–209 (2015)

Lawrence, R.M., Mays, T.J., Rigby, S.P., Walker, P., D'Ayala, D.: Effects of carbonation on the pore structure of non-hydraulic lime mortars. Cem. Concr. Res. **37**, 1059–1069 (2007)

León, C.A.L.: New perspectives in mercury porosimetry. Adv. Colloid Interf. Sci. **76–77**, 341–372 (1998)

Lubelli, B., de Winter, D.A.M., Post, J.A., van Hees, R.P.J., Drury, M.R.: Cryo FIB SEM and MIP study of porosity and pore size distribution of bentonite and kaolin at different moisture contents. Appl. Clay Sci. **80**(81), 358–365 (2013)

Maire, E.: X-ray tomography applied to the characterization of highly porous materials. Annu. Rev. Mater. Res. **42**, 136–178 (2012)

Maire, E., Buffiere, J.Y., Salvo, L., Blandin, J.J., Ludwig, W., Letang, J.M.: On the application of X-ray microtomography in the field of materials science. Adv. Eng. Mater. **3**, 539–546 (2001)

Manger, G.E.: Porosity and bulk density of sedimentary rocks. Geol. Surv. Bull. **114-E** (1963)

Marinello, F., Bariani, P., Savio, E., Horsewell, A., De Chiffre, L.: Critical factors in SEM 3D stereo microscopy. Measur. Sci. Technol. **19** (2008)

Mermut, A.R.: Historical development in soil micromorphological imaging. J. Mountain Sci. **6**(2), 107–112 (2009)

Mostefai, N., Hamzaoui, R., Guessasma, S., Amadou, A.W., Nouri, H.: Microstructure and mechanical performance of modified hemp fibre and shiv mortars: discovering the optimal formulation. Mater. Des. **84**, 359–371 (2015)

Mwaikambo, L.Y., Ansell, M.P.: The determination of porosity and cellulose content of plant fibers by density method. J. Mater. Sci. Lett. **20**, 2095–2096 (2001)

Nimmo, J.R.: Porosity and pore size distribution. Encycl. Soils Environ. **3**, 2935–303 (2004)

Phillipson, M.N., Baker, P.H., Davies, M., Ye, Z., McNaughtan, A., Galbraith, G.H., McLean, R. C.: Moisture measurement in building materials: an overview of current methods and new approaches. Build. Serv. Eng. Res. Technol. **28**(4), 303–316 (2007)

Rachini, A., Le Troedec, M., Peyratout, C., Smith, A.: Chemical modification of hemp fibers by silane coupling agents. J. Appl. Polym. Sci. **123**(1), 601–610 (2012)

Roche, R.C., Abel, R.A., Johnson, K.G., Perry, C.T.: Quantification of porosity in acropora pulchra (brook 1891) using X-ray micro-computed tomography techniques. J. Exp. Mar. Biol. Ecol. **396**, 1–9 (2010)

Rouquerol, F., Rouquerol, J., Sing, K.S.W., Llewellyn, P., Maurin, G.: Adsorption by Powders and Porous Solids Principles, Methodology and Applications. Elsevier, Oxford (2014)

Sassoni, E., Manzi, S., Motori, A., Montecchi, M., Canti, M.: Novel sustainable hemp-based composites for application in the building industry: physical, thermal and mechanical characterization. Energy Build. **77**, 219–226 (2014)

Shen, W., Wan, J., Tokunaga, T.K., Kim, Y., Li, X.: Porosity calculation, pore size distribution and mineral identification within shale rocks: application of scanning electron microscopy and energy dispersive spectroscopy. Electron. J. Geotech. Eng. **20**, 11477–11490 (2015)

Sing, K.S.W.: Reporting physisorption data for gas/solid systems with special reference to the determination of surface area and porosity (Recommendations 1984). Pure Appl. Chem. **57**(4), 603–619 (1985)

Sing, K.: Adsorption methods for the characterisation of porous materials. Adv. Colloid Interf. Sci. **76–77**, 3–11 (1998)

Sing, K.: The use of nitrogen adsorption for the characterisation of porous materials. Colloids Surf. A **187–188**, 3–9 (2001)

Sing, K.: Characterisation of porous materials: past, present and future. Colloids Surf. A **241**, 3–7 (2004)

Stevulova, N., Cigasova, J., Estokova, A., Terpakova, E., Geffert, A., Kacik, F., Singovszka, E., Holub, M.: Properties characterization of chemically modified hemp hurds. Materials **7**, 8131–8150 (2014)

Tamari, S.: Optimum design of the constant volume gas pycnometer for determining the volume of solid particles. Meas. Sci. Technol. **15**, 549–558 (2004)

Tran, L.Q.N., Minh, T.Nguyen, Fuentes, C.A., Chi, T., Truong, Vuure, Van, A.W., Verpoest, I.: Investigation of microstructure and tensile properties of porous natural coir fibre for use in composite materials. Ind. Crops Prod. **65**, 437–445 (2015)

Walker, R., Pavia, S.: Impact of Hydration on the Properties of Hemp Lime Concrete. Civil engineering research in Ireland CERAI, Belfast (2014)

Walker, R., Pavia, S., Mitchell, R.: Mechanical properties and durability of hemp-lime concretes. Constr. Build. Mater. **61**, 340–348 (2014)

Washburn, E.W.: The dynamics of capillary flow. Phys. Rev. **17**, 273–283 (1921)

Webb, P.A.: Introduction to Chemical Adsorption Analytical Techniques and their Applications to Catalysis. Technical publications, Micromeritics instrument corp (2003)

Webber, J.B.W.: Studies of nano-structured liquids in confined geometries and at surfaces. Prog. Nucl. Magn. Reson. Spectrosc. **56**(1), 78–93 (2010)

Williams, J., Lawrence, M., Walker, P.: Thermally modelling bio-composites with respect to an orientated internal structure. Sustainable ecological engineering design for society, Leeds, 17th–18th, September, 2015 (2015)

Yang, B., Wu, A., Miao, X., Liu, J.: 3D characterization and analysis of pore structure of packed ore particle beds based on computed tomography images. Trans. Nonferrous Met. Soc. China **24**, 833–838 (2014)

Yin, J., Song, K., Lu, Y., Zhao, G., Yin, Y.: Comparison of changes in micropores and mesopores in the wood cell walls of sapwood and heartwood. Wood Sci. Technol. **49**, 987–1001 (2015)

Zauer, M., Pfriem, A., Wagenfuhr, A.: Toward improved understanding of the cell wall density and porosity of wood determined by gas pycnometry. Wood Sci. Technol. **47**, 1197–1211 (2013)

Zhao, H., Darwin, D.: Quantitative backscattered electron analysis of cement paste. Cem. Concr. Res. **22**(4), 695–706 (1992)

Ziel, R., Haus, A., Tulke, A.: Quantification of the pore size distribution (porosity profiles) in microfiltration membranes by SEM, TEM and computer image analysis. J. Membr. Sci. **323**, 241–246 (2008)

Chapter 3
Water Absorption of Plant Aggregate

Sofiane Amziane, Vincent Nozahic and Mohammed Sonebi

Abstract In “agro-concretes”, highly-porous plant-based particles are used, and are responsible for massive water absorption.

Keywords Plant aggregate · Water absorption · Wettability · Porous medium · Filtration · Absorption model

3.1 Introduction

In “agro-concretes”, highly-porous plant-based particles are used, and are responsible for massive water absorption. It is often reported (Nguyen et al. 2010; Arnaud and Gourlay 2012) that this hydrophilic nature of the plant materials leads to competition with hydraulic binders, which require at least a certain amount of water in order to form hydrate products and cohesion. Such competition appears to be the reason for the crumbling of hydraulic binders (Cerezo 2005). Binders with high carbonation potential—in particular, aerated-lime-based binders—are often employed to counter this problem, at least on the surface of the agro-concrete (Elfordy et al. 2008; Mounanga et al. 2009; Nguyen et al. 2009a).

Recent studies performed on self-compacting concretes (Kwasny et al. 2012) could help better under the amount of water that needs to be added to the particulates in order to limit competition between the absorption of the plant material and the hydration of the binder (Jacek et al. 2012; Diederich et al. 2010). Indeed, the pre-wetting methods used up until now to provide the water necessary to saturate the granulates and the binder cannot really justify the dosage of water to be used

S. Amziane (✉) · V. Nozahic
Institut Pascal, Clermont Université, Clermont-Ferrand, France
e-mail: sofiane.amziane@univ-bpclermont.fr

M. Sonebi
School of Planning, Architecture and Civil Engineering,
Queen’s University Belfast, Belfast, UK

S. Amziane and F. Collet (eds.), *Bio-aggregates Based Building Materials*,
RILEM State-of-the-Art Reports 23, DOI 10.1007/978-94-024-1031-0_3

(Cerezo 2005; Nguyen et al. 2009a). In addition, the fabrication of the agroconcrete also comes into play, as the method of compaction is used, partially crushing the plant material and thereby releasing some of the mix water (Nguyen et al. 2009b).

3.2 Wetting of Porous, Heterogeneous Surfaces

Whether in the binder, in the lignocellulosic particles or at the interface between those two entities, water transfer by means of capillary phenomena and diffusion in the plant structure play a very important role. However, that transfer can only take place after the *wetting of the plant surface* by the interstitial water contained in the binder.

A surface's *wettability*, where the main characteristic is its surface tension γ_S, is a crucially-important concept when the adhesion of a liquid (which has a surface tension γ_L) to that surface is needed to be calculated. In the particular case of a cement paste applied to a plant particle, the process of adhesion mainly occurs between the interstitial liquid and the surface of the plant.

Wettability, which is necessary for any process of adhesion, can be examined on two different scales:

- *The physical intermolecular bonds*: There are two types of bonds created by the wetting of a plant surface with an aqueous solution: hydrogen bonds (which are OH bonds) and van der Waals bonds (electrostatic forces exerted between the particles).
- *The physical measurements (contact angle and surface tension)*: these measured values physically describe the expression of the bonds within each of the media (surface tension for the liquid and the solid) and at equilibrium between two (surface tensions of the solid and liquid γ_{SL}) or three media (Young–Dupré law defining the contact angle).

3.2.1 Surface Tension and Interface Tension

A molecule of water or of any other liquid, when it is situated within that liquid, is subject to van der Waals-type forces of cohesion, exerted isotropically upon it. On the other hand, when the molecule is at the boundary between the air and the liquid, it is in a slightly higher state of energy, which creates a stress of tension (expressed in N m^{-1} or J m^{-2}). This interfacial tension between a liquid and a gas is called the *surface tension* γ_{LG}. The surface tension of water in air at 20 °C, for example, is 72.7×10^{-3} N m^{-1}. The surface of a solid in contact with a gas is an instability of the same type. It is known as the surface energy of *surface tension* γ_{SG} of the solid.

When a solid and a liquid brought into contact with one another, where each of them has a surface energy with the air or any other surrounding gas, a new interfacial equilibrium is created. That equilibrium is an energy balance between the liquid–gas surface tension γ_{LG}, the solid–gas surface tension γ_{SG} and the *interfacial tension* γ_{SL}. An adhesion energy known as the Dupré energy, E_{DUPRE} (expressed in N m^{-1} or J m^{-2}), is then defined:

$$E_{DUPRE} = \gamma_{LG} + \gamma_{SG} - \gamma_{SL} \quad (3.1)$$

3.2.1.1 Contact Angle and Wettability

General laws

When a drop of liquid is placed on a smooth, solid surface, it undergoes spreading to a greater or lesser degree, which is the resultant of the surface energies. This wettability of the surface can be defined by the so-called *spreading coefficient* *S* (de Gennes et al. 2004):

$$S = \gamma_{SG} - \gamma_{SL} - \gamma_{LG} \quad (3.2)$$

When the value of the coefficient S is greater than 0, the total wetting of the surface can be observed. On the other hand, if S is negative, a *partial wetting can be seen*, leading to the formation of a drop sitting on the solid surface. As a general rule, the lower the liquid's surface tension, the greater the spreading. As water has a high surface tension, the wetting of a solid by water is generally incomplete.

In the case of partial wetting, the *Young–Dupré law* gives the expression of the static *contact angle* θ of a liquid drop placed on a solid substrate, at equilibrium with the vapour phase:

$$\cos\theta = \frac{\gamma_{SG} - \gamma_{SL}}{\gamma_{LG}} \quad (3.3)$$

This relation is obtained by projection of the vectors formed by the interfacial tensions taking the triple point (Fig. 3.1b) as the origin. The following situations may be encountered:

- $\theta > 90°$: The liquid is said to be "non-wetting". In the case of water, the solid is said to be hydrophobic (Fig. 3.1b);
- $\theta < 90°$: The liquid is said to be "wetting". For water, the solid is said to be hydrophilic (Fig. 3.1a).

Wetting of a heterogeneous and porous lignocellulosic plant surface

The surfaces representative of lignocellulosic particles exhibit numerous heterogeneities and significant roughness (Gardner et al. 1991). The wetting of wood and the factors influencing it have been widely studied over last decades (Gardner et al. 1991; Walinder 2000; Bruyne 2008). Generally speaking, the rougher its surface, the larger will be the contact angle formed by a hydrophobic

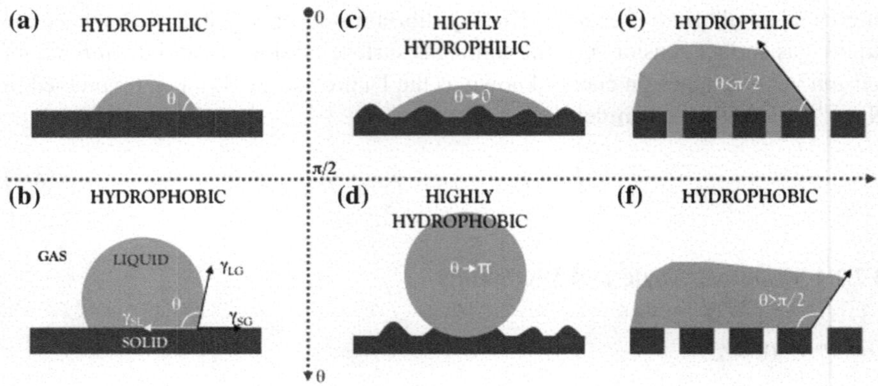

Fig. 3.1 Diagrammatic representation of: a hydrophilic (**a**) and hydrophobic (**b**) smooth surface; a hydrophilic (**c**) and hydrophobic (**d**) rough surface; and a hydrophilic (**e**) and hydrophobic (**f**) porous surface (de Gennes et al. 2004)

material (de Gennes et al. 2004) (Fig. 3.1d). Similarly, the rougher the surface of a *hydrophilic material such as wood*, the more likely it is to be perfectly wettable (Fig. 3.1c).

3.3 Transfer Phenomena in a Porous Medium

3.3.1 Liquid Transfer in the Laminar Regime

3.3.1.1 Capillary Transfer

When the aqueous mineral binder comes into contact with a hydrophilic porous support such as a lignocellulosic particle, its constituting water will wet the surface to which it is applied. Thus, as a function of the size of the pores in the support, a capillary pressure of suction P_C is exerted on the free water in the binder. That pressure P_C [N m^{-2}] is characterised by the *Kelvin–Laplace law*:

$$\mathbf{P_C} = \frac{-2\gamma_{\mathbf{LG}}.\cos\theta}{\mathbf{r}} \tag{3.4}$$

γ_{LG}: surface tension of the liquid (here the free water in the binder) on contact with the surrounding gas [N m^{-1}]

θ: contact angle which the liquid forms with the porous surface

ρ: radius of the capillary pore in question [m]

Thus, the height by which the liquid rises in the capillary h_L defined by *Jurin's law* can be calculated:

$$\mathbf{h_L} = \frac{-\mathbf{P_C}}{\boldsymbol{\rho}_\mathbf{L} \cdot \mathbf{g}} \tag{3.5}$$

g: acceleration due to gravity [m s^{-2}]
ρ_L: density of the liquid (here the free water in the binder) [kg m^{-3}]

The dynamics of that ascension is defined by *Washburn's law*, which establishes a relation of *proportionality* between the capillary rise height h_L *and the square root of the elapsed time.* In particular, it is influenced by the fluid's dynamic viscosity μ:

$$\mathbf{h_L^2} = \frac{\mathbf{r}.\gamma_\mathbf{LG}.\mathbf{cos}\theta}{2\mu}.\mathbf{t} \tag{3.6}$$

μ: Dynamic viscosity [Pa s]

3.3.1.2 Filtration Transfer in a Saturated Porous Medium

In the complex porous medium represented by agroconcretes, the pressure generated by capillary water absorption triggers a second phenomenon at the interface between the binder and the particles. Indeed, this absorption engenders a *phenomenon known as filtration* of water through the large mineral granulates (>50 μm) which make up the binder. This aqueous movement in a saturated medium, which is predominant at the interface, may bring fine particles and ions suspended in the interstitial liquid into the plant structure. The filtration process is similar to the laminar flow of a viscous interstitial fluid, undergoing friction as it passes over the granular skeleton of the binder. It is defined by *Darcy's law*:

$$\vec{\mathbf{u}} = \frac{-\mathbf{k}}{\mu}(\overrightarrow{P_C} - \rho\vec{\mathbf{g}}) \tag{3.7}$$

$\vec{u}$: Rate of filtration [m/s]
P_C: Pressure [Pa]
ρ: Density of the fluid [kg m^{-3}]
$\vec{g}$: Vector of acceleration due to gravity [m s^{-2}]
k: Intrinsic permeability [m^2]

3.3.2 Transfer of Water Vapour, Particles or Ions by Diffusion

There are many different types of diffusion phenomena, which occur in numerous media. Agroconcretes, though, are a good example, because from the moment they are made, they are subject to numerous diffusive transfers, such as:

- the *diffusion of liquid water* into the plant cell walls from the moment of mixing to an age of several days;
- the diffusion of mineral particles from the binder and plant extractible materials into the mix water—particularly at the binder/plant interface;
- the *diffusion of water in vapour form as the concrete dries*, over the course of 1–3 months, but also during the use of the material.

All these phenomena have an influence on the short- and long-term adhesion between a lignocellulosic particle and a mineral binder. Thus, the diffusion of liquid water into plant cell walls may, potentially, destroy the hydrogen bonds established with the binder (Coutts and Kightly 1984; Vick 1999) and cause the plant tissues either to swell or to retract (Rowell 2005).

3.3.2.1 Fick's Law

The numerous diffusive phenomena observed in nature are described by *Fick's laws*, and can be characterised by their *diffusion coefficient D*. These are very common transport phenomena, usually engendered by the creation of a gradient of chemical potential, hydric gradient or concentration gradient. Diffusion tends to even out that gradient, in accordance with Fick's first law:

$$\mathbf{J} = \mathbf{D}.\frac{\partial \boldsymbol{C}}{\partial \boldsymbol{x}} \tag{3.8}$$

J: Diffusive flow [mol m^{-2} s^{-1}]
D: Diffusion coefficient [m^{-2} s^{-1}]
C: Concentration of molecular species [mol m^{-3}]
X: Spatial coordinate where diffusion is observed [m]

Consider the end of a wood core sample, stabilise in an atmosphere of relative humidity RH_0, brought into contact with an atmosphere of HR_1 at a time t = 0. A gradient of humidity concentration C(x,t) is instantaneously created with the wood tissues, giving rise to a diffusion front whose depth is d(t). The Brownian nature of the motion accounts for the *progress of the diffusion front d proportional to the square root of elapsed time.*

3.3.2.2 Einsteinian Laws

The diffusion of the extractible materials and mineral particles in the interstitial liquid medium is responsible for problems with the setting of the mineral binder in contact with a plant particle. This migration of species is governed by the Einstein–Smoluchowski law (Einstein 1905), which defines the Brownian motion of the particles.

$$\mathbf{D} = \mu_0 . \mathbf{K_B} . \mathbf{T} \tag{3.9}$$

D: diffusion coefficient
K_B: Boltzmann's constant
T: Temperature of the medium [K]
μ_0: Mobility of the particle, of the ion, etc.

An important special case of this law defines the diffusion of spherical particles in a medium with a low Reynolds number (laminar regime): this is the Stokes–Einstein law (Bentz et al. 2009):

$$\mathbf{D} = \frac{\mathbf{K_B} . \mathbf{T}}{6\pi . \mu . \rho} \tag{3.10}$$

μ: Dynamic viscosity of the medium [Pa s]
ρ: Radius of the particle diffusing in the medium [m]

This relation is applicable in the capillaries of lignocellulosic stalks and the pores of cell walls, due to their very small diameter, meaning that their Reynolds number with water is less than 2400.

3.4 Analogy with Adhesion of Mortars to a Porous Support

A comparison can be drawn between the lignocellulosic wood granulates of "hempcrete" and a porous substrate (adherent) to which a mortar or flagging (adhesive) is applied. In order to illustrate this, firstly, there is need to set out the initial *hypotheses* that the *granulate is inert and dimensionally stable*. It is on these hypotheses that our discussion in this section is founded.

3.4.1 Capillary Absorbency of a Porous Support

With regard to the establishment of a short-term physical bond, liquid transfers by means of the *forces of capillarity* are of great importance. Numerous studies have demonstrated the influence of the capillary absorption properties of the porous substrate on the development of bonding forces at the mortar/substrate interface (Groot and Larbi 1999; Courard 2000; Sugo et al. 2001).

Courard (2000) defines these capillary exchanges the *absorbency* of the substrate for the interstitial liquid in the mortar. If the absorbency is not high enough, the segregation of water on the surface of the substrate is observed which led to an increase in the final porosity of the interface.

Groot and Larbi (1999) describe the existence of an *optimal capillary absorption coefficient* with which maximal properties are achieved. They introduced the concept of the initial rate of absorption (IRA [kg m^2 min^{-1}]), which is defined as the capacity for capillary absorption per minute of the porous substrate placed in 3 mm of water. The optimum IRA is defined as being that which delivers the highest bond strength between the porous support and the mortar (2). Given that the IRA depends upon the volume and pore size, the authors stress that it is not the only influential factor. This observation is also reported by Sugo et al. (2001).

Sugo et al. (2001) described the *suction potential* existing between a relatively dry porous substrate and a water-saturated mortar, which tends to balance out. The moment the materials are brought into contact, the substrate's great suction potential causes the transport of the interstitial fluid from the mortar toward the substrate/mortar interface, and then into the actual pores of the substrate. In the view of those authors, over the course of these exchanges, gradients of humidity and of suction potential are established at the interface. As the binder becomes hydrated, the gradients evolve and may, depending on the size and interconnectivity of the porous structure, cause a backward flow of liquid from the support to the mortar, facilitated by large pores in the support (Groot and Larbi 1999; Sugo et al. 2001). This backflow may give rise to a regain in hydration (Fig. 3.2).

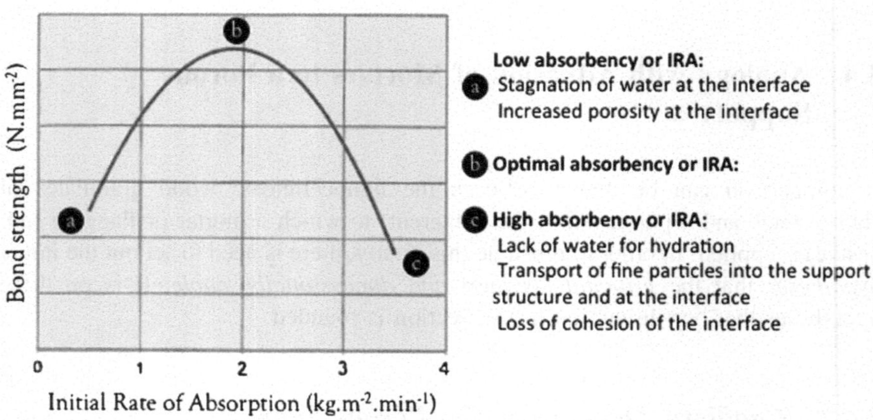

Fig. 3.2 Illustration of the optimum initial rate of absorption (Groot and Larbi 1999)

3.4.2 *Transport of Particles During Filtration*

Many authors reported the point that the processes of capillary absorption are responsible for the transport of fine particles in suspension from the mortar toward the surface and the porous structure of the substrate (Groot and Larbi 1999; Coutts and Kightly 1984; Sugo et al. 2001). Sugo et al. (2001) introduced the concept of a cluster of fine particles from the binder, which stagnate along the brick/mortar interface, providing continuity between the two materials. Excessive transfer would, in their view, lead to the fragilisation of the area of mortar situated before the interface, notably rendering it more porous. Conversely, an insufficient quantity would create an interface which was fragile because of poor interpenetration.

Groot and Larbi (1999) reported that these movements of fine particles from the mortar and their densification at the mortar/support interface are likely to substantially alter the capillary pressures exerted by the mortar.

3.5 Overview of the Processes of Binder/Wood Adhesion

The complexity of wood as a material—one which is highly hydrophilic, porous, subject to swelling, aniso-tropic and heterogeneous—accounts for the multitude of phenomena needing to be taken into account in order to understand its interaction with an adhesive (Coutts and Kightly 1984). The main interactions which take place in the specific case of the short-term adhesion of a mineral binder to wood are illustrated in Fig. 3.3. The important points which must be borne in mind are as follows:

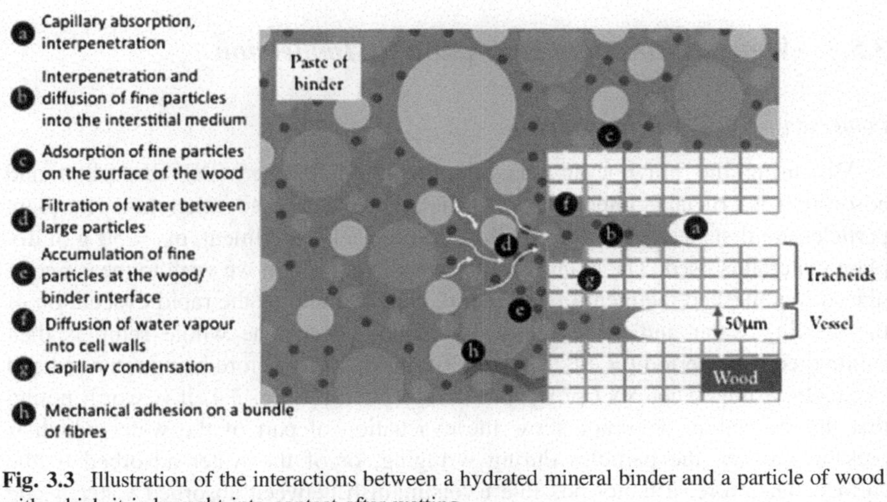

Fig. 3.3 Illustration of the interactions between a hydrated mineral binder and a particle of wood with which it is brought into contact after wetting

- Ensure optimal wetting which facilitates the interpenetration of the binder into the rough surface and the internal pores of the particles (vessels);
- Encourage as many physical bonds as possible, and interpenetration;
- Tend toward hydric transfers which are neither too great (lack of water for hydration) nor too slight (stagnation of water at the interface), by regulating the capillarity or filtration;
- Limit the filtration of fine particles from the binder toward the interface so as to prevent the buildup of particles;
- Limit the diffusion of particles from the binder into the interstitial liquid so as to limit concentration gradients;
- Limit the effects of the variations in the volume of the wood so as to ensure the interface is durable

3.5.1 Hygroscopic Behaviour When Immersed

The hygroscopic behaviour of lignocellulosic plant materials is due, in no small part, to their hydrophilicity. Their complex architecture is marked by multi-scale porosity, designed to carry the fluids necessary for them to flourish (sap and water). Even after cutting and transformation, this porosity continues to play its part, and thus is the primary channel of water absorption in accordance with Laplace's laws. This absorption takes place primarily in the conductive vessels or tracheids before the water propagates to the rest of the cells by diffusion across blockages and cell walls.

3.5.2 Water Absorption/Adsorption by Immersion

Immersion of dispersed granulates

Measuring the gravimetric rate and the rate of water (W) absorption and adsorption is crucially important in formulating plant-based concretes. The plant particles are first steamed at 60 °C for 48 h. For each experiment, m_0 = 50 g of dry plant material is used. The plant particles whose absorption we wish to measure are placed in a meshed recipient (a net, a grid, etc.) that allows the rapid evacuation of the soaking water and some of the interstitial water. The whole setup is then submerged for a period t = {1, 2, 5, 10, 30 min or 48 h} before being removed and manually wrung. The exact procedure is illustrated by Fig. 3.4. It is worth noting that this procedure does not allow the evacuation of part of the water which is present between the particles during wringing, or of the water adsorbed to the surface. Therefore, it is not possible to distinguish between absorbed water inside the capillaries and the water adsorbed to the surface. In addition, the manual

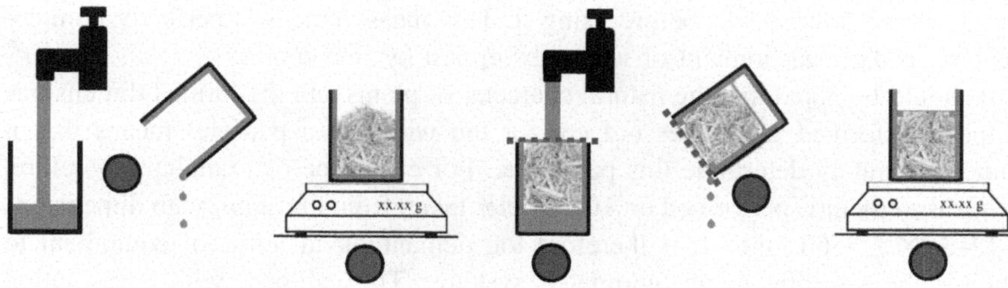

1. Fill the recipient with water and drain it;
2. After taring the balance, weigh out $m_0 = 20$ g of dry granulated material (60 °C, 48 h) into the damp recipient;
3. Cover the recipient with a sieve and fill with water through the sieve;
4. After a given period of immersion, drain the recipient through the sieve, taking care to shake it and achieve maximum evacuation of the water;
5. Wipe the walls of the recipient and weigh it on the balance whose tare has not been altered (*m(t)*).

Fig. 3.4 Step-by-step method for measuring the absorption/adsorption of plant particles by immersion

wringing engenders a not-insignificant degree of uncertainty, mainly when the experiment is of short duration.

$$W(t) = \frac{m(t) - m_0}{m_0}.100 \tag{3.11}$$

For each given duration, the measurement is repeated three times—in total, 18 tests—giving an absorption curve W(t) for a given particulate material. The maximum absorption rate WSAT is defined at 48 h.

Immersion of chips of known dimensions

The water absorption of chips of dimensions [2–3] × 7 × 60 mm^3, was analysed in parallel to that carried out on granulates. In each experiment, 5 chips are weighed and measured in the dry state before being submerged in water. Their mass is measured using a balance with precision of (±0.1 mg) after 5 min, and after 4, 14, 24, 38 and 48 h.

Tangential swelling of chips

The tangential swelling of immersed plant particles gives an idea of their capacity for deformation when brought into contact with water. Lignocellulosic plants, indeed, have the property of being able to integrate molecules of water or any other solvent into their very structure—notably due to the creation of hydrogen bonds (Rowell 2005). It is for this reason that we must allow freshly-cut wood to dry (and therefore shrink) before using it. This measurement is perfectly complementary to the measurement of water absorption by immersion.

It should be noted that the natural character of plants and the limited dimensions of the transformed granulates (<1 cm for the width of a particle) means that it cannot accurately determine this parameter. For each type of granulate, therefore, the measurement is performed on 10 particles taken from the stem, with dimensions of [2–3] × 7 × 60 mm^3. It is therefore too demanding in terms of equipment to monitor the swelling using automated systems. The callipers with a resolution of ± 0.01 mm is selected. With untreated hemp and sunflower stalks, the measurements were taken at regular intervals up to 48 h.

The tangential swelling is expressed simply:

$$G_T(t) = \frac{l(t) - l_0}{l_0}.100 \tag{3.12}$$

Maximum swelling GT MAX is considered to have been achieved after 48 h.

Wettability

The measurement of the contact angles seems to partly contradict the IRTF analyses presented in the previous section (Fig. 3.5a). Thus, the surface that proves most hydrophilic at the moment of deposition of the drop is the sunflower epidermis (76°), with the internal surface being initially quasi-hydrophobic (86°). This is attributable to a very rough surface on the internal face of the sunflower which, for a short period of time, keeps the drop spherical (Fig. 3.5c). For defibred hemp plant granulate, few differences are visible at the moment the drop is deposited: both faces are initially hydrophobic ($\theta > 90°$).

The major differences in terms of the wetting behaviour can be seen after 60 s. The surfaces of the epidermal faces lead to a gradual and limited spreading of the deposited drop (Fig. 3.5b). The behaviour of the internal surfaces is very different because, very quickly, the drop undergoes significant spreading and is then absorbed into the structure of the particle (Fig. 3.5c). This observation can be linked to the internal faces' high content of hemicellulose, which is extremely hydrophilic. In order for the plant to survive, the primary cell walls situated inside of the stem—particularly in the marrow—are composed of hydrophilic substances (cellulose, hemicellulose).

3.5.3 Behaviour in Terms of Water Adsorption/Absorption

Adsorption/absorption and swelling of chips

Before truly turning our attention to the behaviour of the transformed granulates when immersed, preliminary information can be gleaned by examining a single chip. With this goal in mind, chips of hemp and sunflower ([2–3] × 7 × 60 mm^3)

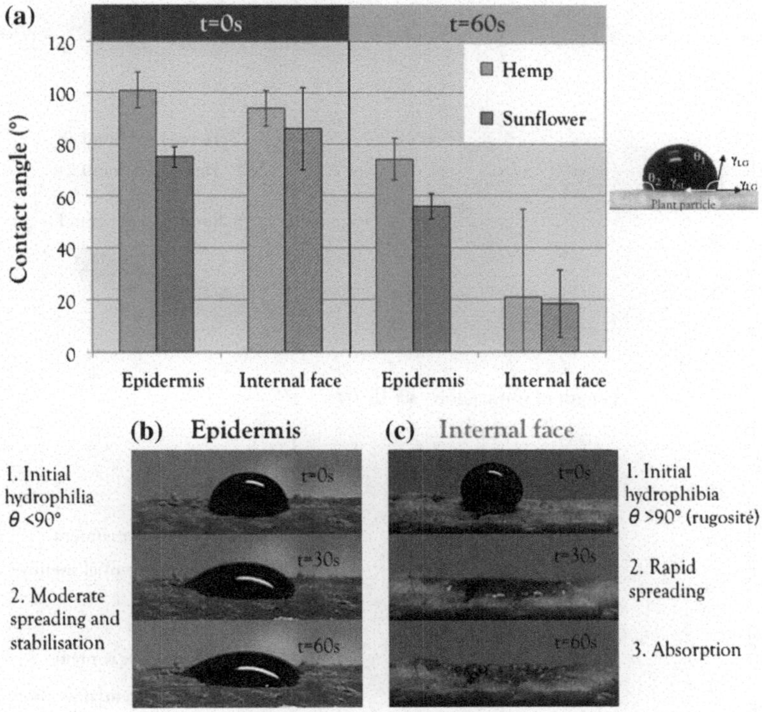

Fig. 3.5 Contact angles formed by a drop of water on the epidermal internal faces of hemp and sunflower particles after 0, 30 and 60 s (**a**). Observation of the characteristic spreading of the drop between 0 and 60 s on the epidermal face (**b**) and the internal face (**c**) of the sunflower

were immersed in mains water ($\gamma_{LG} = 72$ mN m^{-1}). The absorption/adsorption curves obtained (Fig. 3.6a) show three successive phases:

- Adsorption of water to the surface of the particle;
- Absorption of water into the internal structure of the particle;
- Gradual diffusion of the trapped air.

The *first phase of surface adsorption* leads to a *near-instantaneous weight gain* W_0 of the chip. This highlights the rapid nature of the wetting of hemp- and sunflower particles, which we saw earlier when we measured the contact angles. Note that the adsorptive weight gain is directly linked to the ratio between the surface area of the granulate and its mass, but also to the surface tension of the liquid in which it is immersed.

The curves themselves (Fig. 3.6a) show, in a *second phase, the absorption in the plant structure*, for which the kinetics *depends on the square root of the elapsed time*. This demonstrates the *diffusive behaviour* of this propagation of water in the structure up to 14 or 24 h. An inflection of the curve relating to the slowing of that diffusion illustrates the influence of the “finite” dimension of the chips. It should be

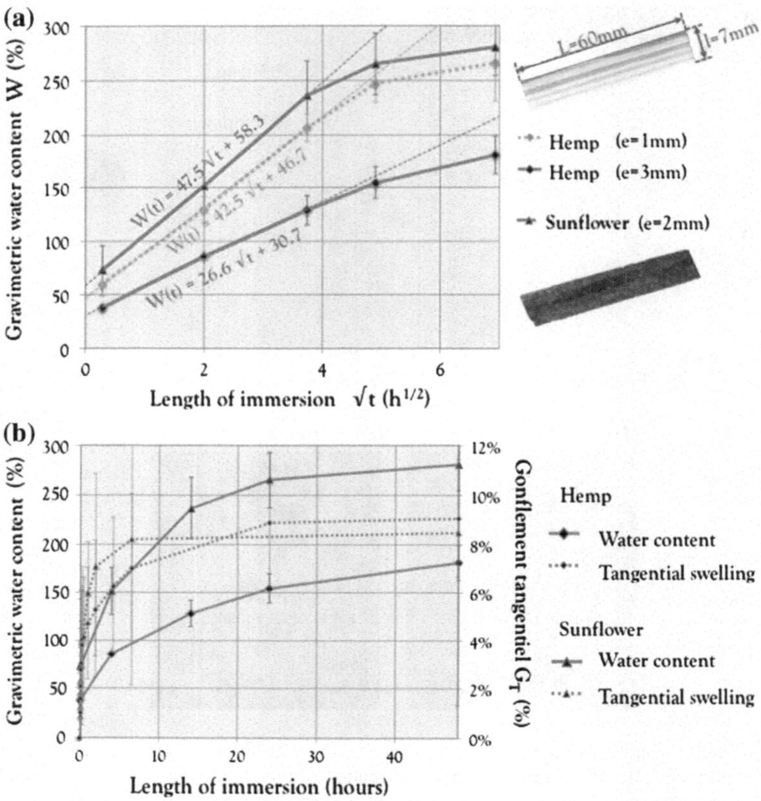

Fig. 3.6 Water adsorption/absorption curve by the immersion of chips (60 × 7 × (1–3) mm^3) of hemp and of sunflower as a function of the square root of elapsed time (**a**). Comparison between the gravimetric water adsorption/absorption and the tangential swelling during immersion (**b**)

noted that the thickness of the chip has a very significant influence on when that slowing starts. A less thick chip tends toward equilibrium more quickly. Hence, diffusion takes place not only in a longitudinal but also in a radial direction.

The *third phase* of the behaviour begins at the start of the inflection on the curve which marks the filling of the particle by the liquid water. This phase corresponds to the *diffusion of the air trapped in the particle* toward the outside, and is extremely slow (AFNOR 1999b).

The phenomenon of adsorption/absorption in the chips before the start of the third phase can therefore be characterised by a simple relation:

$$W(t) = C_A\sqrt{t} + W_0 \tag{3.13}$$

W_O: Initial water adsorption on the surface of the chip [kg kg^{-1}];
C_A: Coefficient of water absorption by the chip [kg s$^{-1/2}$].

Comparing the results between the hemp and sunflower chips highlights the significant capacity for adsorption and absorption of the sunflower granulate. The greater adsorption can be attributed to the rough internal face of the sunflower particles. The available wettable surface is also increased by the multitude of the surfaces provided by the cellulosic marrow, which is shredded during the transformation of the plant material.

The bringing of the plant particles into contact with water, leading to its integration by adsorption/absorption, is responsible for structural swelling. This swelling takes place when water diffuses into the cell walls and forms hydroxide bonds with the hydrophilic compounds such as hemicellulose and cellulose. Figure 3.6b shows that the tangential swelling $G_{T\text{-}SAT}$ is around 9% for pre-dried hemp and sunflower chips. This process is quicker that water absorption, because it takes around 20 h for hemp granulate and 7 h for sunflower granulate to become dimensionally stable. This stabilisation is the sign that the cell walls have reach their fibre saturation point (FSP), which generally lies between 40 and 50% of the initial mass (Rowell 2005).

Adsorption/absorption of divided granulates

The curves of immersion in water for divided granulates of hemp and sunflower (Fig. 3.7a) are of a similar shape to those found in previous works on hemp material (Cerezo 2005; Nguyen et al. 2009a; Arnaud and Gourlay 2012). As is the case with the chips, a first phase of adsorption and a second phase of internal absorption can be seen.

The wetting phase can be considered to have finished after 1 min. The initial adsorption W_0, therefore, will be defined for that duration of immersion. In terms of that value, a significant difference can be seen between the hemp granulate ($W_0 = 214.1\%$) and the sunflower material ($W_0 = 362.3\%$). This too can be attributed, as in the previous section, to the significant roughness of the internal face of the sunflower particles. However, there is an extra factor which comes into play. The divided granulates of hemp and sunflower, have respective specific surfaces estimated at 195 and 226 $cm^2\ g^{-1}$. This larger specific surface of the sunflower granulate also leads to increased initial adsorption. This result is clearly visible when we compare the initial adsorption of that granulate ($W_0 = 362.3\%$) to that of a second sunflower granulate with a smaller specific surface ($S_{SPE} = 157\ cm^2\ g^{-1}$). The W_0 then drops by 40%, ending up at the value of 216% because of the specific surface that is 30% lesser.

From Fig. 3.7, it is evident that the absorption of intra-granular water obeys a logarithmic law. This behaviour is due to the particle-size distribution of the granulates, which, as we saw, also obeys a logarithmic distribution. During this absorption phase, the granules become saturated, one by one, starting with the finest and ending with the coarsest. Thus, we can define the following relation, which is valid until the granulates reach saturation:

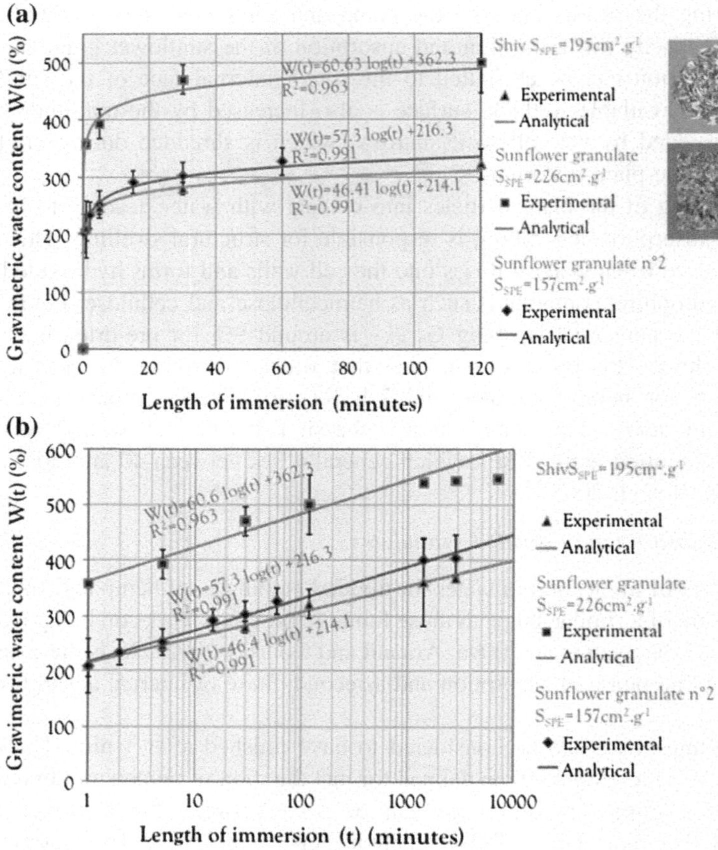

Fig. 3.7 Water adsorption/absorption curves for immersion of hemp and sunflower granulates on a classic timescale (**a**) and on a logarithmic timescale (**b**)

$$W(t) = C_A . \log(t) + W_0 \tag{3.14}$$

W_0: Initial water adsorption on the surface of the granulates;
C_A: Water absorption coefficient of the granulates.

The analytical absorption laws governing the behaviour of granulates report an absorption coefficient that is 30% higher in a sunflower (C_A = 60.6) than a hemp plant (C_A = 46.4). It should be noted that, unlike with surface adsorption, the specific surface has a limited impact on absorption. The sunflower granulate n°2, which has 30% less of a specific surface, thus has absorption kinetics reduced by only 5% (C_A = 57.3).

3.6 Conclusion

In view of the problems relating to water absorption by plant particles, this property has an important factor on the performance of agro-concrete.

It is crucial to understand that the creation of an interface worthy of the name in a composite material, whatever its nature, involves the creation of forces of adhesion between its components.

To date, limited studies have been published on the adhesion between a mineral binder and a lignocellulosic material.

In this chapter, a summary of the adhesion processes and the problems engendered by the combination of a mineral binder and porous lignocellulosic particles is given and the the process of the water as the main problem that needs to be solved.

In this chapter, an overview of the physical laws governing water absorption/adsorption by a plant particle was highlighted. The link with the process of adhesion of a binder to the plant matter was then reported.

The methods used to measure the absorption rate and of the tools used to model it was also described, followed by an analysis of the effect of absorption on the dimensional stability of a particle.

References

AFNOR (b): Méthodes d'essai pour pierres naturelles—Détermination du coefficient d'absorption d'eau par capillarité, AFNOR (1999)

Arnaud, L., Gourlay, E.: Experimental study of parameters influencing mechanical properties of hemp concretes. Constr. Build. Mater. **28**, 50–56 (2012)

Bentz, D.P., Peltz, M.A., Snyder, K.A., Davis, J.M.: VERDiCT: viscosity enhancers reducing diffusion in concrete technology. Concr. Int. **31**, 31–36 (2009)

Bruyne, L.E.: Aspects on wettability and surface composition of modified wood, Doctoral thesis, Manufacturing systems, KTH-Royal Institute of Technology, Stockholm (2008)

Cerezo, V.: Propriétés mécaniques, thermiques et acoustiques d'un matériau à base de particules végétales, Doctoral thesis, INSA de Lyon (2005). http://docinsa.insa-lyon.fr/these/pont.php?id=cerezo, 14 Sept 2010

Coutts, R.S.P., Kightly, P.: Bonding in wood fibre-cement composites. J. Mater. Sci. **19**, 3355–3359 (1984)

Courard, L.: Parametric study for the creation of the interface between concrete and repair products. Mater. Struct. **33**, 65–72 (2000)

De Gennes, P.G., Brochard-Wyard, F., Quéré, D., Reisinger, A.: Capillarity and wetting phenomena: drops, bubbles, pearls, waves, p. 291p. Springer, New York (2004)

Diederich, P., Mouret, M., Ponchon, F.: Design of self-compacting concrete according to the nature of the limestone filler, Actes du congrès SCI 2010, Montreal (Canada) pp. 137–147 (2010)

Einstein, A.: Über die von der molekularkinetischen Theorie der Wärme geforderte Bewegung von in ruhenden Flüssigkeiten suspendierten Teilchen. Ann. Phys. **322**, 549–560 (1905)

Elfordy, S., Lucas, F., Tancret, F., Scudeller, Y., Goudet, L.: Mechanical and thermal properties of lime and hemp concrete ("hempcrete") manufactured by a projection process. Constr. Build. Mater. **22**, 2116–2123 (2008)

Gardner, D.J., Generalla, N.C., Gunnells, D.W., Wolcott, M.P.: Dynamic wettability of wood, Langmuir, vol. 7, pp. 2498–2502, 1 Nov 1991

Groot, C., Larbi, J.: The influence of water flow (reversal) on bond strength development in young masonry. Heron **44**, 63–77 (1999)

Kwasny, J., Sonebi, M., Taylor, S., Bai, Y., Owens, K., Doherty, W.: Influence of the type of coarse lightweight aggregate on properties of semi-lightweight self-consolidating concrete. ASCE Mater. J. Civil Eng. **24**(12), 1474–1483 (2012)

Mounanga, P., Poullain, P., Bastian, G., Glouannec, P., Khelifi, H.: Influence de la composition et du mode de mise en oeuvre sur le développement des propriétés mécaniques du béton de chanvre, Actes du 27ème congrès de l'AUGC, Saint-Malo, 3–5 June 2009. http://www.augc09.univ-rennes1.fr/, 14 Sept 2010

Nguyen, T.T.: Contribution à l'étude de la formulation et du procédé de fabrication d'éléments de construction en béton de chanvre, Doctoral thesis, Université Bretagne Sud (2009a). http://web.univ-ubs.fr/limatb/lab/index.php?option=com_docman&task=doc_download&gid=67, 14 Sept 2010

Nguyen, T.T., Picandet, V., Amziane, S., Baley, C.: Influence of compactness and hemp hurd characteristics on the mechanical properties of lime and hemp concrete. Eur. J. Environ. Civil Eng. **13**, 1039–1050 (2009b)

Nguyen, T.T., Picandet, V., Carre, P., Lecompte, T., Amziane, S., Baley, C.: Effect of compaction on mechanical and thermal properties of hemp concrete. Eur. J. Environ. Civil Eng. **14**, 545–560 (2010)

Rowell, R.: Moisture properties. In: Rowell, R. (ed.) Handbook of Wood Chemistry and Wood Composites, CRC Press, 21p (2005)

Sugo, H.O., Page, A.W., Lawrence, S.J.: The development of mortar/unit bond. In: Proceedings of 9th Canadian Masonry Symposium, Fredericton, 4–6 June 2001

Vick, C.: Adhesive bonding of wood materials. In: Wood Handbook: Wood as an Engineering Material. vol. 113, W. U. F. S. Madison, Ed.: General technical report FPL, pp. 9.1–9.24 (1999)

Walinder, M.: Wetting phenomena on wood—factors influencing measurements of wood wettability, Doctoral thesis, Manufacturing Systems, KTH-Royal Institute of Technology, Stockholm (2000)

Chapter 4
Particle Size Distribution

Vincent Picandet

Abstract In this chapter, a state of the art of Particle Size Distribution (PSD) measurement of bio-based aggregates and characterization methods is presented. Shiv particles coming from the stem of plants cultivated either for their fibers (hemp, flax, etc.) or for their seeds (oleaginous flax, sunflower, etc.) are very different from the mineral aggregates typically used in concretes. Owing to the structure of the stem of the plant they are made from, such aggregates are generally malleable, elongated and highly porous with a low apparent density. Irregular shape are generally observed, especially in case of shiv coming from fiber plant due to the shredding action of the decortication process. Such ground bio-mass lead usually to uni-modal size distribution that can be efficiently characterized using basic distribution models with two parameters. Starting from the standardized tools and techniques developed for mineral aggregates, other technics using image processing are investigated and discussed in the global perspectives of the effect of the PSD on the properties of the in-service building material.

Keywords Shiv · Image-Analysis · Size Distribution · Sieving · Fibre plant · Comminution

4.1 Introduction

At present, no norm exists to cover the PSD of bio-sourced aggregates. They are different in many respects from the mineral aggregates traditionally employed in hydraulic concretes, which rounder, very unyielding with low porosity and considerably denser, for which methods of characterization, mainly by sieving, have been defined and are employed in the published standards. Yet the industrial

V. Picandet (✉)
IRDL—CNRS FRE 3744—Université Bretagne Sud,
Rue de Saint-Maudé, 56100 Lorient, France
e-mail: vincent.picandet@univ-ubs.fr

S. Amziane and F. Collet (eds.), *Bio-aggregates Based Building Materials*,
RILEM State-of-the-Art Reports 23, DOI 10.1007/978-94-024-1031-0_4

implementation either on-site or in a precast factory necessitates a better characterization of these aggregates to stay abreast of the quality of the finished materials. This chapter gives the state of the art of previous studies dealing with various grounded bio-mass. Results lead to suggest some PSD measurement recommendations and to focus on some relevant parameters to characterize efficiently the size distribution of shiv particles.

4.2 General Characteristics of Shiv Particles

Straw of bast fibre plants is composed of very long and not heavily lignified cortical fibres surrounding a woody part at the centre of the stem, made of very heavily lignified short fibres (Crônier et al. 2005). This woody part carried the sap while the plant was growing. The cortical fibre, rich in cellulose, represents the main value of this agricultural product (Bouloc et al. 2006). In order to separate the flexible fibres from the separate the stiff and more brittle ligneous woody parts, the straw is usually mechanically processed (Sponner et al. 2000).

During the process of decortication, the straw is ground, usually using a hammer mills (Munder et al. 2005). The woody part is detached from the fibres, and shredded into small pieces to form hemp or flax shiv. In the case of hemp, 100 kg of straw, when ground, yield around 30 kg of fibre, 60 kg of hemp shiv and 10 kg of dust (Bouloc et al. 2006; Bevan and Woolley 2008; De Bruijn et al. 2009). Although it is the main constituent, hemp shiv is merely a co-product of the exploitation of hemp. Its main use up until now has been in animal litter or horticultural straw.

The term “hemp shiv” is currently used to denote aggregates from the stem of the hemp plant which may be very varied, as they come from agricultural products that are subject to weather hazards and obtained using various post-harvest processes. Flax shiv, for its part, present still more disparate characteristics, particularly due to the larger variety of species that can be grown.

The preferential orientation of the cell walls along the stem axis lead to a strong orientation of the porosity inside shiv particles, with significant anisotropic properties, as observed in wood (Phạm 2014). As a consequence, the process of decortication usually produces elongated particles from the woody part, elongated along the preferential orientation of the microstructure or the porosity, i.e. along the stem axis where they come from. In case of hemp shiv, these pores have a diameter that essentially varies between 10 and 50 µm and a length of around 80 µm (Picandet 2013). In addition, the implementation processes employed generally tend to induce a preferential orientation of these particles which are usually elongated, and therefore—depending on the overall shape of the particles—give rise to a tangible anisotropy of the material, particularly in terms of its thermal characteristics (Elfordy et al. 2008; Nguyen et al. 2010b).

4.2.1 Fibre Contents

When the plant is felled, it may be left on the ground for a variable length of time so that retting will facilitate the process of decortication. The advance of this process, the dampness of the straw and the regulations of the grinders used affect the size of the particles obtained when the straw is ground (Mani et al. 2004; Miao et al. 2011). Also, the decortication performed may be more or less vigorous, and the fibre content in the hemp shiv or the flax shiv may be variable.

Studies based on two types of hemp shiv: a hemp shiv gained from an advanced process of decortication with very few residual fibres, and another gained from a partial process of decortication, containing a significant amount of short cortical fibres (shorter than 4 cm) show a significant effect of residual fibre on the bulk density of the hemp shiv. The remaining fibres, usually short fibres attached to the particles induces an expansion of the apparent bulk volume. When the cortical fibre mass content is higher than 10%, the presence of fibre has a significant effect on bulk shiv and it can be distinguish with the naked eye (Picandet 2013).

Any standard methods exist to evaluate the fibre content in hemp or flax shiv. When fibres are present, the longest fibres can be easily separated manually in a few grams dry sample. Sieving can be used to complete the separation of the fibres from the hemp shiv. Indeed, the fibres tend to form pellets in the first sieves—see Fig. 4.1. Also, when the fibre and the hemp shiv are still linked, the vibrations of the sieve are enough to detach them. Generally, the amount of fibres in the hemp shiv tested varies between 1 and 15% of the mass (Picandet 2013). This amount of fibre should not be taken into account in the PSD in the cumulative size distributions as shown in Fig. 4.2. Such flexible fibres are not consider as aggregate. They have a higher specific weight and represent a very small part of the overall particle volume. Yet even in the case of hemp shiv obtained by an advanced process of decortication, a detailed examination of the aggregates shows that a small amount of short fibres may still be found attached to a few shiv particles, even after the stages of sieving performed, for at least 30 min (Fig. 4.1).

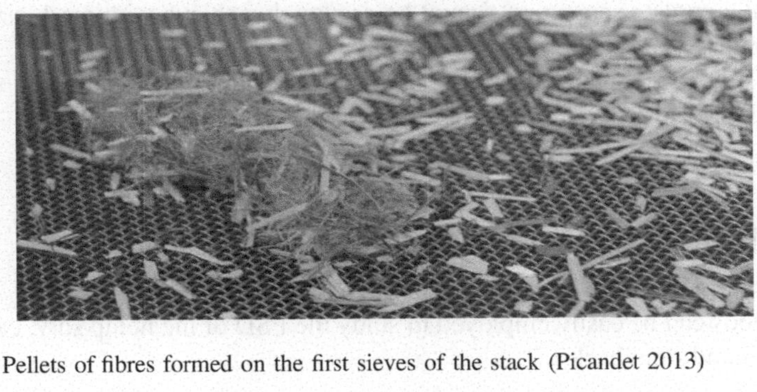

Fig. 4.1 Pellets of fibres formed on the first sieves of the stack (Picandet 2013)

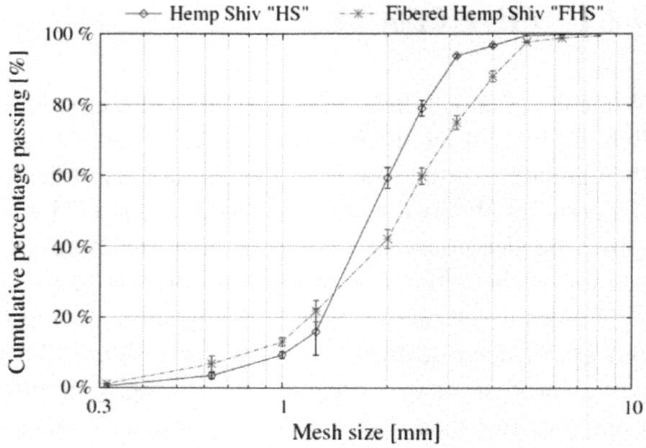

Fig. 4.2 Cumulative size distribution obtained by sieving (Picandet 2013)

4.2.2 Dust Content

According to authors, dust represent all the finest particle, with a mineral or organic composition. The shiv particles, even they are bio-based material, are made of insoluble mineral matter that can be determined by a number of methods. The quantity of ash in fibrous plants is generally low, whereas in grasses it may be significantly higher—particularly in rice or wheat straw, whose silica (SiO_2) content is higher. Separation of organic and mineral compounds will not lead to relevant results regarding the dust content.

Many studies considers that dust represent all particles with a size lower than 0.5 mm (Bevan and Woolley 2008), (De Bruijn et al. 2009). However, as mentioned bellow, this method let suppose a sieving procedure using corresponding mesh size. Is such a case, the size would represent the width of the particle and lower mesh size would be more appropriate to separate particles considered as dust. Sieving method with shiv particle have to be performed with dry material in order to separate the finest particles from the others as far as possible. Nevertheless, a part of the finest particle could remain attached or stuck to the coarser ones. Experience shows that complete separation cannot be accurately achieve with mesh size lower than 0.1 mm.

Hemp shiv also contains dust of organic or mineral origin, usually less than 2% of the weight of the particles passing through the 0.315 mm sieve.

4.2.3 Methods to Measure PSD

Two methods can be easily employed to study the PSD of the hemp shiv, each with advantages and drawbacks:

- The conventional method of sieving with dry particles enables us to take measurements directly on a sample of a few hundred grams. The limitations of this method are due to the elongated shape of hemp shiv particles, and their low density, which render sieving less appropriate and unreliable (Igathinathane et al. 2009a).
- 2D image analysis of particles spread out over a flat surface gives us access to more information. Thereby, the width and length of each particle detected can be measured. However, this method is more complex and requires samples no larger than a few grams. The precision of the results produced is therefore limited by the representativeness of the sample and by the only dimensions obtained for the aggregates by projection onto a plane.

4.3 Sieving Methods

The shiv particles have to be dried until constant weight, usually in a ventilated oven, at 60 °C or 50 °C depending on the running procedures in the laboratories. Sieves with standardized square mesh are used, as is a mechanical siever for the study of soils and mineral aggregates (NF ISO 3310.1–ASTM E-11-95). The devices used induce a vertical motion which is created by a tapping impulse. In order to distribute the sample amount over the whole sieving surface, the vertical throwing motion is overlaid with a slight horizontal circular motion. This procedure has been selected since most of the laboratory involved already use such devices to characterize mineral aggregates.

However, it should be noted that other devices have been especially designed for needle-shaped, flat, long or fibrous samples such as shiv particles. For instance, horizontal sieve shaker (only horizontal circles motion in a plane) with sieves made of perforated slick steel plates can be used, as recommended by ASABE (2006), see Fig. 4.3. The preferential orientation of particles laid on the screen means that only a few disoriented particles pass through the sieve when their length are lower than the diagonal sieve aperture (Bitra et al. 2009a).

In order to obtain repetitive results, the vibration time has to be extended to 30 min at least for a 100-g sample and for 5 consecutive sieves. The apertures of the sieves used can range from 10 to 0.315 mm.

PSD analysis by sieving assumes that all the particles are practically spherical in shape, and pass through a square aperture when their diameter is less than the side of the square. For flat or elongated particles, such as hemp shiv particles, this point is developed in further detail in Sect. 3.1. The particles may either pass through the sieve in the direction of their length (see Fig. 4.8) or retained if they are positioned across the aperture. In the latter case, these particles may also block the passage of particles located above them.

Generally speaking, increasing the time taken over sieving helps to reduce the relative differences of the refused particles obtained for each sieve. On the basis of

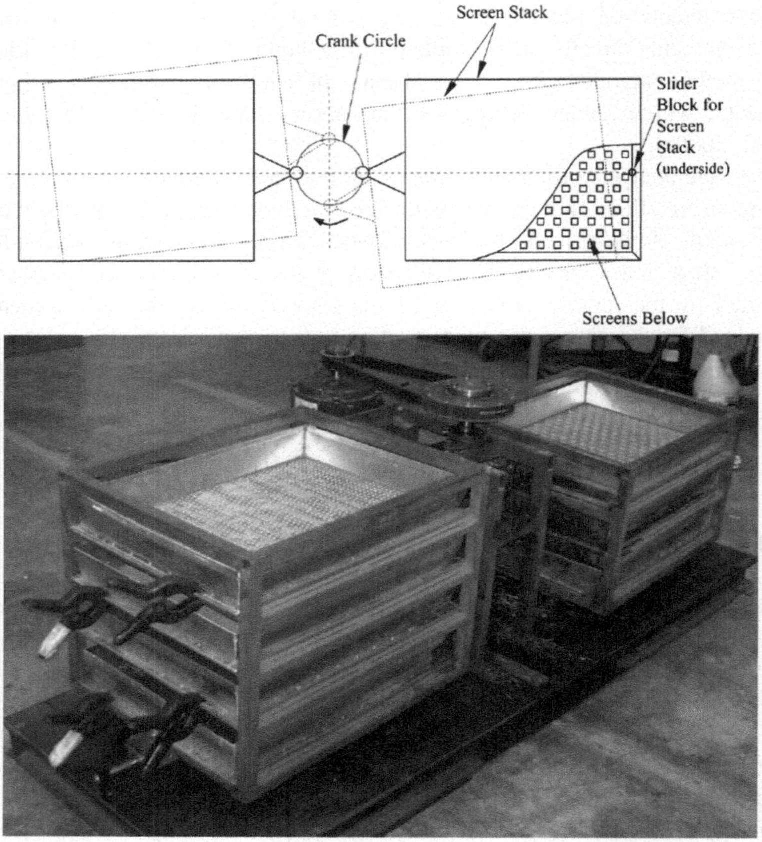

Fig. 4.3 Sectional plan view of ASABE sieve analyser and picture of American Society of Agricultural and Biological Engineers (ASABE) sieve analyser (Bitra et al. 2009b)

several tests, the precision of the results, i.e. the evaluation of the particles retained in each sieve, was evaluated at ±15% with the series of sieves used. These uncertainties are illustrated in Fig. 4.2 with two type of hemp shiv: one with a low fibre.

4.4 Image-Processing Methods

This method requires good-quality sampling, in that only a finite amount of material can reasonably be analysed. A classic method of quartering can be applied and repeated as many times as need be. The representativeness of the sample selected is the key element guaranteeing the relevance of the results produced since only few grams can be reasonably analysed.

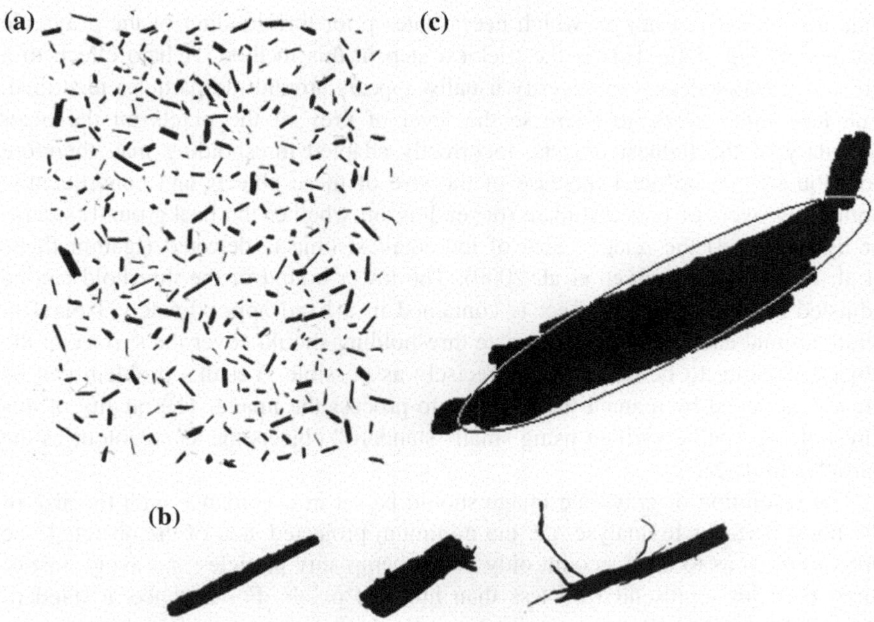

Fig. 4.4 **a** Binarized image of singulated arrangement of hemp shiv particles over an A4 size area. **b** Hemp shiv particles after binarization of the image, classified on the basis of biases in the analysis which will cause protuberances due to tearing and to remaining connected fibres. **c** Adjusted ellipse on a binarized particle to evaluate its length and width

Image analysis is based on a two-dimensional observation of particles spread out over a flat surface. Images can be obtained with a conventional scanner generally used to digitize documents. This technique offers the advantage of avoiding any distortion of the image which might occur if a camera were used (Picandet 2013). The scanner can acquire a colour image or an image converted into 8-bit grayscale that can be processing by an image-analysis program such as "ImageJ", which is an open source program freely accessible online (Ferreira et al. 2012).

The particles are spread out so that they do not overlap or touch, see Fig. 4.4a. This requires very detailed attention when the particles are fine and in these conditions it is difficult to avoid some overlap (Igathinathane et al. 2009a, b). With the aim of simplifying the procedure and increasing the precision of our measurements, rather dispersed distribution of particles reduces touching or overlap bias is recommended even it requires a greater number of frames to be analysed. This method refers to a "static image analysis method" according to the international standard ISO 13322-1:2014(E). Other "dynamic image analysis" methods exist ISO 133222:2006(E), and use image capture of moving particle dispersed on a conveyor or falling particles dispersed in the air.

As the shiv under investigation could be light in colour, a dark background should be used in order to obtain a maximum degree of contrast. Image analysis requires a binarized image, which necessitates prior thresholding of the grayscale image, see Fig. 4.4a. This is the trickiest step in this method. A halo effect, to a greater or lesser degree of severity usually appears around the particles identified. The halo effect tends to decrease the level of grey of the pixels on the outer boundary of the lightest objects. Incorrectly adapted thresholding may therefore contribute to an artificial increase of the size of these objects and consequently, noticeably over- or underestimate (depending on whether the background is dark- or light-colored) the relative size of the smallest objects detected (Igathinathane et al. 2009a, b, c; Nguyen et al. 2009). The lower bound of the threshold can be adjusted so that this halo effect is contained in a band approximately 1 pixel in width around each object. Appropriate thresholding should cover the surface of the objects needing to be detected as precisely as possible. Yet this problem can be greatly assuaged by manual thresholding to process the image. The quality of this thresholding can be verified using small "standard" objects the same colour as the particles to detect.

The resolution of grayscale image should be set in accordance with the size of the finest particles to analyse, i.e. the minimum projected area of the objects to be processed, so as to take account only of the hemp shiv particles and avoid dust or finest particles represented by less than hundred pixels. For instance, a standard resolution of 600 DPI (dots per inch) on both the vertical and horizontal axes corresponds to a constant scale factor of 0.04233 mm per pixel. This scale factor can be verified by calibration. The precision of the measurements can further be improved by increasing the resolution. For instance, with a 600 DPI resolution, only the particles whose area is greater than 0.2 mm^2 (i.e. objects represented by at least 110 pixels) can be taken into account for the PSD analysis. Such detection threshold, allow to take into account most of the finest hemp shiv as the coarser dust particle while it fulfil the recommendations of ISO 9276-6:2008 setting the limit to 100 pixels belong each analysed object.

Other processing operations can also be carried out with a view to preparing the images before analysis. One such operation, which involves producing an erosion of a given number of pixels followed by an operation of expansion of the object by adding the same number of pixels onto its boundary, can help eliminate dust particles and fibres which are not representative of the hemp shiv particles needing to be identified. This operation is referred to as an opening operation, because it may lead to the de-compartmentalization of cavities, separated by thin boundaries, which may be contained in the objects (Ferreira et al. 2012) but some tests with hemp shiv do not show significant effects (Picandet 2013)

4.5 Image-Analysis

Image analysis can go further in particle characterization and gives us access to far more information than does sieving. For each particle detected, its projected area and the perimeter of that projected area are directly measured and recorded. Shape as well as size parameters can be evaluated, and multiclass analysis of bio-based materials can be achieved. The aim of this section is to provide information that could help characterization of different commercial products.

4.5.1 Size Estimation

For each particle detected, its projected area and the perimeter of that projected area can be directly measured and recorded. Other more elaborate parameters, such as the minimum convex hull area surrounding the object, can be used to define different formal parameters. Of these, the convexity ratio, χ (also called "solidity") (Mora and Kwan 2000), defined as the object's projected area over the convex area surrounding that object, reveals the form of the particles. Perfectly convex particles have a convexity ratio of 1.

Observation of different types of ground-up straw shows that the resulting particles have irregular and angular forms due to the microstructure of the plant and to the shredding action of the decortication process, see Fig. 4.4b. In this scenario, the shapes of the finest particles therefore tend to be polygonal and convex, whereas the shapes of the coarsest particles tend to diversify to include non-convex particles (Bitra et al. 2009a). Particles with too low convexity ratio, less than 2/3 for instance, can be removed from analysis devoted to convex particles. Actually, selection before analysis does not lead to a significant difference in case of hemp shiv particles (Picandet 2013). Overall, in the case of convex and non-convex particles gleaned from ground straw, the method for determining the length based on the diameter of the smallest enclosing circle or maximal calliper (which some writers also refer to as the Féret diameter) is fairly representative of the length of the object (Igathinathane et al. 2009a, b, c). Hence, the length can be directly quantified using this maximum diameter, defining the major axis of the projected area (see Fig. 4.4b).

The measurement of the length and width of these particles may be subject to different definitions, depending on the representativeness of these dimensions in the case of the type of object needing to be analysed.

The width can be first defined as the minimal Féret diameter or minimal calliper, i.e. the minimum distance between two parallel straight lines (or planes) encompassing the object, or indeed as the width of the narrowest rectangle (or parallelepiped) containing the object, see Fig. 4.5. This method lead to suppose that the estimation of the width is correct in the case of rectangular particles. However, in the case of hemp shiv, the short fibres still connected and the shredded particles give rise to outcrops from the projected areas and ultimately cause an overestimation of the widths obtained by way of this method.

In order to iron out some of the outgrowth of the objects studied (Fig. 4.4b) and analyse them using geometric forms deemed to be representative, other methods

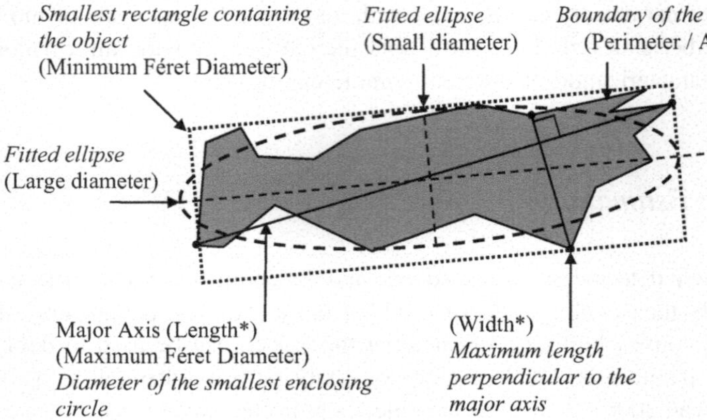

Fig. 4.5 Lengths and widths analysed (Picandet 2013)

exist. Such methods consist of adjusting the basic geometric shapes (rectangles, ellipses, triangles, polygons, etc.) to the objects detected (Bitra et al. 2009a; Igathinathane et al. 2008) so as to determine their length, and above all their representative width. Of these, in the case of hemp shiv, an ellipse can be adjusted so that its centre of gravity corresponds to that of the object and its projected area is identical to that of the object. The lengths and widths of the object are therefore defined respectively in accordance with the large and small radii of the adjusted ellipses (see Fig. 4.5).

It should be noted that with rectangular shapes, the adjustment of an ellipse also leads to an overestimation of the lengths and widths in identical proportions, so that the elongation of the particles, ε, again remains unchanged.

For around 2600 analysed particles contained in a 4 g sample hemp shiv, Fig. 4.6 gives a view of the logarithmic scale of the widths in comparison to the lengths analysed. The elongation of the particles, ε (the ratio of length to width of

Fig. 4.6 Lengths and widths deduced from 2600 analysed hemp shiv particles

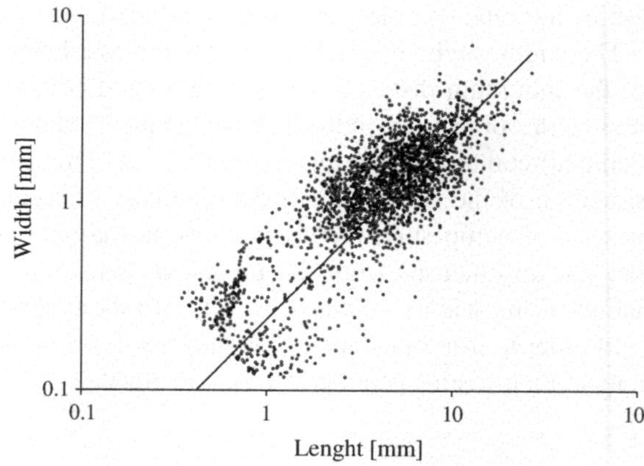

the particles). Of these point clouds, two general categories of particles may appear: the particles representative of hemp shiv, for which the corresponding point cloud is centred on a width of around 2 mm and dust or micro-fibres, for which the point cloud seems to be truncated by the threshold selected.

Of the various characteristics that are measured, the elongation seems critically important, because it will condition the orientation of the granular arrangement and the anisotropy of the finished materials. The point cloud is primarily clustered around a straight line denoting a constant length-to-width ratio and let suppose an homothetic shape of the hemp shiv particle analysed (Picandet 2013).

4.5.2 *Distributions*

Various types of distribution, or probability density, of size of particles may be deduced from these tests, depending on whether they are distributed in accordance with their number or with their projected area. However, so as to be meaningfully representative of potential granular packing and to be comparable with the analyses obtained by sieving, this analysis should be performed on the basis of a distribution of the volume of the particles.

4.5.2.1 Frequency Distribution

Distributions on the basis of the number of particles can be performed directly from the raw data from the image analysis step. However, this type of distribution is very sensitive to the number of the smallest particles considered. Such a distribution based on the relative cumulative number of particles passing through a sieve of a given size, denoted as N% in Fig. 4.7, therefore depends heavily on the detection threshold. Moreover, such distribution is not relevant in case of aggregates to be incorporated into a mix since a large number of smallest particle represent a minor part of the volume of the whole particles.

4.5.2.2 Area Fraction Distribution (Projected Area)

Image analysis also gives us access to the projected area of each particle detected: A_i. The influence of the finest particles on the PSD can therefore be weighted by this criterion, so as to consider a distribution by the cumulative projected area of the particles whose considered size is less than a given value. The distribution of cumulative passing can be directly calculated on the basis of the sum of the projected areas of n particles, arranged in order of increasing size, for a total number of N particles detected, whose total area is A_T. The cumulative distribution by

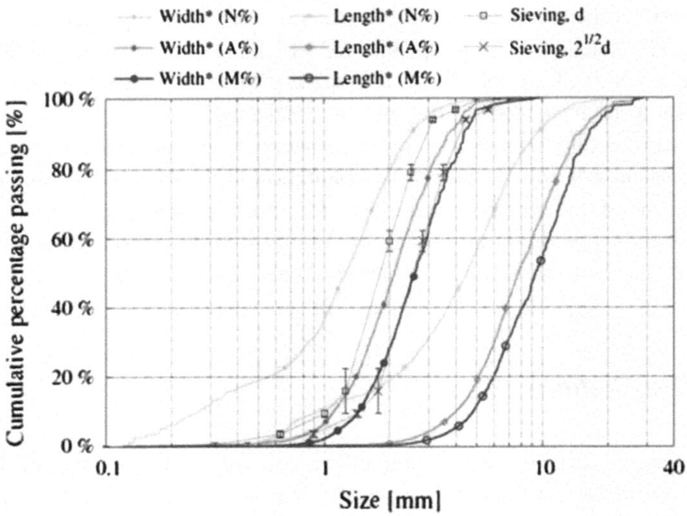

Fig. 4.7 Lengths and widths deduced from 2600 analysed hemp shiv particles

increasing size, $P_A(X \leq x_n)$, similar to the "cumulative passing" obtained by sieving (see Fig. 4.2), can then be written as:

$$P_A(X \leq x_n) = \frac{\sum_{i=1}^{n} A_i}{\sum_{i=1}^{N} A_i} = \frac{1}{A_T} \sum_{i=1}^{n} A_i \tag{4.1}$$

where X is the considered size of the particles and $P_A(X \leq x_n)$ the proportion of the projected area accounted for by particles smaller than the nth particle of size x_n. The cumulative distributions $P_A(X \leq x_n)$ and the distributions of size based on the area of the particles are therefore annotated ($A\%$) in the figures.

4.5.2.3 Mass Fraction Distribution

The PSD curves are usually traced on the basis of the results obtained by sieving, i.e. on the basis of a mass distribution. If M_i is the mass of the particle i, the distribution $P_M(X \leq x_n)$ is directly calculated from the cumulative mass of the n smallest particles passing through a sieve of given size out of a total of N particles of mass M_T:

$$P_M(X \leq x_n) = \frac{1}{M_T} \sum_{i=1}^{n} M_i \tag{4.2}$$

where X is the considered size of the particles and $P_M(X \leq x_n)$ the proportion of the mass of particles smaller than the nth particle of size x_n.

In order to compare sieving result with image analysis, a cubic weighting with respect to the particle size (width or length) of the particles is required. In case of spherical particles, a thousand 100 μm-diameter particles have the same volume than a single 1 mm particle.

If the apparent density of the particles is independent of their dimensions, $P_M(X \leq x_n)$ can also be written in accordance with the volume V_i of each particle and their total volume V_T. In order to deduce a volume distribution from a projected area distribution some assumptions need to be introduced. Based on many observation on hemp shiv, the assumption of quasi-homothetic particles can be naturally introduced. It considers that the particles are similar in shape, e_i denotes the average thickness over the whole of the projected area of each particle, i.e. The volume-based cumulative distribution is obtained by weighing the area, A_i, with the width, e_i, of each particle considering $V_i = A_i e_i$.

$$P_M(X \leq x_n) = \frac{1}{V_T} \sum_{i=1}^{n} V_i \cong \frac{\sum_{i=1}^{n} e_i.A_i}{\sum_{i=1}^{N} e_i.A_i} \tag{4.3}$$

Many complementary tests can be performed; yet it is not easy to approximate the thickness e_i of each particle. It can only be supposed that since the particles are simply and freely spread out over a plane, this average thickness is less than the width of each particle. The thickness of the woody part in the stem varies noticeably, depending on the climatic conditions, the date of harvesting and the density of the plantation (Schäfer and Honermeier 2006) on the one hand, and then depending on the height of the particular section within the stem (Rahman Khan et al. 2010). However, the process of decortication is applied to all the cropped straw, giving rise to multidirectional grinding. As observed for different types of ground straw, the general shape of the particles does not seem to be affected by the diameter of the stems being ground (Igathinathane et al. 2009a, b, c; Nguyen et al. 2010) whereas the size of the particles produced depends essentially on the process of grinding itself and on the settings used (Bitra et al. 2009a). In the presented case, the average elongation ratio of the particles is, overall, independent of projected area. Extension of this assumption into the third dimension lead to consider the average ratio of width to average thickness, $\bar{\Phi} = e_i/l_i$, constant as well.

If the density of the particles is identical, with a similar shape irrespective of their size, i.e. if they are generally homothetic, the mass distribution of the particles can be deduced from the projected area A_i and the width l_i of the particles by the following relation:

$$P_M(X \leq x_n) \cong \frac{\sum_{i=1}^{n} l_i.A_i}{\sum_{i=1}^{N} l_i.A_i} \left(\text{if } \frac{e_i}{l_i} \cong \bar{\Phi}(\text{const.}) \right) \tag{4.4}$$

It is in view of this hypothesis that the cumulative distributions $P_M(X \leq x_n)$ and distributions of size based on the mass of the particles are annotated as (*M%*) in Fig. 4.7 hereafter.

It should be noted that the cumulative distribution $P_M(X \leq x_n)$ depends neither on the value of the apparent density of the particles, ρ_a, nor on $\bar{\Phi}$. Because $P_M(X \leq x_n)$ is sensitive to the largest particles, this distribution may be considerably different from $P_A(X \leq x_n)$ in the case of a spread distribution. It is interesting to note that if all the particles had the same thickness, the approximation of the cumulative distribution by mass, $P_M(X \leq x_n)$, defined in equation (Eq. 4.4), would be equivalent to $P_A(X \leq x_n)$ defined in equation (Eq. 4.1). Figure 4.7 illustrate the difference observed between the cumulative distributions based on the number of particles, the projected area and the supposed mass of the particles, for both the width and length of hemp shiv particles.

4.5.2.4 Average Flatness Estimation

The average ratio of the thickness to the width, $\bar{\Phi}$, can be considered as the flakiness of the particles. As mentioned in the previous section, it is not needed to deduce the volume or mass fraction distribution from the area distribution. It can be evaluated if the mass or total volume of the particles being analysed is known (Kwan et al. 1999).

$$\bar{\Phi} = \frac{V_T}{\sum_{i=1}^{N} l_i . A_i} = \frac{M_L}{\rho_a \sum_{i=1}^{N} l_i . A_i} \tag{4.5}$$

In our case, the mass of the sample is known, and the apparent density of the particles ρ_a can be estimated (Ceyte 2008; Nguyen et al. 2009; Nguyen et al. 2010). Introducing the apparent density ρ_a of the particles approximately equal to 300 kg/m^3, in the case at hand; $\bar{\Phi} \cong 1/3$ if the dimensions are estimated on the basis of adjusted ellipses.

4.5.3 Comparison with the Results Obtained by Sieving

As presented in Fig. 4.7, the cumulative distributions of the particles are, overall, closer to the curve of cumulative passing obtained by sieving, considering the width, rather than the length. This implies that the particles are able to pass through the sieves in a direction perpendicular to the meshes, and that the sorting of the hemp shiv when the particles are sieved is done primarily on the basis of the width rather than the length of the particles.

Assuming ellipsoidal particles, elongated and flaky, when these particles can pass lengthways through the sieves, their width may be oriented along the diagonals of the square holes (see Fig. 4.8). In this case, only those particles whose width is greater than $2^{1/2}d$ are retained by a sieve whose mesh size measures d.

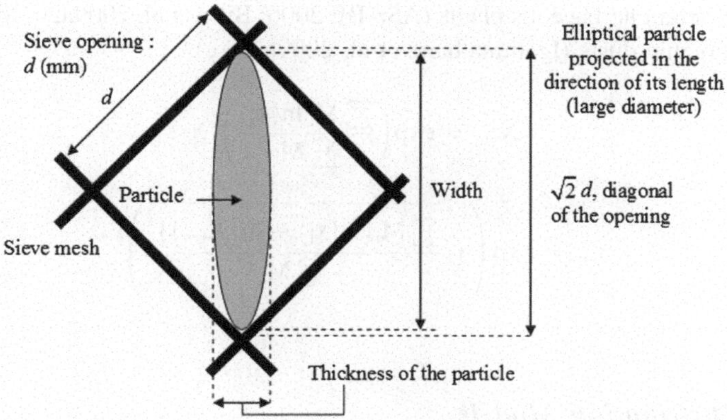

Fig. 4.8 Cross-section of the largest elliptical particle able to pass through the sieve, in the direction of its length and along the diagonal of the square hole

Figure 4.8 shows that the distribution of the cumulated mass widths (M%) is very similar to the distribution obtained by considering the diagonal of the square opening in the sieves, $2^{1/2}d$ mm instead of the mesh size. As Fig. 4.7 confirms, the size-per-mass distribution of the particles is also that which corresponds most closely to the results obtained by sieving. In addition, the standard distributions of particle size presented shows uni-modal nature of the area fraction (*A*%) and mass fraction (*M*%) distributions for the width and length of the particles that can be characterized using standard distributions laws.

4.6 Characterization of the PSD

4.6.1 Means and Standard Deviations

The arithmetic mean of the size (width or length) of the particles, weighted by their area or mass, and the associated standard deviation, can be used to gain an overall characterization of the PSD. The standardized degrees of skewness and kurtosis relative to the 3rd- and 4th-order moments of distribution can also be calculated.

However, in that the distributions can be approximated by a normal law according to a scale of logarithmic size, the weighted geometric mean X_{gm} and its associated standard deviation σ_{gm}, defined in the two equations below, appear to be more relevant when seeking to characterize the distribution, as in the case of numerous distributions of particle sizes obtained from the comminution of the stems of different herbaceous plants (ASABE 2006; Bitra et al. 2009a, b; Miao et al. 2011) or organic dusts (Igathinathane et al. 2009a, b, c).

$$X_{gm} = \exp\left(\frac{\sum M_i \ln(x_i)}{\sum M_i}\right) \quad (4.6)$$

$$\sigma_{gm} = \exp\left(\sqrt{\frac{\sum M_i (\ln(x_i) - \ln(X_{gm}))^2}{\sum M_i}}\right) \quad (4.7)$$

4.6.2 Distribution Models

The maximum amount of information can be obtained if a distribution model is adjusted to the PSD under investigation (Djamarani and Clark 1997). Different laws exist that are based on semi-infinite variables. Usually, basic models with two parameters—one relating to the average size and the other to the spreading of the distribution—are used.

4.6.2.1 Log-Normal Distribution

Given the uni-modal nature of the distributions presented and apparently symmetrical according to a logarithmic scale of size, the Log-Normal law seems obvious as a first approach (Limpert et al. 2001). Its distribution function of the lengths X is written as follows:

$$P_{Log.N}(X \leq x) = \frac{1}{2}\left[1 + \text{erf}\left(\frac{\ln(x) - \mu}{\sigma\sqrt{2}}\right)\right] \quad (4.8)$$

where μ and σ are the parameters needing to be identified, and erf(x) denotes the Gauss error function. It should be noted that e^{μ} and e^{σ} reciprocally represent the weighted geometric mean and the associated standard deviation, which can both also be calculated on the basis of the whole dataset including each particle identified, and usually denoted as X_{gm} and σ_{gm}.

4.6.2.2 Rosin-Rammler Distribution

When the size distribution relates to particles or fragments obtained by grinding, one of the other models used most frequently in the existing body of literature is the Rosin-Rammler model (Djamarani and Clark 1997), particularly in the case of biomass, (Allaire and Parent 2003, 2004; Bitra et al. 2009b, 2011). The distribution function of this model, identical to the Weibull distribution, can better represent skewed distributions (Rosin and Rammler 1933):

$$P_{RR}(X \le x) = 1 - \exp\left[-\left(\frac{x}{\lambda}\right)^k\right] \tag{4.9}$$

where λ and k are constants relating respectively to the dimension of the 63.2th percentile of the distribution function and to the tightening of the distribution. This distribution function has the advantage of exhibiting a reciprocal function which can be used to directly calculate the dimensions corresponding to a given cumulative fraction.

$$x = \lambda\left[-\ln(1 - P_{RR}(X \le x))^{1/k}\right] \tag{4.10}$$

The median size of the particles $D_{RR}(50)$ corresponds to $P(X \le x) = 50\%$ and is equal to $\lambda.\ln(2)^{1/k}$ mm.

4.6.3 *Fitting of the Distribution Laws*

The aforementioned two distribution models can easily be adjusted to the area fraction and volume fraction distributions. Figure 4.9, shows the fitted models encompass each of the distributions, in width or length. Overall, the distributions are better represented by a log-normal law, particularly for the smallest values.

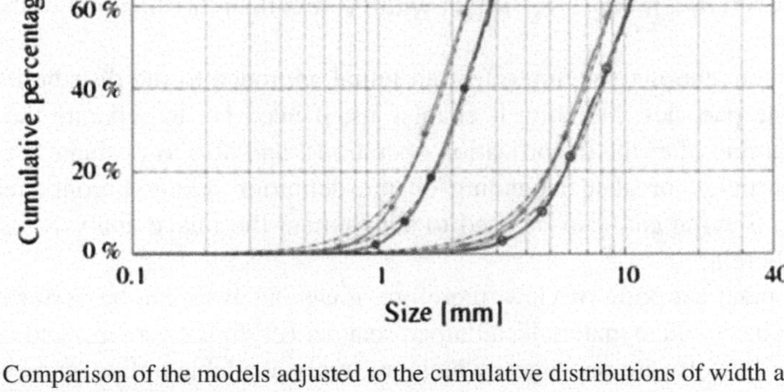

Fig. 4.9 Comparison of the models adjusted to the cumulative distributions of width and length in area fraction and mass fraction

These two models can also be adjusted for data obtained by sieving, but with far fewer points (see Fig. 4.2). Using a common "least-squares" method, as mentioned in international standard ISO 9276-3:2008 about the adjustment of an experimental curve to reference model of PSD, the correlation coefficient R^2 is lower than 0.999 over 2600 identified particles. The values deduced from the parameters of these models can thus be used to give a fairly accurate overall characterization of the PSD under study, as in the case of various biomass coming from the comminution of the stem (Allaire and Parent 2003, 2004; Bitra et al. 2009b, 2011). The Rosin-Rammler model can be better at describing the PSD obtained by sieving of the finer particles, because it takes account of greater skewness toward fine particles. The log-normal law, for its part, is better at describing the distributions obtained by image analysis and corresponds more closely to the modes observed.

4.7 Conclusions

The image analysis method, based solely on 2D observations, can be used to gain a precise measurement of the length distribution of the hemp shiv particles and, to a lesser extent, when these particles retain connected fibres, of their width. Various image analysis algorithms can be used to determine their width, such as the minimum Féret diameter or the measurement of the small diameter of an adjusted ellipse. Many different shape parameters are available, and a large number of particles can be identified to deliver statistically robust results.

A near-constancy of the average elongation of the particles for the different surface intervals is observed. In addition, according to this basic hypothesis, image analysis allows us access to the average elongation $\bar{\varepsilon}$ and the average flatness $\bar{\Phi}$ of the particles assuming homothetic shape in 3D, i.e. a quasi-constant average flakiness. These parameters cause significant anisotropy in the materials due to casting process of LHC. The comparison between sieving and image analysis shows that the square-mesh sieves conventionally used for mineral aggregates separate the hemp shiv particles, elongated and flat, essentially on the basis of their width. Furthermore, given how thin they are in comparison to their width, the hemp-shiv particles pass through the sieves if that width is less than the diagonal of the square holes.

Hence, conventional sieving offers an initial approach to the distribution of the width of the particles. Notably, it enables us, if need be, to separate the cortical fibres remaining after the decortication operations, and also to evaluate the amount of finest particles, or dust depending on the definition adopted, contained in the hemp shiv. Sieving can also be used to supplement the image analysis techniques presented herein.

From a practical point of view, therefore, image analysis can be performed very easily with basic office materials, a flatbed scanner (or digital camera) and a modern computer. The software packages, which are freely available online, can be used in their basic, default configurations. Experience shows us, in the case of the hemp shiv under investigation, that a detection threshold for the particles set around 0.2 mm^2 (or 100 pixels at least), with no special discrimination between the particles in view of their convexity, yields similar results: with no significant differences in terms of the distributions either of area fraction or of mass fraction. In addition, it is sufficient to sample 4 g of material obtained by successive stages of quartering to provide reliable and reproducible results.

The distributions obtained by sieving and image analysis, for their width or length, cumulative in terms of area or mass, are usually uni-modal and can be accurately approximated using a classic log-normal law. The weighted geometric mean and the associated standard deviation therefore constitute the main representative parameters with regard to the granular packing and to the overall properties induced in the materials, to characterize this type of PSD.

References

Allaire, S.E., Parent, L.E.: Size guide number and Rosin-Rammler approaches to describe particle size distribution of granular organic-based fertilisers. Biosyst. Eng. **86**, 503–509 (2003)

Allaire, S.E., Parent, L.E.: Physical properties of granular organic-based fertilisers, part 1: static properties. Biosyst. Eng. **87**, 79–87 (2004)

ASABE.: Standards. Method of determining and expressing particle size of chopped forage materials by screening ANSI/ASAE S424.1. St. Joseph, MI, pp. 619–662 (2006)

Bevan, R., Woolley, T.: Hemp lime construction, a guide to building with hemp lime composites. HIS Bre Press (2008)

Bitra, V.S.P., Womac, A.R., Chevanan, N., Miu, P.I., Igathinathane, C., Sokhansanj, S., Smith, D. R.: Direct mechanical energy measures of hammer mill comminution of switchgrass, wheat straw, and corn stover and analysis of their particle size distributions. Powder Technol. **193**, 32–45 (2009a)

Bitra, V.S.P., Womac, A.R., Chevanan, N., Yang, Y.T., Miu, P.I., Igathinathane, C., Sokhansanj, S.: Mathematical model parameters for describing the particle size spectra of knife-milled corn stover. Biosyst. Eng. **104**, 369–383 (2009b)

Bitra, V.S.P., Womac, A.R., Yang, Y.T., Miu, P.I., Igathinathane, C., Chevanan, N., Sokhansanj, S.: Characterization of wheat straw particle size distributions as affected by knife mill operating factors. Biomass Bioenergy **35**(8), 3674–3686 (2011)

Bouloc, P., Allegret, S., Arnaud, L.: Le chanvre industriel: production et utilisations. France agricole, Paris (2006)

Ceyte, I.: Béton de chanvre, définition des caractéristiques mécaniques de la chènevotte. Travail de Fin d'Etudes, ENTPE (2008)

Crônier, D., Monties, B., Chabbert, B.: Structure and chemical composition of bast fibers isolated from developing hemp stem. Journal of agricultural and food chemistry, no **53**, 8279–8289 (2005)

De Bruijn, P.B., Jeppsson, K.-H., Sandin, K., Nilsson, C.: Mechanical properties of lime-hemp concrete containing shives and fibres. Biosyst. Eng. **103**, 474–479 (2009)

Djamarani, K.M., Clark, I.M.: Characterization of particle size based on fine and coarse fractions. Powder Technol. **93**, 101–108 (1997)

Elfordy, S., Lucas, F., Tancret, F., Scudeller, Y., Goudet, L.: Mechanical and thermal properties of lime and hemp concrete ("hempcrete") manufactured by a projection process. Constr. Build. Mater. **22**, 2116–2123 (2008)

Ferreira, T., Rasband, W.: ImageJ User Guide 1.46r, U.S. National Institues of Health, Bethesda, Maryland, USA (2012)
Igathinathane, C., Pordesimo, L.O., Columbus, E.P., Batchelor, W.D., Methuku, S.R.: Shape identification and particles size distribution from basic shape parameters using ImageJ. Comput. Electron. Agric. **63**, 168–182 (2008)
Igathinathane, C., Pordesimo, L.O., Columbus, E.P., Batchelor, W.D., Sokhansanj, S.: Sieveless particle size distribution analysis of particulate materials through computer vision. Comput. Electron. Agric. **66**, 147–158 (2009a)
Igathinathane, C., Melin, S., Sokhansanj, S., Bi, X., Lim, C.J., Pordesimo, L.O., Columbus, E.P.: Machine vision based particle size and size distribution determination of airborne dust particles of wood and bark pellets. Powder Technol. **196**, 202–212 (2009b)
Igathinathane, C., Pordesimo, L.O., Batchelor, W.D.: Major orthogonal dimensions measurement of food grains by machine vision using ImageJ. Food Res. Int. **42**, 76–84 (2009c)
Kwan, A.K.H., Mora, C.F., Chan, H.C.: Particle shape analysis of coarse aggregate using digital image processing. Cem. Concr. Res. **29**, 1403–1410 (1999)
Limpert, E., Stahel, W.A., Abbt, M.: Log-normal Distributions across the sciences: keys and clues. Bioscience **51**(5), 341–352 (2001)
Mani, S., Tabil, L.G., Sokhansanj, S.: Grinding performance and physical properties of wheat and barley straws, corn stover and switchgrass. Biomass Bioenergy **27**, 339–352 (2004)
Miao, Z., Grift, T.E., Hansen, A.C., Ting, K.C.: Energy requirement for comminution of biomass in relation to particle physical properties. Ind. Crops Prod. **33**, 504–513 (2011)
Mora, C.F., Kwan, A.K.H.: Sphericity, shape factor, and convexity measurement of coarse aggregate for concrete using digital image processing. Cem. Concr. Res. **30**, 351–358 (2000)
Munder, F., Fürll, C., Hempel, H.: Processing of bast fiber plants for industrial application (Chap. 3). In: Mohanty A.K, Misra M., Drzal L.T. (eds.) Natural Fibers, Biopolymers, and Biocomposites, Taylor & Francis (2005)
Nguyen, T.T., Picandet, V., Amziane, S., Baley, C.: Influence of compactness and hemp hurd characteristics on the mechanical properties of lime and hemp concrete. Eur. J. Environ. Civil Eng. **13**, 1039–1050 (2009) p133 and p139.
Nguyen, T.T.: Contribution à l'étude de la formulation et du procédé de fabrication d'éléments de construction en béton de chanvre. Doctoral Thesis, Université de Bretagne Sud, Ecole Doctorale SICMA, Lorient (2010a). Avaiable online: http://web.univ-ubs.fr/limatb/lab/
Nguyen, T.T., Picandet, V., Carré, P., Lecompte, T., Amziane, S., Baley, C.: Effect of compaction on mechanical and thermal properties of hemp concrete. Eur. J. Environ. Civil Eng. **14**, 545–560 (2010b)
Phạm, T.-H.: Modélisation multi-échelles des propriétés thermiques et élastiques de composites chaux-chanvre, PhD thesis, Université de Bretagne-Sud (2014)
Picandet, V.: Characterization of plant-based aggregates (Chap. 2). In: Amziane S., Arnaud, L. (eds) Bio-Aggregate-Based Building Materials: Applications to Hemp Concretes, Wiley, pp. 27–74 (2013)
Rahman Khan, Md. M., Chen, Y., Laguë, C., Landry, H., Peng. Q., Zhong, W.: Compressive properties of hemp (Canabis Sativa L.) stalks. Biosyst. Eng. **106**, 315–323 (2010)
Rosin, P., Rammler, E.: The laws governing the fineness of powdered coal. J. Instrum. Fuel **7**, 29–36 (1933)
Schäfer, T., Honermeier, B.: Effect of sowing date and plant density on the cell morphology of hemp (Cannabis sativa L.). Ind. Crops Prod. **23**, 88–98 (2006)
Sponner, J., Toth, L., Cziger, S., Franck, R.R.: Hemp (Chap. 4), pp. 176–206. In: Franck R.R. (ed.) Bast and other plant fibres, Woodhead Publishing Limited in association with The Textile Institute Abington Hall, Abington Cambridge CB1 6AH England (2000)

Chapter 5
Bulk Density and Compressibility

Vincent Picandet

Abstract Hemp is made of highly deformable particles. Depending on the water content, on the particle size distribution, and on many other material parameters such as initial porosity and retting of the processed stalks, the mechanical behaviour of shiv in bulk can change significantly. In a compaction process, the mass per volume of the raw material increases with the applied stress and some creep or relaxation effects occur as observed in wood based materials. Hence, the mechanical properties of the bulk impact the packaging of the raw material as the shiv density inside the final mix and the in-service properties of the composite material. In this way, the bulk compressibility is primarily useful to manage the building processes, from transport of the raw material, to mixing and casting.

Keywords Shiv · Density · Bulk · Compressibility · Mechanical properties · Casting process

5.1 Introduction

Unlike mineral aggregates, usually employed in hydraulic concretes, bio-based aggregates made from plant particles are highly deformable and cannot provide a rigid skeleton to the composite material. Their mechanical behaviors impact significantly the building processes, especially in case of pre-casting where highest compaction can be achieved on the fresh mixes to optimize their mechanical resistance (Nguyen et al. 2009, 2010; Nguyen 2010), and finally strongly influence the cast concrete and its in-service behavior (Tronet et al. 2014). In case of building materials, only few studies deal with the mechanical characterization of the bio-based aggregate in bulk. This chapter presents the test methods used to characterize the bulk

---

V. Picandet (✉)
Centre de Recherche, IRDL—FRE CNRS 3744—Université Bretagne Sud, 56100 Lorient, France
e-mail: vincent.picandet@univ-ubs.fr

S. Amziane and F. Collet (eds.), *Bio-aggregates Based Building Materials*, RILEM State-of-the-Art Reports 23, DOI 10.1007/978-94-024-1031-0_5

compressibility and the results obtained with the same hemp shiv. Some analysis are also presented, especially in case of compression in a die where friction effects have to be simultaneously analyzed.

5.2 Density and Porosity, Case of Hemp Shiv

The densities and porosities measured on two types of hemp shiv (Nguyen 2010), named as "HS" and "FHS" with respectively less than 2% and more than 12% of cortical fibers by weight are summarized in Table 5.1.

The apparent density when loose and dry is measured on the basis of a cylindrical volume 160 mm in diameter and 320 mm in height, in which the loose dry hemp shiv is poured. Many test have been proceeded in the frame of second round robin test of this TC Rilem 236 to evaluate the effect of the pouring procedure and the effect of the size of the specimens on the results.

Otherwise, the apparent density of the particles can be measured on the basis of a straight section of stem, the area of which was determined by image analysis and the measured height. This figure, given as an indicative value, is underestimated in that the hemp shiv particles probably have a greater density (Ceyte 2008; Cerezo et al. 2005), owing to the stresses undergone during the decortication process and the stress of confinement when they were conditioned and kept in a 20 kg sack. A value of 300 kg m^{-3} would, at first glance, seem to constitute a more meaningful density of the particles used. It would represent a decrease of the intra-granular porosity but a slight increase in the inter-granular porosity as reported in Table 5.1.

The apparent density of the solid phase is determined by a pycnometer, using toluene as filling fluid.

From these measurements, we deduce the following porosity values:

- total porosity: $\phi_{total} = 1 - \rho_V / \rho_S$
- intra-granular porosity: $\phi_{intra} = 1 - \rho_P / \rho_S$
- inter-granular porosity: $\phi_{inter} = 1 - \rho_P / \rho_V$

Table 5.1 Densities and porosity of the hemp shiv under examination (Nguyen et al. 2009)

	HS	FHS
ρ_L apparent density loose and dry (kg m^{-3})	112	71
ρ_P apparent density of the dry particles (kg m^{-3})	256	256
ρ_S apparent density of the solid phase (kg m^{-3})	1460	1440
ϕ_{total}, total porosity (%)	92	95
ϕ_{intra}, intra-granular porosity (%)	82	82
ϕ_{inter}, inter-granular porosity (%)	56	72

It should be noted that when cortical fibers are present in hemp shiv, these fibers make up a significant proportion of the loose volume, and largely contribute to the increased inter-granular porosity.

Although the apparent density of the dry particles is probably underestimated, the intra-granular porosity proves to be very high, which accounts for the excellent absorbent quality of this material.

5.3 Bulk Compressibility

5.3.1 Low Stress Compression in a Die

The apparent density ρ_L of loose dry hemp shiv (see Table 5.1), can be significantly increased by the application of confining stress. The application of such stress is sometimes necessary for storage and/or transport. Applied upon implementation or casting, it can help greatly reduce the porosity in the elaborated material and increase its mechanical resistance (Nguyen et al. 2009).

A compacting test is performed in a steel cylinder with an internal diameter of 160 mm and a height of 320 mm. The stress is applied axially by a piston sliding into the cylinder. Compacting is performed by a hydraulic press with a capacity of 500 kN. The movement of the piston and the force applied are recorded. The force-displacement curve for both types of aggregates is recorded and expressed as Fig. 5.1, plotting axial applied stress against equivalent bulk density of dry hemp shiv.

For this test, the hemp shiv was humidified so that its water content was 100%, so that the test would be representative of the mechanical behavior of the hemp shiv

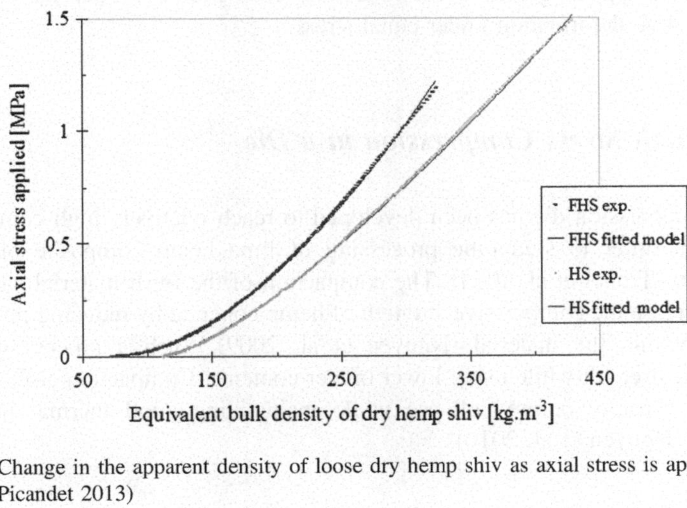

Fig. 5.1 Change in the apparent density of loose dry hemp shiv as axial stress is applied in the cylinder (Picandet 2013)

Table 5.2 Parameters of the compressibility model to the cases under investigation

	ρ_L [kg m^{-3}]	σ_o [MPa]	k (compressibility) [−]
HS	112	0.38	1.3
FHS	71	0.13	1.8

at the time when it is cast. This in fact corresponds to its water content at the moment of mixing during the manufacture of blocks destined for prefabrication by controlled compacting in the fresh state of hemp concrete (Nguyen et al. 2009; Nguyen 2010).

Figure 5.1 demonstrates that the apparent density of loose hemp shiv becomes greater than that of the uncompressed shiv once stress of 0.7 MPa is applied. The inter-granular porosity is then probably reduced, but, in view of the deformability of the shiv particles, no information can be gleaned from these tests about the inter-granular porosity in the compressed bulk.

In order to characterize the compressibility of hemp shiv in bulk, a basic model with two parameters, σ_o and k, (Picandet 2013) described by Eq. 5.1, inspired by models frequently used in the study of compacting of different biomasses (Emami and Tabil 2007), was used. It directly links the stress applied to the hemp shiv, σ, to the equivalent bulk density of dry hemp shiv in its initial state, ρ_{Vo}, and to that in the stressed state ρ_L.

$$\sigma = \sigma_o \left(\frac{\rho - \rho_L}{\rho_L} \right)^k \tag{5.1}$$

This model fairly accurately describes the experimental behavior observed within the range of stresses applied. The values of the parameters σ_o and k are recapped in Table 5.2. Here, σ_o is the compression stress required to double the initial bulk density, ρ_L. The parameter k, which relates to compressibility (Jones 1960), indeed appears greater in the case of fibrous hemp, FHS, because it exhibits greater relative deformation under equal stress.

5.3.2 *High Stress Compression in a Die*

A rigid compression die has been developed to reach relatively high compression pressure in order to study the processing of lime hemp composite blocks by compaction (Tronet et al. 2014). The compaction of the fresh material can significantly increase the compressive strength of hemp concrete by reducing the volume of voids within the material (Nguyen et al. 2009). Such a process improves mechanical strength while using lower binder contents. Compaction also increases the strain capacity before collapse while maintaining good thermal insulation properties (Nguyen et al. 2010).

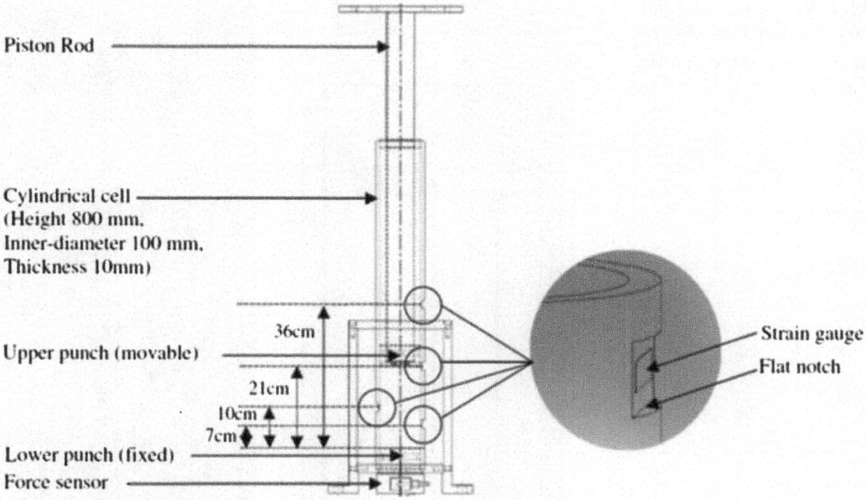

Fig. 5.2 The compression device, fitted with strain gauges (Tronet et al. 2014)

However, (Nguyen et al. 2010) observe that LHC compacted blocks are not homogeneous; they have a higher density, mechanical strength and conductivity in the upper area of the block than in the lower area. This effect is due to friction along the die walls during compression. An instrumented die has been developed to identify the cause of heterogeneity induced during moulding. This die was designed to provide information on the effective stress state undergone by LHC during compression, and particularly on the friction effects at the walls. It consists of a steel cylinder cell and two punches. The upper punch can move and is driven by the piston rod. This piston rod is mounted in a 250 kN compression test machine equipped with force and displacement transducers. The lower punch is also mounted on a force sensor. The thickness of the cylinder wall is 10 mm thus ensuring negligible strain up to a radial pressure of 10 MPa.

A mean radial pressure up to 10 MPa can be reached, and several flat spots were machined on the outside wall of the cell to ensure good accuracy and local measurements of radial stress, see Fig. 5.2. The strain gauge calibration was carried out with water. The stress state has been firstly evaluated for several mixes (dry or wet shiv without lime) to understand compressibility, friction and stress transmission mechanisms.

5.3.2.1 Measurements, Boundary Conditions and Stress

The cylinder, containing the mixture to be compacted, is circular cross section with a die radius R. The compacted height is h, see Fig. 5.3. Considering a rigid die, it is assumed that any radial displacement occurs along the sample height.

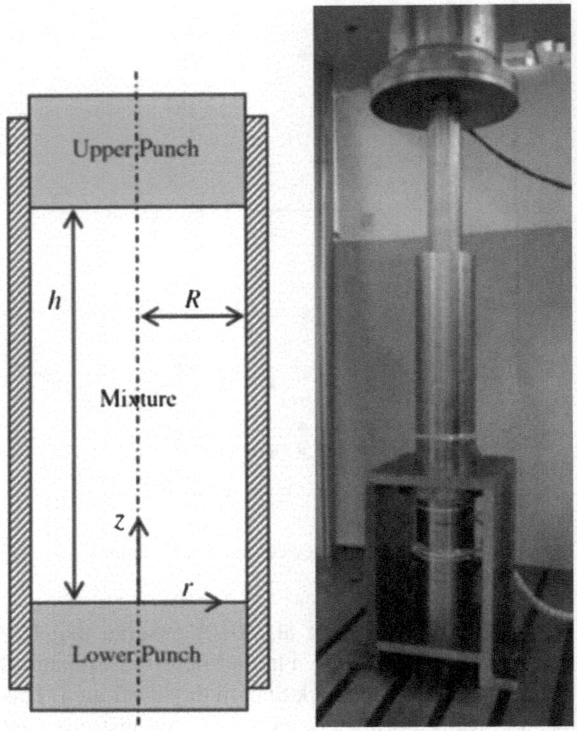

Fig. 5.3 Geometry of the device and picture of the experimental setup (Tronet et al. 2014)

The axial mean stress on the upper punch, σ_{zUP}, is controlled as $F_{UP} = \pi R^2 \sigma_{zUP}$.

The axial mean stress on the lower punch, σ_{zLP}, is deduced from the force measured from the transducer F_{LP} as $F_{LP} = \pi R^2 \sigma_{zLP}$.

The force F_{LP} is not a controlled parameter; it is dependent on σz_{UP}, on friction and on the stress transmission of the mixture from an axial to a radial direction. Generally, as proposed first by Janssen for stresses, in silos (Janssen 1895), and elaborated by Jacky in soil mechanics (Jacky 1944), and by many authors working on powder compaction (Doremus et al. 2001; Modnet 2000; Chtourou et al. 2002; Sinka et al. 2003; Jonsén and Häggblad 2005), or extrusion process of granular pastes (Perrot et al. 2006), this stress transmission consists of a proportionality law between the radial stress at the wall σ_r and the longitudinal stress along the cylinder σ_z as $\sigma_r/\sigma_z = \alpha$, where α is the transmission ratio. Moreover, tangential stress at wall τ_z and radial stress σ_r are linked by a Coulomb law (wall/sample friction coefficient μ_f), such as $\tau_z\ (z) = \mu_f \sigma_r\ (z)$ neglecting any cohesion in case of granular material.

Considering radial stress independent from the radial abscissa, i.e. constant in an ortho-axial plan, transmission ratio α and the friction coefficient μ_f, constant along the specimen. These two coefficients must be considered as mean values and are consequently computed from the values of σ_{zm} and σ_{rm}, i.e. the mean values of axial and radial stresses respectively. Mean radial stress σ_{rm}, is deduced from the radial

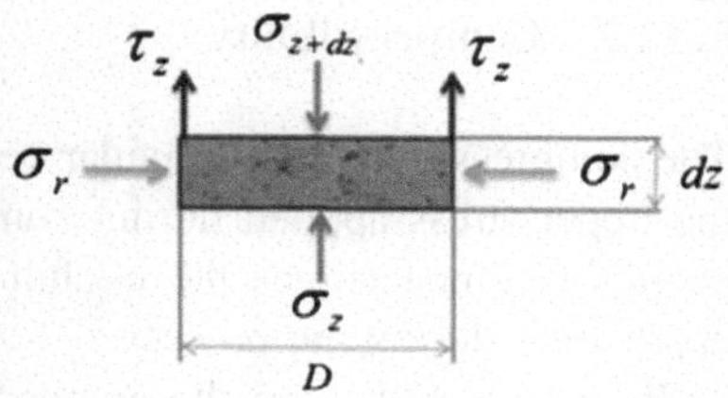

Fig. 5.4 Stress balance on a virtual slice of compacted material in the die

strain gauge measurements. In general terms, the literature considers that σ_{rm} is very close to radial stress at half-height of the specimen:

$$\sigma_{zm} \cong \frac{1}{h}\int_0^h \sigma_z(z)dz \text{ and } \sigma_{rm} \cong \sigma_r\left(\frac{h}{2}\right) = \alpha\sigma_z\left(\frac{h}{2}\right) \tag{5.2}$$

Neglecting the gravity effect, the balance of the stresses acting on an horizontal disk of material within the die, see Fig. 5.4, leads to express the vertical stress gradient as $d\sigma_z/dz = 2\tau_z/R$. Introducing the coefficients α and μ_f in order to express τ_z as a function of σ_z, this axial stress can be also expressed as $d\sigma_z/dz = 2\alpha\mu_f\sigma_z/R$. The solution of this first order differential equation leads to the following expression of σ_z according to z:

$$\sigma_z(z) = \sigma_{zLP}\exp(2\alpha\mu_f z/R) \tag{5.3}$$

Since the mean upper stress on the upper punch σ_{zUP} is known, the forces balance on the whole specimen gives for a given compression state and a given height h of the sample:

$$\mu_f(h)\sigma_{rm}(h) = \frac{R}{2h}(\sigma_{zUP} - \sigma_{zLP}) \tag{5.4}$$

where R is the inner radius of the cell, h is the specimen height, σ_{zUP} and σz_{LP} are the pressures deduced from the upper and lower force transducer measurements, and σ_{rm} is the mean radial stress, deduced from strain gauge measurements. The evolution of stress transmission coefficient α and friction coefficient μ_f can be then written as:

$$\alpha = \frac{\sigma_{rm}}{(\sigma_{zUP} - \sigma_{zLP})}\ln\left(\frac{\sigma_{zUP}}{\sigma_{zLP}}\right) \text{ et } \mu_f = \frac{R}{2h}\frac{(\sigma_{zUP} - \sigma_{zLP})}{\sigma_{rm}} \tag{5.5}$$

In the following, the use of a compressibility model such as Cooper-Eaton model allow to analyse the evolution of this coefficients according to the applied stress during the compression test.

5.3.2.2 Compressibility

The compressibility of a granular bed characterizes the evolution of density versus the upper stress applied during compression (Leuenberger 1982) while the compactibility characterizes the evolution of strength versus maximum compression application during the process.

Based on studies on the compaction of biomass resources, several models of compressibility were tested for hemp mixtures through granular materials (Chevanan et al. 2010; Mani et al. 2004; Emami and Tabil 2007) or powder compressibility (Walker 1923; Jones 1960; Barbosa-Cànovas et al. 1987; Cooper and Eaton 1962; Kawakita and Lüdde 1971). These models generally link volume or density to upper pressure σ_{zUP}. Compactness, C, is the most significant parameter to characterize the compaction state. In this study, compactness corresponds to the sum of solid volume fraction and liquid volume fraction, i.e. the ratio of liquid and solid volumes to the total volume of the sample. The liquid phase is included because it is assumed to be physically linked to shiv when compaction starts. This assumption is valid as soon as the water content is low, ensuring that no free and draining water issue from the sample. During experiments, no loss of mass was recorded.

The mechanisms of compaction are generally distinguished in three steps, as first described by Seeling and Wülff (1946): (1) packing of particles, (2) elastic and plastic deformation, (3) fragmentation.

This is the reason why Cooper and Eaton considers that void filling depends on pore size, i.e. when rearrangement and filling of small voids occur or when plastic deformation and fragmentation occur. In Cooper and Eaton's model, P_r and P_d are coefficients related to the pressure needed to induce the process that has the greatest probability of occurring.

$$\frac{1 - C_0/C}{1 - C_0} = a_1 \exp(-P_r/\sigma_{zUP}) + a_2 \exp(-P_d/\sigma_{zUP}) \tag{5.6}$$

For shiv and LHC mixtures, P_d values are ten to twenty times higher than P_r, as observed in the compaction of an alumina powder but these values are very low compared with metal or mineral powder beds; similarly to straw particles, wet or dry, which are very flexible and compressible. Both the rearrangement and deformation phase are easier to discern for biomass resource particles than for rigid and frictional ones. Wet shiv are easier to compact than dry mixtures: the shiv aggregates absorb water and become heavier, more flexible and more compressible (at a local scale). These factors facilitate both rearrangement and deformation of the granular bed. But water content tends also to increase contact surface at the wall and between particles. Actually, the product $\alpha\mu_f$, see Eq. 5.4 and 5.5, is relative to the portion of the vertical compression stress exerted at the wall that induces heterogeneity inside specimens. In case of highly densified shiv, this product $\alpha\mu_f$ is minimum for dry state and usually reaches a maximum around 15% water content in case of "HS" hemp shiv type (Tronet 2014; Tronet et al. 2014).

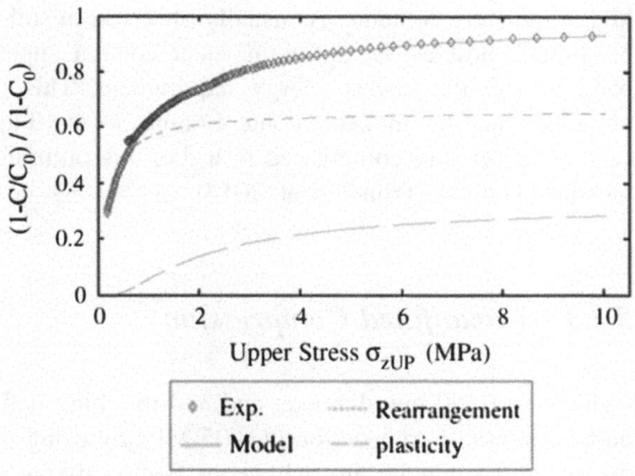

Fig. 5.5 Cooper and Eaton's model versus experimental results with rearrangement and plasticity contributions to the compactness

As a consequence, for biomass resources, the sum ($a_1 + a_2$) of the first and second parameters of Cooper and Eaton's Model tend towards the theoretical compactness at infinite pressure, which is ideally the unity (Mani et al. 2004), see Fig. 5.5. Therefore in the computations, the parameter a_2 was always taken as being equal to $1 - a_1$. The parameter 'a_1' makes it possible to quantify compactness at the end of rearrangement and/or at the beginning of the plasticity phenomenon.

If the pressure increases towards an infinite value and the first term of the model is only considered, as shown in Fig. 5.7, a basic relationship between C, C_0 and a_1, is found to compute a rearrangement compactness C_r:

$$C_r = \frac{C_0}{1 - a_1(1 - C_0)} \tag{5.7}$$

where C_0 is the initial (bulk) compactness.

The values of a_1 as the rearrangement pressure P_r and deformation pressure P_d, can therefore be computed by fitting experimental curves with Cooper and Eaton's model, see Fig. 5.5. Examining the coefficient of determination, R^2 obtained with hemp shiv coming from "Chanvrières de l'Aube, LCDA" the three-parameters fitting Cooper and Eaton's models (1962) gives the best results, see Fig. 5.5.

Otherwise, mong two-parameters models found in literature, the Kawakita and Lüdde's (1971) gives the best fit (Tronet et al. 2014).

5.3.2.3 Effect of the Moisture Content

Compactness is generally higher with wet shiv than with dry shiv: water makes shiv heavier and easier to deform, but moisture has the opposite effect below and above 20% moisture content: a first optimum state is observed around 10% for initial compactness, and for characteristic pressures P_r and P_d. Actually, moisture could act as an adhesive up to a given moisture (around 15%), and as a lubricant for higher moisture contents. As usually observed in soil compaction, in case of many bio-based particles, an optimum water content appears to achieve a given compaction with the lowest energy requirement. This maximum density is usually obtained when the moisture content ranges from 10 to 45% (O'Dogherty 1989). In case of hemp shiv compressed in a die, this optimum is reached around 15% of moisture content (Tronet et al. 2014).

5.3.3 Unconfined Compression

Cylinders of 160 mm diameter and 320 mm high is filled with 50 mm thick layers under a stress of compaction of 0.05 MPa by using only dry hemp shiv similar as the one tested in a die. In such a case, hemp shiv particles exhibit enough cohesion to make test specimen that can stand alone, without any applied stress, when the cylindrical mold is removed, see Fig. 5.6. Even with dried particles, (dried at 60 °C until constant weight) the 160 mm diameter and 320 mm diameter cast specimens can be directly loaded with uniaxial compressions (Cerezo et al. 2005; Arnaud and Gourlay 2012).

The stress–strain curves resulting from simple and cyclic compressions are perfectly superimposed which ensures the reproducibility of the tests carried out.

Fig. 5.6 Tested 320 mm high specimen made of hemp shiv only. The initial particle cohesion is obtained with a slight compaction in a 50 mm layers in a 160 mm inner diameter cylinder (Cerezo et al. 2005)

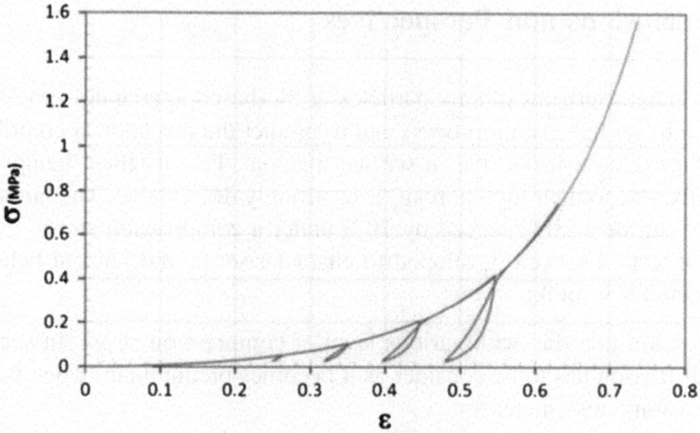

Fig. 5.7 Simple and cyclic compression tests on 320 mm high and 160 mm diameter specimens made of dry hemp particles without binder (Chevanan et al. 2010)

Hemp particles in bulk are very strongly compressible: the cylinders can indeed sustain strain levels higher than 70% before the particles heap collapses (Arnaud and Gourlay 2012). Figure 5.7 shows that the stiffness of the sample continuously increases while its height decreases due to compression. At the beginning of the test, the apparent stiffness is about 0.25 MPa while it reaches 4.5 MPa at the end, under a load of 1.5 MPa.

Moreover, load cycle exhibits significant hysteresis especially when the strain range, or load range increases. The unloading branch with a delayed stress response shows a viscosity effects that can be due to the air flows inside the highly porous deformable material. The radial strain measurement allow to deduce that the volume of the sample decreases significantly during compression, with Poisson coefficient, ν, lower than 0.05 (Cerezo et al. 2005). The air drained during compaction induces an incoming flow during unloading that delays the stress response of the gradually unloaded specimen.

Significant residual strain is observed at the end of each loading cycle. Rearrangements of particle should be the main effect leading to such large irreversible strain (Cerezo et al. 2005). This rearrangement, especially in case of unconfined specimens, leads to the filling of small voids. Specimen density, or compactness, increases irreversibly due to a first load cycle, as assumed in the Cooper and Eaton model providing the best fitting of the experimental results (Tronet et al. 2014), see previous section.

5.4 Conclusions and Perspectives

Mechanical characteristics of shiv particles or bio-based aggregates provides many information to set the casting process and to predict the mechanical contribution of the granular skeleton inside the in-service material. For instance, hemp shiv particles in bulk, due to their high porosity, are strongly deformable. The initial volume of the bulk can be easily reduced by 70% under a compression stress. Up to day, two kind of tests has been developed to characterise the mechanical behaviour of the shiv particles in bulk:

- Compression in a die, with various level of compression stress. In such a case, the wall friction has to be consider as it becomes predominant when the applied compression stress increases.
- Unconfined compression on initially compacted shiv. Prior testing, compaction of successive shiv layer in a rigid cylinder is required in order to obtain enough cohesion to make an initial cylindrical heap of particles. This test needs diametrical measurement to be correctly analysed. In case of an initial aspect ratio (high to diameter) of 2, the compression stress is limited to few MPa before the particles heap collapses.

Otherwise, a confined compression test in a triaxial cell would provide more information with negligible friction effects at the wall and without initial compaction. Preferential orientation of the particle due to initial compression before testing may introduce a bias in the measurement. In this way of investigation, few unpublished results have been obtained with hemp shiv and should be consolidated to provide more information on the mechanical behaviour of such bulk particles.

The creep and relaxation effects should also be investigated, according to the stress level and the moisture content, as they are representative of the aggregates skeleton inside the in-service material.

Finally, correlations between particle size distribution (PSD), see previous chapter, and initial density or compressibility of such highly deformable particles have to be investigated. In such bio-based building materials, as in the case of mineral aggregates in concrete, the PSD should be set to optimize the mechanical and/or thermal properties of the in-service material.

References

Arnaud, L., Gourlay, E.: Experimental study of parameters influencing mechanical properties of hemp concretes. Constr. Build. Mater. **28**, 50–56 (2012)

Barbosa-Cànovas, G.V., Malavé, J., Peleg, M.: Density and compressibility of the selected food powder mixtures. J. Food Process Eng. **10**(1), 1–19 (1987)

Ceyte, I.: Béton de chanvre, définition des caractéristiques mécaniques de la chènevotte. Travail de Fin d'Etudes, ENTPE (2008)

Cerezo, V.: Propriétés mécaniques, thermiques et acoustiques d'un matériau à base de particules végétales: approche expérimentale et modélisation théorique, Ph.D. Thesis, ENTPE - INSA de Lyon (2005)

Chevanan, N., Womac, A.R., Bitra, V.S.P., Igathinathane, C., Yang, Y.T., Miu, P.I., Sokhansanj, S.: Bulk density and compaction behaviour of knife mill chopped switchgrass, wheat straw, and corn stover. Bioresour. Technol. **101**, 207–214 (2010)

Chtourou, H., Guillot, M., Gakwaya, A.: Modelling of metal powder compaction process using the cap model. Part I. Experimental material characterisation and validation. Int. J. Solids Struct. **39**, 1059–1075 (2002)

Cooper, A.R., Eaton, L.E.: Compaction behaviour of several ceramic powders. J. Am. Ceram. Soc. **45**(3), 97–101 (1962)

Doremus, P., Toussaint F., Alvain O.: Simple tests standard procedure for the characterisation of green compacted powder, In: Proceedings of the Nato Advanced Research Workshop on Recent Developments in Computer Modelling of Powder Metallurgy Processes, pp. 29–41, Kiev (2001)

Emami, S., Tabil, L.G.: Friction and compression characteristics of chickpea flour and components. Powder Tech. **175**, 14–21 (2007)

Jacky, J.: The coefficient of earth pressure at rest, J. Soc. Hung. Archit. Eng. pp. 355–358 (1944)

Janssen, H.A.: Versuche über Getreidedruck in Silozellen. Z. Ver. Dtsch. Ing. **39**, 1045–1049 (1895)

Jones, W.D.: Fundamental Principle of Powder Metallurgy. Edward Arnold Publisher Ltd., London, UK (1960)

Jonsén, P., Häggblad, H.A.: Modelling and numerical investigation of the residual stress state in a green metal powder body. Powder Technol. **155**(3), 196–208 (2005)

Kawakita, K., Lüdde, K.H.: Some considerations on powder compression equations. Powder Technol. **4**, 61–68 (1971)

Leuenberger, H.: The compressibility and compactibility of powder systems. Int. J. Pharm. **12**(1), 41–55 (1982)

Mani, S., Tabil, L.G., Sokhansanj, S.: Grinding performance and physical properties of wheat and barley straws, corn stover and switchgrass. Biomass Bioenergy **27**, 339–352 (2004)

Modnet, P.M.: Methods and measurements group. Measurement of friction for powder compaction modelling. Powder Metall. **43**(4), 364–374 (2000)

Nguyen, T.T., Picandet, V., Amziane, S., Baley, C.: Influence of compactness and hemp hurd characteristics on the mechanical properties of lime and hemp concrete. Eur. J. of Env. Civil Eng. **13**, 1039–1050 (2009)

Nguyen, T.T.: Contribution à l'étude de la formulation et du procédé de fabrication d'éléments de construction en béton de chanvre. Doctoral Thesis, Université de Bretagne Sud, Ecole Doctorale SICMA, Lorient. Avaiable online: http://web.univ-ubs.fr/limatb/lab/ (2010)

Nguyen, T.T., Picandet, V., Carré, P., Lecompte, T., Amziane, S., Baley, C.: Effect of compaction on mechanical and thermal properties of hemp concrete. E. J. of Env. Civil Eng. **14**, 545–560 (2010)

O'Dogherty, M.J.: A review of the mechanical behaviour of straw when compressed to high densities. J. Agric. Eng. Res. **44**, 241–265 (1989)

Perrot A., Lanos C., Estellé P., Melinge Y.: Ram extrusion force for a frictional plastic material: model prediction and application to cement paste. Rheol Acta **45**(4), 457–467 (2006)

Picandet, V.: Characterization of plant-based aggregates (Chap. 2). In: Amziane, S., Arnaud, L., (eds.) Bio-Aggregate-Based Building Materials: Applications to Hemp Concretes, pp. 27–74. Wiley Inc., London (2013)

Seeling, P., Wulff, J.: Pressing operation in fabrication of articles by powder metallurgy. Trans. Am. Inst. Min. Metall. Eng. **166**, 492–505 (1946)

Sinka, I.C., Cunningham, J.C., Zavaliangos, A.: The effect of wall friction in the compaction of pharmaceutical tablets with curved faces: a validation study of the Drucker-Prager Cap model. Powder Technol. **133**, 33–43 (2003)

Tronet, P.: Contribution à l'étude des matériaux chaux-chanvre: influence du compactage sur les propriétés, Ph.D. Thesis, Université de Bretagne-Sud (2014)

Tronet, P., Lecompte, T., Picandet, V., Baley, C.: Study of lime hemp composite precasting by compaction of fresh mix - An instrumented die to measure friction and stress state, Powder Tech. **258**(14), 285–296 (2014)

Walker, E.E.: The properties of powders—part VI: the compressibility of powders. Trans. Faraday Soc. **19**(1), 73–82 (1923)

Chapter 6
Hygric and Thermal Properties of Bio-aggregate Based Building Materials

Florence Collet

Abstract This chapter gives the state of the art of previous studies on hygric and thermal properties of bio-aggregate based building materials. Firstly, hygric characteristics such as sorption isotherms, water vapor permeability and moisture diffusivity are given. The ability of bio-aggregate based building materials to moderate ambient relative humidity may be valued using moisture buffer value. Then thermal properties (thermal conductivity, thermal diffusivity conductivity and specific heat capacity) are reported. Finally, concluding remarks on hygrothermal behavior with simultaneous heat and mass transfer are provided, they underline that considering only thermal conductivity and specific heat capacity is not sufficient to evaluate the energy performance of bio-aggregate based building materials. The results found in bibliography mainly concern wood-based and hemp-based materials.

Keywords Thermal conductivity · Thermal diffusivity · Specific heat capacity · Sorption isotherm · Water vapour permeabilty · Moisture buffer value

6.1 Introduction

Bio-aggregate based building materials (BBM) are made from various binders and bio-aggregates. The more efficient formulation parameter is the aggregate to binder ratio. These materials are highly porous. Their porosity includes a wide range of pore sizes: macropores due to the imperfect arrangement of bio-aggregates, mesopores within aggregate and binder, and micropores in the binder. This porosity is open and interconnected and is thus the place of heat and mass transfer and moisture storage.

F. Collet (✉)
Laboratoire de Génie Civil et Génie Mécanique de Rennes, équipe Matériaux-Thermo-Rhéologie, Université de Rennes 1, Rennes, France
e-mail: florence.collet@univ-rennes1.fr

F. Collet
IUT Génie Civil Construction Durable, 3, Rue Du Clos Courtel, BP 90422, 35704 Rennes Cedex, France

S. Amziane and F. Collet (eds.), *Bio-aggregates Based Building Materials*, RILEM State-of-the-Art Reports 23, DOI 10.1007/978-94-024-1031-0_6

This chapter gives the state of the art of previous studies on hygrothermal behavior of bio-aggregate based building materials. Firstly, hygric characteristics such as sorption isotherms, water vapour permeability and moisture diffusivity are given. Their ability to moderate ambient relative humidity is also quantified by moisture buffer value. Then thermal properties (conductivity and specific heat capacity) are investigated. Finally, concluding remarks on hygrothermal behavior with simultaneous heat and mass transfer are provided.

6.2 Hygric Properties

6.2.1 Moisture Storage: Sorption Isotherm

Sorption isotherm relates the equilibrium moisture content of the material to the ambient relative humidity for a given temperature. From dry state to humid state, the water uptake occurs following three stages. Firstly molecules of water are adsorbed on the internal pore wall surfaces forming a monolayer: this is the monomolecular adsorption. Then water molecules adhere to the monolayer: this is the polymolecular adsorption. Finally, molecules of water lead to a liquid bridge and fill the pore with the formation of liquid menisci: this is the capillary condensation. Depending on the porous structure of the material, these phenomena may occur successively or simultaneously (the first and the following layers are thus adsorbed simultaneously). Besides, a hysteresis may appear between adsorption and desorption branches. This hysteresis is often explained qualitatively by capillary condensation, by existence of ink-bottle shaped pores or, more generally, interconnected pores spaces (Naono and Hakuman 1993). The IUPAC gives a classification of physisorption isotherms in six types with reference to pore size and of hysteresis loops in four types related with pore structures (IUPAC 1986). Sorption isotherms can be measured according to continuous methods (under quasi-equilibrium) or discontinuous methods (at successive stages). Several models have been developed to describe sorption curve. The GAB model is often met as, despite it is valid when there is no capillary condensation; it is convenient to fit experimental adsorption data all over the relative humidity range (Guggenheim 1966; Anderson 1946; Anderson and Hall 1948; de Boer 1953).

Bio-aggregate based building materials are strongly hygroscopic. Their water uptake is much higher than in other building materials, as illustrated Fig. 6.1 by comparing hemp concrete with Aerated Autoclaved Concrete and with Vertical Perforated Bricks (Amziane and Arnaud 2013). As well for wood concrete as for hemp concretes, the water content at equilibrium at very high relative humidity (resp. 99.9 and 95%RH) is much lower than the water content at saturation (Bouguerra et al. 1999; de Bruijn and Johansson 2013). This is linked to the macroporosity due to the bio-based aggregate. Moreover, their sorption curves show hysteresis which extends all over the range of relative humidity, as well for

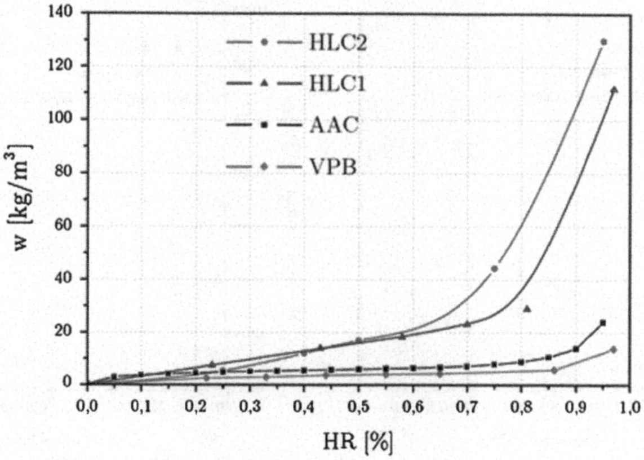

Fig. 6.1 Adsorption isotherms of materials. *HLC* Hemp-Lime Concrete, *AAC* Aerat-ed Autoclaved Concrete, *VPB* Vertical Perforated Bricks (Amziane and Arnaud 2013)

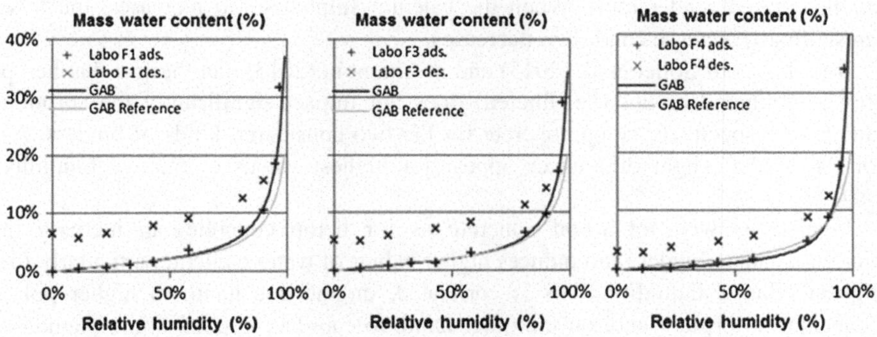

Fig. 6.2 Sorption isotherms of hemp concretes with increasing substitution of lime-based binder for calcium sulphate-based binder. *F1* 1/3 sulphate-based binder + 2/3 lime-based binder; *F3* 1/2 + 1/2; *F4* 2/3 + 1/3 (Chamoin 2013)

wood concrete as for hemp concretes (Figs. 6.2 and 6.3) (Bouguerra et al. 1999; Cerezo 2005; Chamoin 2013; Collet 2004; Collet et al. 2008, 2013).

Comparing sorption curve of hemp concrete with sorption curve of hemp shiv (Zaknoune 2011; Collet et al. 2013) underline that hemp concrete shows lower sorption curve than hemp shiv but exhibits larger hysteresis. So, the porosity of the binder widely influences the hygroscopic behavior of the mix. Actually, the binder develops specific surface area and is the place of capillary condensation in the inter-particle pores. The kind of binder slightly impacts the sorption curve. In Chamoin (2013), substituting lime-based binder for calcium sulphate-based binder leads to similar values of water content as the reference material, excepted at highest relative humidity (>80%RH), where capillary condensation takes place. The hysteresis is also slightly impacted, but the effect depends on the calcium

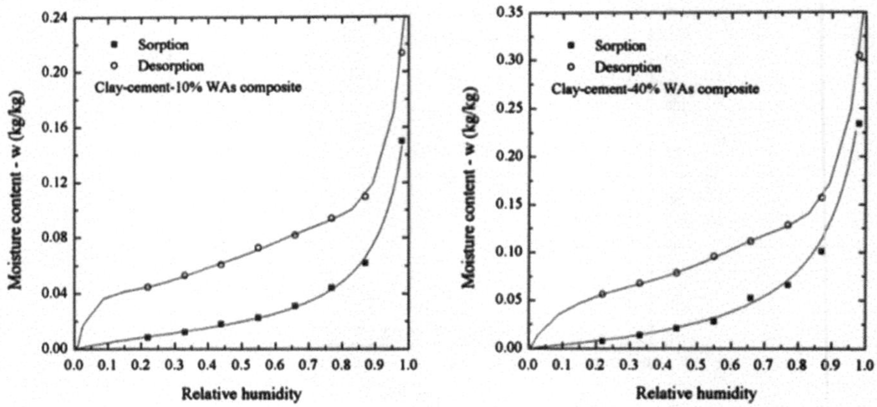

Fig. 6.3 Sorption-desorption isotherms of various mixtures obtained by substituting clay with wood aggregates (Bouguerra et al. 1999)

sulphate-based binder ratio. When the calcium sulphate-ratio increases, the hysteresis firstly increases and then decreases.

It is shown in Collet et al. (2013) and in Chamoin (2013) that the fact that hemp shiv is defibered or not (i.e. fibered) does not impact significantly the sorption curves of respectively hemp concrete for the two considered kinds of binders. As for the binder, slight differences appear for highest values of relative humidity (>80%RH).

Finally, as well for wood concrete as for hemp concrete, an increase in bio-aggregate to binder ratio induces higher values of water content, particularly for highest relative humidity. This is correlated, on the one hand, to higher polymolecular adsorption and, on the other hand, to capillary condensation. Actually, Bouguerra et al. remind that raising the aggregate to binder ratio increases macroporosity and the specific surface area is thus the highest for highest aggregate to binder ratio (Bouguerra et al. 1999). Collet et al. point out that, in the range of relative humidity 81–97%RH, the capillary condensation occurs in pores of width from 0.005 to 0.040 μm. This size of pore is met in the skin of the cell wall of hemp shiv (Collet et al. 2013). Increasing the hemp to binder ratio increases these phenomena.

6.2.2 *Moisture Transfer: Water Vapor Permeability, Capillarity, Moisture Diffusivity*

Moisture transfer takes place under vapor or/and liquid transport. The driving force for vapor diffusion is the vapor pressure gradient. The liquid transport occurs by surface diffusion or capillary flow, in both cases the driving force is the capillary pressure.

Water vapor permeability characterizes the ability of a material to transfer moisture under a vapor pressure gradient once the steady state is reached. The commonly called "vapor permeability" includes (i) vapor transfer by diffusion (transport by collision of water molecules with each other), (ii) vapor transfer by effusion (transport by collision of water molecules with walls of pores) and (iii) liquid transfer (connected with capillary condensation). For hygroscopic materials, the water vapor permeability increases with the water content of the material. The measurement of water vapor permeability is often performed following the well-known cup method.

The mainly open high porosity of bio-aggregate based building material gives them high moisture vapor permeability (i.e. low water vapor diffusion resistance). The water vapor diffusion resistance, at dry point, is 15 for wood shaving concrete (Amziane and Arnaud 2013), it ranges from 5 to 12 for hemp concrete (Collet 2004; Evrard 2008; Walker and Pavia 2014) while it is equal to 130 for solid concrete, 50 for light weight aggregate (natural pumice stone) concrete, and 10 for aerated autoclaved concrete (ThU 2005). Besides, for hygroscopic materials, such as bio-aggregate based building materials, the water vapor permeability is a strong function of the local relative humidity. Figure 6.4 gives the variation of water vapor permeability with relative humidity for two hemp concretes (Collet et al. 2013) and for clay brick and concrete brick (Kumaran et al. 2002). This figure highlights the high water vapor permeability of hemp concrete and its increase with relative humidity.

As well as in Chamoin (2013) as in Walker and Pavia (2014), the type of binder does not have a significant effect on the water vapor permeability of hemp concrete. In Chamoin (2013) hemp concretes with fibered hemp shiv show similar values as hemp concrete with defibered hemp shiv in the hygroscopic domain. For highest values of relative humidity, fibered hemp shiv leads to lower values than defibered hemp shiv. The macropores between hemp particles have a greater influence on the water vapor permeability than the micropores in binder. However, the use of water retainer in Walker and Pavia (2014) reduces the water vapor permeability of hemp concrete.

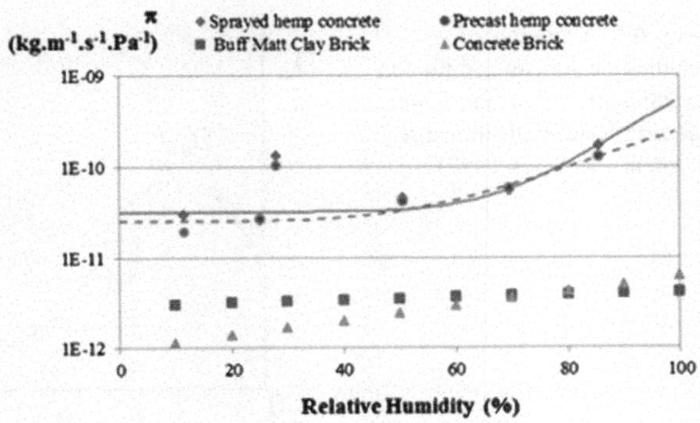

Fig. 6.4 Variation of moisture permeability versus ambient relative humidity for several building material: sprayed and precast hemp concrete (Collet et al. 2013), clay brick and concrete brick (Kumaran et al. 2002)

The capillary action characterizes the spontaneous flow of a liquid into a porous structure. The water absorption coefficient gives the amount of water absorbed par unit area as function of the square root of time.

The water absorption coefficient by capillarity (free water uptake) ranges from 2.40 to 4.42 kg/(m^2 $h^{1/2}$) for hemp concretes (Walker and Pavia 2014; Evrard 2008; Evrard and de Herde 2010). A higher value of water absorption coefficient [9 kg/(m^2 $h^{1/2}$)] is found in de Bruijn et al. (2009) for higher density of hemp concrete. These values are in the range of values found in Nielsen (1976) where the water absorption coefficient are 2.52 kg/(m^2 $h^{1/2}$) for sand-lime brick, 4.92 kg/(m^2 $h^{1/2}$) for cellular concrete, 10.5 and 13.98 kg/(m^2 $h^{1/2}$) for brick with respective density of 1775 and 2025 kg/m^3. Bouguerra et al. observe by Magnetic Resonance Imaging the nonsorption effect of wood aggregates on the capillary absorption of the wood composite. The macrostructure of wood aggregate, embedded in a clay-cement matrix which shows essentially micropores, tend to slow down the capillary invasion in the material. Finally, the presence of wood aggregates leads to a significant reduction in the sorptivity (Bouguerra et al. 1999).

The moisture diffusivity of a material characterizes the rate of change of its moisture content under transient moisture evolutions. De Vries has shown that the variation of moisture diffusivity versus water content evolves according to three phenomena (de Vries, 1958) (Fig. 6.5): (i) for low water content ($u < u_k$), moisture transfer is essentially due to vapor transport (diffusion and effusion), the condensed phase exists in the form of an adsorbed film or in small islands of water; (ii) when water content increases, small islands of water increase in number and in size, the area for vapor flux decreases and moisture transfer occurs by mechanisms of condensation and evaporation at vapor–liquid interfaces, as soon as the continuity of the liquid phase is reached ($u = u_k$) the liquid transfer increases; (iii) then, for high water content, liquid transfer is predominant.

The isothermal moisture diffusivity calculated from the derivative of the sorption curve and from the vapor permeability ranges from 10^{-10} to 10^{-7} m^2 s^{-1} (Collet 2004; Amziane and Arnaud, 2013; Collet et al. 2013). For comparison, the value given by Peuhkuri (2003) for aerated autoclaved concrete is 8.10^{-9} m^2 s^{-1} for relative humidity ranging between 40 and 60%. However, as underline in Rode

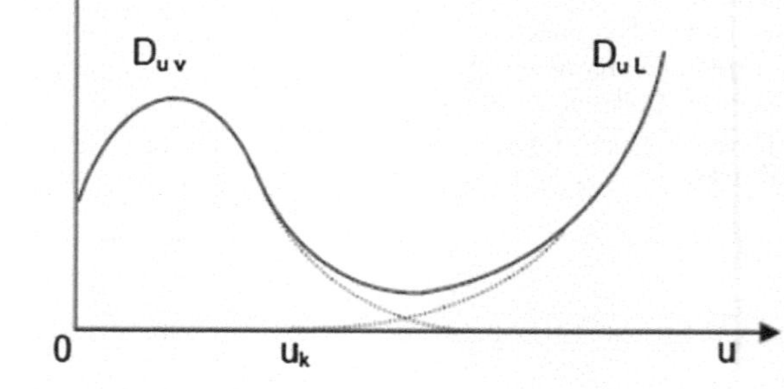

Fig. 6.5 Variation of isothermal moisture diffusion coefficient (vapor phase and liquid phase) with moisture content (de Vries 1958)

(2005), there may be some discrepancy between the basic material properties depending on whether they have been determined under steady state or dynamic conditions. It is shown in Collet and Pretot (2012a) that the moisture effusivity is slightly higher when it is calculated from steady-state data than when it is from dynamic measurements. It is suggested that it is due to a non-fickian behavior of sprayed hemp concrete. New method to determine the mass diffusivity in materials, was developed by Perré et al. (2015). This method allows non-Fickian behavior to be detected, and can still be used in such case with dual-scale model in the identification procedure. No data are found with such method on bio-aggregate based building materials.

The moisture diffusivity calculated from capillary test ranges from 10^{-10} to 10^{-8} $m^2\ s^{-1}$ for wood concrete (Bouguerra et al. 1999) and from 10^{-10} to 10^{-8} $m^2\ s^{-1}$ for cellular concrete (Kumaran et al. 2002) (Fig. 6.6).

In line with previous comments, the formulation of binder does not have a significant effect on moisture diffusivity. It is shown in Chamoin (2013) that by increasing the substitution rate of lime-based binder for calcium sulphate-based binder, the moisture diffusivity is slightly but not significantly amplified (Fig. 6.7).

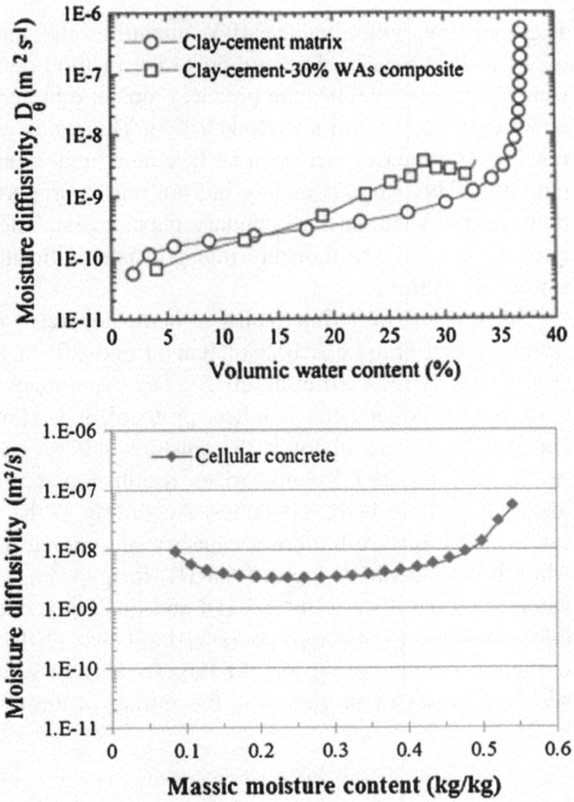

Fig. 6.6 Moisture diffusivity of materials from capillary test—*top* clay-cement matrix and clay-cement-30% of wood aggregates composite (Bouguerra et al. 1999); *bottom* cellular con-crete (Kumaran et al. 2002)

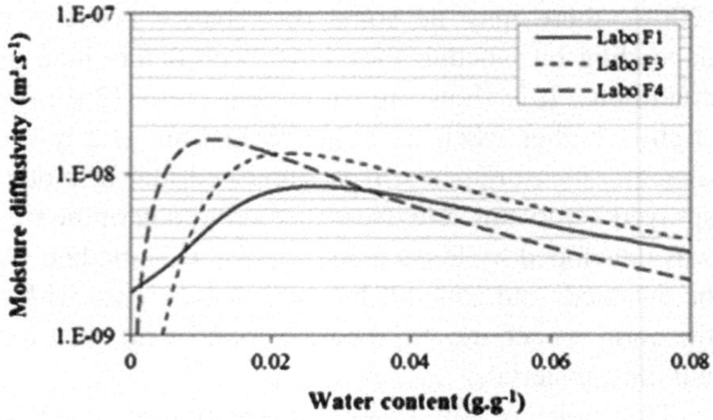

Fig. 6.7 Moisture diffusivity of hemp concrete with increasing substitution of lime-based binder for sulphate-based binder. *F1*: 1/3 sulphate-based binder + 2/3 lime-based binder; *F2* 1/2 + 1/2; *F3* 2/3 + 1/3 (Chamoin 2013)

Moreover, Bouguerra et al. (1999) underline that the wood aggregates do not seem to participate to the liquid phase transfer because of their macroporous structure. Thus, the liquid transfer occurs mainly in the microporous matrix.

6.2.3 *Moisture Buffering: Moisture Buffer Value*

The moisture buffer value MBV quantifies the moisture buffering ability of a material. It is measured according to the method defined in the NORDTEST project. This project defines the practical moisture buffer value of materials, measured under dynamic conditions (Rode 2005). This value relates the amount of moisture uptake (and release), per open surface area, under daily cyclic variation of relative humidity. This value is mainly but not only a property of the material as the mass transfer coefficient at the boundary plays a role. Though, for many materials, the internal resistance to moisture transport is significantly larger than the convective surface resistance.

The moisture buffering quality of hemp concretes was studied by several authors (Evrard 2006; Tran Le 2010, Collet and Pretot 2012a, b, c; Collet et al. 2013; Dubois et al. 2012, 2013; Latif et al. 2015). The experimental investigations are generally performed following the Nordtest protocol. It is shown that hemp concrete is an excellent hygric regulator with moisture buffer values globally higher than 2 g/(m^2 %RH). Figure 6.8 summarizes results found in literature for usual building materials and for hemp concretes. According to the NORDTEST project classification, the moisture buffering capacity of concrete is limited [<0.5 g/(m^2 %RH)] while it is moderate [<1 g/(m^2 %RH)] for gypsum, good for cellular concrete and wood fibreboard [<2 g/(m^2 %RH)] and finally excellent [>2 g/(m^2 %RH)] for cellulose insulation and hemp concretes. Latif et al. (2015) found higher values of MBV for hemp concrete [4.3 g/(m^2 %RH)]. This higher value is due to higher air velocity which increases mass transfer at the surface of the material.

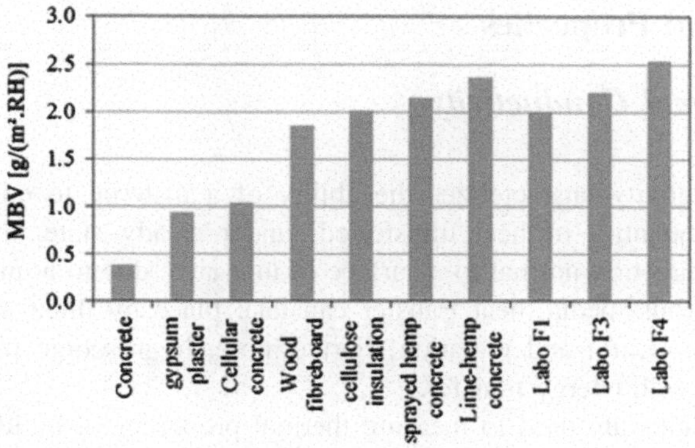

Fig. 6.8 Review of moisture buffer value of building materials: concrete and cellular concrete from (Rode 2005), gypsum plaster, wood fibreboard and cellulose insulation from (Janssen and Roels 2008), sprayed hemp concrete from (Collet and Pretot 2012), lime-hemp concrete from (Dubois et al. 2012), hemp lime concretes with increasing substitution of lime-based binder for sulphate-based binder. *F1* 1/3 sulphate-based binder + 2/3 lime-based binder; *F3* 1/2 + 1/2; *F4* 2/3 + 1/3 (Collet and Pretot 2012b)

The composition and the manufacturing method impact the moisture buffering ability of hemp concretes (Collet et al. 2013). It is shown in Collet and Pretot (2012b) that the composition of binder slightly affects the MBV. Actually, the MBV of hemp concrete made of mix binder increases with increasing substitution of lime-based binder for calcium sulphate-based binder. This phenomenon is correlated with an increase in moisture effusivity.

Finally, it should be underlined that this ability is impacted when hemp concrete is coated. Collet and Pretot (2012c) show that the MBV is more or less reduced depending on the kind of coating added on its surface. For closed coating (sand-lime plaster), the MBV is nearly half the value of neck hemp concrete [respectively 1.08 and 2.14 g/(m^2 %RH)]. For open coating (hemp-lime plaster), the value is much less reduced [from 2.14 to 1.82 g/(m^2 %RH). Similar results are found in Latif et al. (2015). The wall assemblies with breather membrane show the lower moisture buffer potential.

Dubois and Lebeau (2013) performed inverse modelling from MBV test in order to estimate hygric parameters such as vapor permeability and hygric capacity. Boundary and initial conditions are also optimized: vapor diffusion resistance factor at exchange surface and initial equilibrium relative humidity in the specimen. It is shown that the vapor resistance factor of the material (or its vapor permeability) and its moisture capacity are strongly correlated. Actually, these two parameters can be combined in one: the moisture diffusivity.

6.3 Thermal Properties

6.3.1 Thermal Conductivity

Thermal conductivity characterizes the ability of a material to conduct heat. It quantifies the quantity of heat transferred, under steady state, through a unit thickness in a direction normal to a surface of unit area, due to a unit temperature gradient. In porous media, heat transfer can take place by three modes such as conduction, convection and radiation. Furthermore, hygroscopic phenomena are also associated with energy transfer.

Several methods are used to measure thermal properties of building materials. Steady-state methods like guarded hot plate, heat flow meter and guarded hot boxes and transient methods like hot wire and line source give thermal conductivity value. Other transient methods such as flash method, transient plane source or hot disk allow simultaneous measurement of conductivity and diffusivity and/or heat capacity. Small discrepancies may appear between the results from these different methods.

Several authors use the self-consistent homogenization to compute thermal conductivity of bio-aggregate based building materials. This homogenization is performed at dry state versus the formulation and the density of the material or at humid state, taking into account the water content of the material (Arnaud 2000; Collet 2004; Cerezo 2005; Bederina et al. 2007).

6.3.1.1 Thermal Conductivity of Bio-aggregate Based Building Materials

Figure 6.9 gives examples of thermal conductivity versus density found in literature for several bio-aggregate based building materials BBM (made of wood aggregate or hemp shiv) (Agoua et al. 2013; Aigbomian and Fan 2013; Al Rim et al. 1999; Bederina et al. 2007; Pretot et al. 2009; Sassoni et al. 2014; Taoukil et al. 2013; Walker and Pavia 2014; Magniont et al. 2012). The thermal conductivity of BBM ranges from very low (about 0.04 W m^{-1} K^{-1}), allowing their use as insulating material, to medium (1–2 W m^{-1} K^{-1}). In a general way, BBM are lightweight materials and therefore good, not exceptional, thermal insulator; with thermal conductivity about 0.1–0.3 W m^{-1} K^{-1}. Their thermal conductivity is equivalent to that of other building materials with similar density. Actually, cellular concrete shows a thermal conductivity of 0.115 W m^{-1} K^{-1} for a density of 400 kg m^{-3} (resp. 0.162 W m^{-1} K^{-1} for 600 kg m^{-3}) (Gawin et al. 2004).

Thermal conductivity of bio-aggregate based building materials is thus related with density. It also depends on several parameters such as formulation (binder, aggregate, aggregate to binder ratio…), production method and water content. The density is induced by the formulation and the production method. The effect of the formulation data will be detailed further.

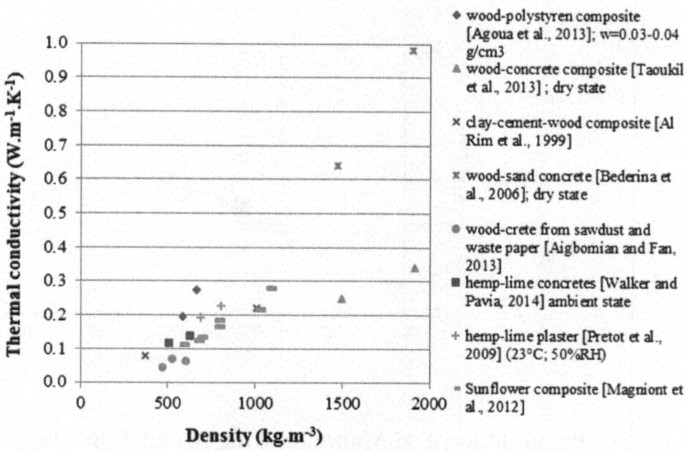

Fig. 6.9 Thermal conductivity of bio-aggregate based building composites (made of wood or hemp shiv)

Fig. 6.10 Evolution of thermal conductivity of hemp concrete depending on density (Cerezo 2005)

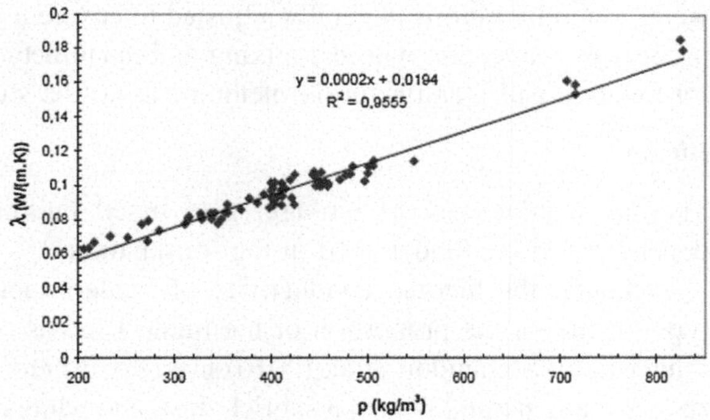

Cerezo gives a linear relationship between thermal conductivity and density of hemp concrete (Cerezo 2005) (Fig. 6.10). This curve fit with high accuracy the experimental data for density ranging from 300 to 520 kg m^{-3}. Actually these points correspond to hemp concrete with similar formulation. For higher and lower densities, the hemp-binder-water ratios differ. For a given formulation Cerezo models the variation of thermal conductivity with density from self-consistent homogenization (Cerezo 2005) (Fig. 6.11). There is a high correlation between experimental data and modeling. Similar results are found by Collet and Pretot on sprayed hemp concrete (Collet and Pretot 2014). The variation of thermal conductivity with density is measured and modeled with self-consistent homogenization as the spraying process induces a variation of density (Elfordy et al. 2008). For a given formulation (the one used to build wall), the variation of thermal conductivity versus density fit a linear regression curve. At dry state, the thermal conductivity rises by 109% while density increases from 250 to 600 kg m^{-3}.

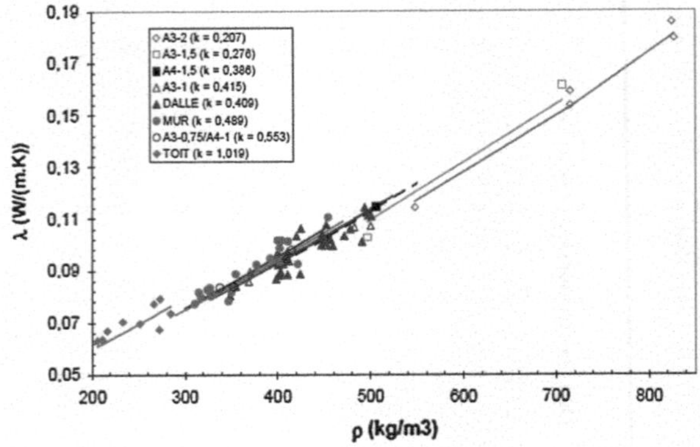

Fig. 6.11 Comparison of experimental data to self-consistent homogenization of thermal conductivity (Cerezo 2005)

6.3.1.2 Effect of Formulation and Manufacturing Method on Thermal Conductivity of Bio-Aggregate Based Building Materials

The formulation embeds several data: the type of binder, the type of aggregate, the aggregate to binder ratio, the water to binder ratio and the use of water retainer. The water to binder ratio is generally adjusted to ensure a good workability of the fresh product in connection with the mixing and manufacturing method. So, the effect of formulation and manufacturing method can not be studied separately.

Binder

Thermal conductivity of bio-aggregate based building materials is significantly dependent on the binder used in the formulation.

Actually, the thermal conductivity of binders themselves is dependent on the type of binder, the proportion of the binders in case of mixture, and the water to binder ratio. Stefanidou et al. (2010) used six different binders (traditional binders such as lime, natural pozzolans, brick dust, and white cement) to produce 15 pastes. The thermal conductivities of studied materials range from 0.16 (mixture of air lime and pozzolan) to 0.39 W m^{-1} K^{-1} (white cement). They show that, with similar proportion by mass, traditional materials such as lime are less conductive than modern materials like cements. They also underline that adding white cement increases the thermal conductivity of mixtures and that adding pozzolanic materials reduces the thermal conductivity in comparison with pure lime. Finally, they also show that an increase in the water to binder ratio in the mixture of binder increases the porosity and thus reduces the thermal conductivity of the binder.

Furthermore, the physico-chemical interaction occurring between the bio-aggregate and the binder affects the microstructure of the composite and thus its thermal conductivity. Gourlay and Arnaud (2010) give the thermal conductivity of hemp concretes made with three kinds of binder. They show that hemp concretes made with cement based binder are less conductive than hemp concretes made of lime based binder (respective thermal conductivities of 0.06 and 0.08 W m^{-1} K^{-1}). They underline that, with the same hemp-binder-water ratio, the density of hemp

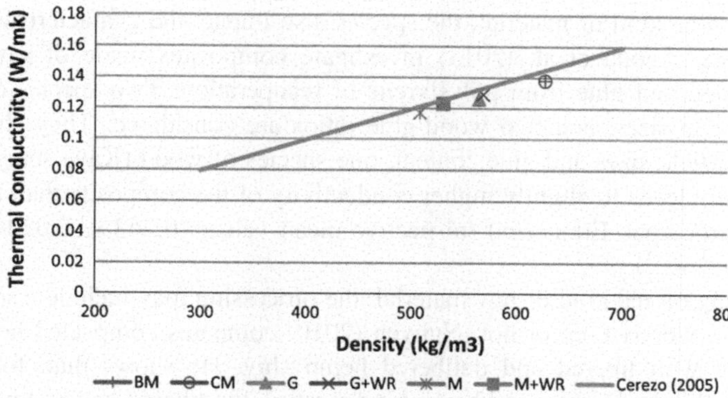

Fig. 6.12 Relationship between density and thermal conductivity of hemp concrete walls and the relationship established by Cerezo (2005). *BM* builder's mix, *CM* commercial mix, *G* lime: Ground Granulated Blastfurnace Slag (GGBS) binder, *G + WR* lime: GGBS binder with water retainer, *M* lime: metakaolin binder, *M + WR* lime: metakaolin binder with water retainer (Walker and Pavia 2014)

concretes is lower with cement based binder than with lime-based binder (resp. 297/393 kg m^{-3}) even if the density of cement-based binder is higher than the density of lime-based binder (resp. 1166/843 kg m^{-3}). They conclude that the physico-chemical interaction between hemp aggregates and binder affects the microstructure of the material by increasing its porosity and this induces a decrease in thermal conductivity. Walker and Pavia (2014) also show that increasing the binder's hydraulic content in hemp concretes slightly reduces thermal conductivity. The decrease in thermal conductivity occurs simultaneously with a decrease in density (Fig. 6.12).

Bio-aggregate

Thermal conductivity of bio-aggregate based building materials also depends on the characteristics of the bio-aggregate. The lower the thermal conductivity of inclusions, the more the material is insulating. In addition, the increase in porosity decreases the density of the composite and consequently its thermal conductivity. The characteristics of bio-aggregate rely on the kind of material, its origin, its species, and its processing (treatment, separation of woody core from fiber, particle size…).

The morphological characteristics of bio-aggregates depend on their growing conditions (land, weather…). Stevulova et al. compared lightweight composites made with technical hemp (cannabis sativa) from two origins (Stevulova et al. 2013). They show that Hungarian hemp leads to higher values of thermal conductivity of concrete than Netherlands hemp (respective thermal conductivity: 0.111 and 0.069 W m^{-1} K^{-1} for same components ratios), while giving higher strength parameter. They suggest that this fact is probably due to creating a stronger structure during the growing, with smaller pores.

For the same kind of material, the species also impact the characteristics of the bio-aggregate. Agoua et al. (2013) investigate composites made of sawdust of recycled wood and glue from polystyrene of recuperation. Two species of wood, several particle sizes, and two wood-glue ratios are considered. They show that, whatever particle sizes and glue content, one species of wood (Kaya senegalensis) systematically leads to slightly higher conductivity of the composite than the other species (Pterocarpus Erinaceus) (respective mean values: 0.263 ± 0.022/0.242 ± 0.022 W m^{-1} K^{-1}).

Depending on the kind of raw material, the processing may include a separation of fiber from woody core or not. Nguyen (2010) compares compacted hemp concretes made with fibered and defibered hemp shiv. He shows that, for similar formulation (kind of binder and hemp-binder ratio), the fibered hemp shiv leads to lower conductivity of the composite than defibered hemp shiv. According to the author, the fibers induce a higher inter-particular porosity, for the same density of the composite, leading to lower thermal conductivity. It would be interesting to corroborate this result for other processing methods.

The particle size distribution, resulting from the processing, also impacts the thermal conductivity of composites. Agoua et al. show that the more the composites contain thin elements of sawdust of wood, the more they drive heat (Agoua et al. 2013). Similar results are found with hemp-based materials. Stevulova et al. show that decreasing the mean particle length in hemp composites induces a stronger structure with smaller pores and also leads to higher thermal conductivity (Stevulova et al. 2013). Pretot et al. compare two hemp-lime plasters (Pretot et al. 2009). The hemp with the finest particle size distribution leads to slightly higher thermal conductivity (0.227 vs. 0.193 W m^{-1} K^{-1}: +17.6%), simultaneously with higher density (809 vs. 692 kg m^{-3}: +16.9%).

Bio-aggregate to binder ratio

Bio-aggregate to binder ratio impacts the thermal properties of bio-based building materials. Studies performed on wood-aggregate based composites and on hemp-aggregate based composites show that by increasing the bio-aggregate to binder ratio, the thermal conductivity of the composite decreases (Agoua et al. 2013; Al Rim et al. 1999; Bederina et al. 2007; Benfratello et al. 2013; Bouguerra 1999; Collet and Pretot 2014; de Bruijn and Johansson 2013; Ledhem et al. 2000; Taoukil et al. 2013). The increase in bio-aggregate to binder ratio induces a decrease in density of the composite due to a low density of the bio-aggregate. Furthermore the bio-aggregate induces porosity in the matrix which appears to be more porous when the aggregate ratio increases (Al Rim et al. 1999). Bederina et al. underline that the decrease in thermal conductivity is not linearly proportional to the increase of bio-aggregate in wood-based composite (Bederina et al. 2007) (Fig. 6.13]. Similar results are found on hemp concretes (Benfratello et al. 2013; Collet and Pretot 2014).

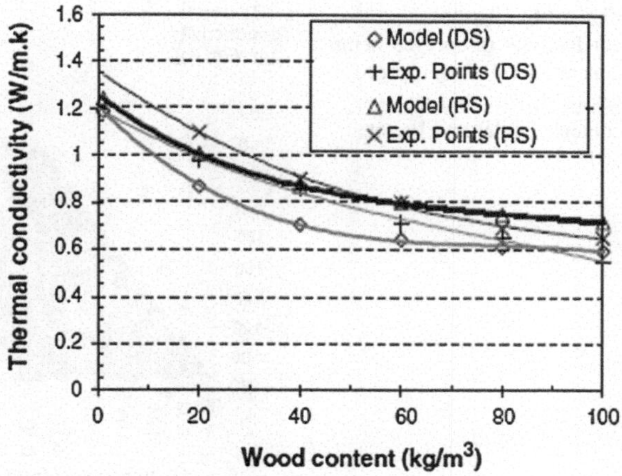

Fig. 6.13 Thermal conductivity according to the content of wood: confrontation of the experimental results to the results obtained by auto-coherent model (Bederina et al. 2007)

Anisotropy

The geometry and the capillary structure of bio-aggregate make them anisotropic. It is shown in Carré and Le Gall (1990) and in Suleiman et al. (1999) that the longitudinal thermal conductivity of wood (parallel to capillaries network) is higher than the transversal one.

Depending on the production method, bio-aggregate based building materials can also show anisotropy. Nguyen studied the effect of anisotropy on thermal conductivity of compacted hemp concretes (Nguyen 2010). He shows that, depending on the formulation, the thermal conductivity in longitudinal axis was 1.2–1.8 higher than the thermal conductivity in the transversal axis. Such results were also found in Pierre et al. (2013) where the longitudinal thermal conductivity was higher than the transversal conductivity of hemp concrete and in Dinh et al. (2015) where the ratio between thermal conductivity measured according to parallel direction and perpendicular direction of compaction was from 1.13 to 1.19 for dry hempcretes and from 1.15 to 1.17 for humid hempcretes.

6.3.1.3 Effect of Water Content on Thermal Properties of Bio-aggregate Based Building Materials

The hygroscopicity of bio-aggregate based building materials impacts their thermal properties.

Actually, an increase in water content induces an increase in thermal conductivity, as water shows much higher thermal conductivity than air ($\lambda_{water} = 0.6$; $\lambda_{air} = 0.026$ W m^{-1} K^{-1}). This increase is more or less important, depending on the density and on the porosity of the material. For wood concrete, the thermal conductivity increases rapidly with water content. The thermal conductivity at saturation is two to three times higher than the value at dry state (Taoukil et al. 2013). It can reach five times this value in Bouguerra (1999). For hemp-based materials,

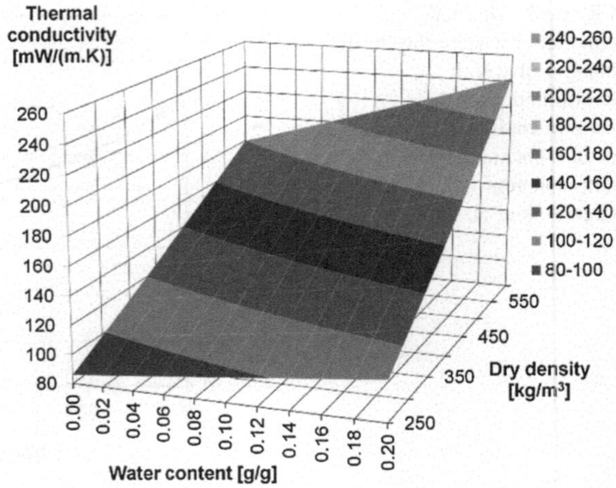

Fig. 6.14 Thermal conductivity of sprayed hemp concrete (wall formulation) versus density and water content (Collet and Pretot 2014)

thermal conductivity also increases with water content. In Amziane and Arnaud (2013), the variation of thermal conductivity with water content is modeled by double homogenization for two hemp concretes. For the HLC1 hemp concrete, thermal conductivity is related to water content following:

$$\lambda = 0.105 + 0.035 \times w \quad (6.1)$$

with λ the thermal conductivity (W m^{-1} K^{-1}) *and* w the mass water content (%).

The values from this equation are in line with experimental data obtained by Cerezo (2005) and are also consistent with the results from Collet (2004) who considers a three phase model with air, water and solid phase. Taking into account adsorption/desorption isotherm, the thermal conductivity rises by 30% when the material has been left in humid environment (75%RH) (Amziane and Arnaud 2013). In Collet and Pretot (2014), it is underlined that the ambient humidity impacts more or less the thermal conductivity of hemp concrete, as, depending on their formulation, they are more or less hygroscopic. From dry to humid state (90%RH), the thermal conductivity of hemp concrete rises by 25% for wall formulation with low density; while, in the same range of relative humidity, it increases by 11% for floor formulation (water content at 90%RH about 0.08 g g^{-1}). Finally, for a given formulation, even if the impact of moisture content is lower than the impact of density (Fig. 6.14), it is not negligible.

6.3.2 *Heat Capacity and Thermal Diffusivity*

Specific heat capacity and volumetric heat capacity characterize the ability of a material to store heat. The specific heat capacity is the amount of heat per unit mass required to raise the temperature by one degree Celsius while the volumetric heat capacity is related to a unit of volume.

Thermal diffusivity describes the rate at which heat flows through a material. Thermal diffusivity is related to thermal conductivity, density and specific heat capacity according:

$$a = \frac{\lambda}{\rho.C_p} \tag{6.2}$$

with a the thermal diffusivity ($m^2\ s^{-1}$), λ the thermal conductivity ($W\ m^{-1}\ K^{-1}$), ρ the density ($kg\ m^{-3}$) and C_p the specific heat capacity ($J.kg^{-1}\ K^{-1}$).

Thermal diffusivity and/or heat capacity are measured with transient methods as flash method, transient plane source or hot disk. Specific heat capacity can also be measured with calorimeter or DSC.

6.3.2.1 Heat Capacity and Thermal Diffusivity of Bio-aggregate Based Building Materials

The specific heat capacity of bio-aggregate based building materials is in the same range as the specific heat capacity of usual building materials. For wood concrete, Taoukil et al. estimate the thermal diffusivity for several wood shaving ratios (Taoukil et al. 2013). The specific heat capacity calculated from these measures ranges from 640 to 1490 J $kg^{-1}\ K^{-1}$, depending on wood shaving ratio. For hemp concrete, the study of Walker and Pavia with six kinds of binder gives specific heat capacities ranging from 1240 to 1350 J $kg^{-1}\ K^{-1}$ (Walker and Pavia 2014). The specific heat capacity of sprayed hemp concrete, calculated from the thermal diffusivity at dry point at 20 °C in Pierre et al. (2013), ranges from 650 to 870 J $kg^{-1}\ K^{-1}$. There is a high discrepancy between these results that may be linked with the measurement methods. Actually, one should underline that the measurement of thermal diffusivity of such materials appears difficult. Taoukil et al. underline that the value depends on the counting model used to estimate the thermal diffusivity from experimental thermogram (Taoukil et al. 2013). On the other hand, the use of calorimetry may also be distorted by chemical reaction producing heat during the measurement.

6.3.2.2 Effect of Formulation and Manufacturing Method on Heat Capacity and Thermal Diffusivity of Bio-Aggregate Based Building Materials

Like thermal conductivity, heat capacity and thermal diffusivity are impacted by formulation of bio-aggregate based building materials.

The effect of the kind of binder is studied in Walker and Pavia (2014). They show that increasing the binder's hydraulic content in hemp concretes slightly increases heat capacity, while it reduces thermal conductivity. More, in their study the use of water retainers increases both heat capacity and conductivity.

The effect of bio-aggregate to binder ratio is investigated on hemp concrete and on wood concrete. De Bruijn and Johansson (2013) find that the amount of hemp in hemp lime concrete has no significant effect on their thermal diffusivity as the measured values overlap. However, they conclude that a larger amount of hemp in hemp lime concrete leads to better thermal properties, especially for wet conditions, with lower thermal conductivity and lower specific heat capacity. Taoukil et al. estimate the thermal diffusivity of wood-concrete for several wood shaving ratios (Taoukil et al. 2013). It is highlighted that increasing the wood-shaving content in composite decreases its thermal diffusivity, in link with the decrease in thermal conductivity. Globally, increasing the bio-aggregate amount in the composite allows improving its thermal insulating performance.

6.3.2.3 Effect of Water Content on Heat Capacity and Thermal Diffusivity of Bio-Aggregate Based Building Materials

Regarding the variation of thermal diffusivity with water content, the study performed by de Bruijn and Johansson on two hemp-lime concretes conditioned to 15 and 65%RH, does not allow concluding (de Bruijn and Johansson 2013). Taoukil et al. highlight that a maximum value, for given water content, can be observed on wood-concrete (Taoukil et al. 2013). This is explained by the fact that in a first step, the thermal conductivity increases faster than volumetric heat capacity, inducing an increase in thermal diffusivity. Then, the thermal conductivity increases slower than volumetric heat capacity and thermal diffusivity thus decreases.

Evrard (2008) uses the thermal balance to calculate the variation of the specific heat capacity of hemp concrete with water content.

$$C(w) = \frac{(\rho_0 \times C_0 + w \times C_w)}{\rho} \tag{6.3}$$

with: ρ the density (kg m^{-3}); C the specific heat capacity (J kg^{-1} K^{-1}); subscript *0*: at dry state; w the water content (kg kg^{-1}); C_w the specific heat capacity of water stored in the porous structure (J kg^{-1} K^{-1}).

The variation of specific heat capacity with ambient relative humidity is thus calculated from this relationship and from sorption curves. Figure 6.15 illustrates the direct increase from thermal capacity of water stored.

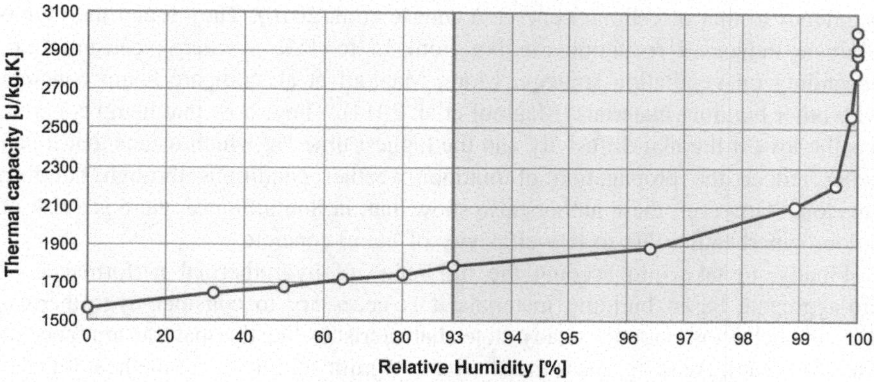

Fig. 6.15 Evolution of thermal capacity of hemp lime material—wall in moist state (Evrard 2008)

6.4 Concluding Remarks on Hygrothermal Behavior of Bio-aggregate Based Building Materials

The hygric and thermal properties of bio-aggregate based building materials give them a particular hygrothermal behavior which allows reducing energy demand of building while maintaining indoor relative humidity (Tran Le et al. 2010; Evrard and de Herde 2010).

It was shown in Evrard and de Herde (2010) and in Lawrence et al. (2013) that hemp concrete wall perform better than traditional wall assemblies on phase shift point of view. For example, in the simulation of Evrard and de Herde (2010) the phase shift was 15 h in the hemp concrete wall while it was 5 h in the mineral wool wall.

These performances are correlated with hygric behavior of hemp concrete. Actually, Lawrence et al. (2013) underline that, after a sudden drop in temperature on one side of the experimental wall, the steady state is reached at approximately 240 h while it is reached within 72 h in simulations ignoring the effects of relative humidity. Thus, relative humidity has a strong effect on the hygrothermal behavior of the wall. Pretot and Collet (2012) studied the response of a hemp concrete test—wall to several hygrothermal solicitations. They show that (i) under isothermal conditions with vapor pressure gradient, homogeneous vapor diffusion occurs; (ii) under constant vapor pressure with a decrease of ambient temperature, huge variations of vapor pressure are observed through the wall, in correlation with adsorption-desorption and/or condensation-evaporation phenomena. The experimental study reported by Arnaud, Samri and Gourlay in Amziane and Arnaud (2013) shows that an internal source (or well) of heat and/or the presence of another flow of heat (notably by convection) exist in hemp concrete and directly impact the heat balance equation.

Numerical studies at building scale show that hemp concrete allows high hygrothermal performances of building. Tran Le et al. compare hemp concrete behaviour to that of cellular concrete (Tran le et al. 2010). They found that hemp concrete induces a reduction ranging from 15 to 45% in energy consumption, depending on ventilation strategy. More, Maalouf et al. compare hemp concrete with other building materials (Maalouf et al. 2014). They show that hemp concrete has the lowest thermal diffusivity and the highest time lag which means that it can better reduce the propagation of outdoor weather conditions through building envelope. However, these authors also show that, in South France, there is a risk of indoor superheating due to low effusivity of hemp concrete.

Finally, to take into account the full range of hygrothermal performance of bio-aggregate based building materials it is necessary to consider hygrothermal dynamic behavior, not only steady-state characteristics like thermal transmittance of wall (U-value). Actually, mass transfer has a significant impact on the heat transfer in bio-aggregate based building materials, in correlation with latent heat of phase-change or sorption heat. It is thus necessary to accurately model mass transfer taking into account the sorption hysteresis of bio-aggregate based building materials. Currently numerical model are developed following this objective (Aït Oumeziane et al. 2014). Besides, it is necessary to develop new experimental benches to highlight the contribution of hygric behavior in the whole energetic performance of these materials. More, investigations are needed to develop measurement methods of thermal diffusivity and/or heat capacity of bio-aggregate based building material as usual methods are not well adapted.

References

Agoua, E., Allognon-Houessou, E., Adjovic, E., Togbedji, B.: Thermal conductivity of composites made of wastes of wood and expanded polystyrene. Constr. Build. Mater. **41**, 557–562 (2013)

Aigbomian, E.P., Fan, M.: Development of wood-crete building materials from sawdust and waste paper. Constr. Build. Mater. **40**, 4361–4366 (2013)

Al Rim, K., Ledhem, A., Douzane, O., Dheilly, R.M., Queneudec, M.: Influence of the proportion of wood on the thermal and mechanical performances of clay-cement-wood composites. Cem. Concr. Compos. **21**, 269–276 (1999)

Amziane, S., Arnaud L. (eds.): Bio-aggregate-based Building Materials. Applications to Hemp Concrete. ISTE Ltd and Wiley, Inc (2013)

Anderson, R.B.: Modifications of the Brunauer, Emmett and Teller. J. Am. Chem. Soc. **68**(4), 686–691 (1946)

Anderson, R.B., Hall, W.K.: Modifications of the Brunauer Emmett and Teller, equation II. J. Am. Chem. Soc. **70**(5), 1727–1734 (1948)

Arnaud, L.: Mechanical and thermal properties of hemp mortars and wools: experimental and theoretical approaches. In: 3rd International Symposium Of Bioresource Hemp (2000)

Bederina, M., Marmoret, L., Mezreb, K., Khenfer, M.M., Bali, A., Quéneudec, M.: Effect of the addition of wood shavings on thermal conductivity of sand concretes: experimental study and modelling. Constr. Build. Mater. **21**, 662–668 (2007)

Benfratello, S., Capitano, C., Peri, G., Rizzo, G., Scaccianoce, G., Sorrentino, G.: Thermal and structural properties of a hemp-lime biocomposite. Constr. Build. Mater. **48**, 745–754 (2013)

Bouguerra, A.: Prediction of effective thermal conductivity of moist wood concrete. J. Phys. D: Appl. Phys. **32**(12), 1407–1414 (1999)

Bouguerra, A., Sallée, H., de Barquin, F., Dheilly, R.M., Quéneudec, M.: Isothermal moisture properties of wood-cementitious composites. Cem. Concr. Res. **29**, 339–347 (1999)

Carré, P., Le Gall, R.: Définition et détermination des conductivités thermiques dans les structures multicouches C.V.R. – balsa. Rev. Gén. Therm. **340**, 211–215 (1990)

Cerezo, V.: Propriétés mécanique, thermique et acoustique d'un matériau à base de particules végétales: approche expérimentale et modélisation théorique. Ph.D. thesis, ENTPE, INSA de Lyon (juin 2005)

Chamoin, J.: Optimisation des propriétés (physiques, hydriques et mécaniques) de bétons de chanvre par la maîtrise de la formulation. Ph.D. thesis, INSA de Rennes (juillet 2013)

Collet, F.: Caractérisation hydrique et thermique de matériaux de génie civil à faibles impacts environnementaux. Ph.D. thesis, INSA de Rennes (décembre 2004)

Collet, F., Pretot, S.: Experimental investigation of moisture buffering capacity of sprayed hemp concrete. Constr. Build. Mater. **36**, 58–65 (2012)

Collet, F., Pretot, S.: Variation de la capacité hydrique tampon de bétons de chanvre en fonction de la formulation, Ecobat Sciences et techniques 1ère édition, 7 et 8, communication orale, Paris (mars 2012)

Collet, F., Pretot, S.: Effect of coating on moisture buffering of hemp concrete. In: 2nd International Conference on Building Energy and Environment, Boulder, Colorado, 1–4 August 2012

Collet, F., Pretot, S.: Thermal conductivity of hemp concretes: variation with formulation, density and water content. Constr. Build. Mater. **65**, 612–619. doi:10.1016/j.conbuildmat.2014.05.039 (2014)

Collet, F., Bart, M., Serres, L., Miriel, J.: Porous structure and water vapour sorption of hemp-based materials. Constr. Build. Mater. **22**, 1271–1280 (2008)

Collet, F., Chamoin, J., Prétot, S., Lanos, C.: Comparison of the hygric behaviour of three hemp concretes. Energy Build. **62**, 294–303 (2013)

De Boer, J.H.: The Dynamical Character of Adsorption. Clarendon Press, Oxford (1953)

de Bruijn, P., Jeppssona, K.H., Sandin, K., Nilssona, C.: Mechanical properties of lime–hemp concrete containing shives and fibres. Biosyst. Eng. **103**(4), 474–479 (2009)

de Bruijn, P., Johansson, P.: Moisture fixation and thermal properties of lime-hemp concrete. Constr. Build. Mater. **47**, 1235–1242 (2013)

de Vries, D.A.: Simultaneous transfer of heat and moisture in porous media. Eos, Trans. Am. Geophys. Union **39**, 909–916 (1958)

Dinh, T.M., Magniont, C., Coutand, M., Escadeillas, G.: Hemp concrete using innovative pozzolanic binder. In: ICBBM 2015, Clermont Ferrand, France, 22–24 June 2015

Dubois, S., Lebeau, F.: Hygrothermal behavior of different lime-hemp concrete mixes. In: First International Conference on Concrete Sustainability (ICCS13), Tokyo, Japan (2013)

Dubois, S., Evrard, A., Lebeau, F.: Hygrothermal modelling of lime-hemp concrete used as building material and indoor climate buffering characterization. In: International Conference of Agricultural Engineering, Valence, Espagne, 8–12 July 2012

Dubois S., Evrard A., Lebeau F.: Modeling the hygrothermal behavior of biobased construction materials. J. Build. Phys. (2013)

Elfordy, S., Lucas, F., Tancret, F., Scudeller, Y., Goudet, L.: Mechanical and thermal 415 properties of lime and hemp concrete ("hempcrete") manufactured by a projection process. Constr. Build. Mater. **22**(10), 2116–2123 (2008)

Evrard, A.: Sorption behaviour of Lime-hemp concrete and its relation to indoor comfort and energy demand. In: 23rd Conference on Passive and Low Energy Architecture, Geneva, Switzerland, 6–8 September 2006

Evrard, A.: Transient hygrothermal behavior of Lime–Hemp materials, Ph.D. thesis, Université catholique de Louvain (2008)

Evrard, A., De Herde, A.: Hygrothermal performance of lime-hemp wall assemblies. J. Build. Phys. **34**, 15–25 (2010)

Gawin D., Kosny J., and Wilkes K.: Thermal conductivity of moist cellular concrete—experimental and numerical study. ASHRAE Build. **IX**, 1–10 (2004)

Gourlay, E., Arnaud, L.: Comportement hygrother-mique des murs de béton de Chanvre. Actes du congrès SFT, Le Touquet, France (2010)

Guggenheim, E.A.: Application of Statistical Mechanics. Clarendon Press, Oxford (1966). (Chapter 11)

International Union of Pure and Applied Chemistry: Reporting physisorption data for gas/solid systems with special reference to the determination of surface area and porosity. Pure Appl. Chem. **57**(4), 603–619 (1986)

Janssen, H., Roels, S.: The dependable characterisation of the moisture buffer potential of interior claddings, Proceedings of the Nordic Symposium on Building Physics, Copenhagen, Denmark, June 2008, pp. 16–18 (2008)

Kumaran, K., Lackey, J., Normandin, N., van Reenen, D., Tariku, F.: Summary Report from Task 3 of MEWS Project at the Institute for Research in Construction—Hygrothermal Properties of Several Building Materials, IRC-RR-110 (2002)

Latif, E., Lawrence, M., Shea, A., Walker, P.: Moisture buffer potential of wall assemblies incorporating hemp-lime. In: ICBBM 2015, Clermont Ferrand, France. 22–24 June 2015

Lawrence, M., Shea, A., Walker, P., De Wilde, P.: Hygrothermal performance of bio-based insulation materials. Proc. Inst. Civil Eng. Constr. Mater. **166**(4), 257–263. ISSN: 1747-650X (2013)

Ledhem, A., Dheilly, R.M., Benmalek, M.L., Queneudec, M.: Properties of wood-based composites formulated with aggregate industry waste. Constr. Build. Mater. **14**, 341–350 (2000)

Maaloufa, C., Tran Le, A.D., Umurigirwaa, S.B., Lachia, M., Douzane, O.: Study of hygrothermal behaviour of a hemp concrete building envelope under summer conditions in France. Energy Build. **77**, 48–57 (2014)

Magniont, C., Escadeillas, G., Coutand, M., Oms-Multon, C.: Use of plant aggregates in building ecomaterials. Eur. J. Environ. Civil Eng. **16**(Supplement 1), s17–s33 (2012)

Naono, H., Hakuman, N.: Analysis of porous texture by means of water vapour adsorption isotherm with particular attention to the lower limit of hysteresis loop. J. Colloid. Interf. Sci. **158**, 19–26 (1993)

Nguyen, T.T.: Contribution à l'étude de la formulation et du procédé de fabrication d'éléments de construction en béton de chanvre. Ph.D. thesis, Université de Bretagne Sud (janvier 2010)

Nielsen, A.F.: Free Water Intake of Cellular Concrete and Bricks Measured by Gamma Ray Attenuation. Thermal Insulation Laboratory, Technical University of Denmark, Report no. 41 (January 1976)

Oumeziane, Y.A., Bart, M., Moissette, S., Lanos, C.: Hysteretic behaviour and moisture buffering of hemp concrete. Transp. Porous Med. (2014). doi:10.1007/s11242-014-0314-7

Perré, P., Pierre, F., Casalinho, J., Ayouz, M.: Determination of the mass diffusion coefficient based on the relative humidity measured at the back face of the sample during unsteady regimes. Drying Technol. Int. J. doi:10.1080/07373937.2014.982253 (2015)

Peuhkuri, R.: Moisture dynamics in building envelopes. Doctoral thesis, Department of Civil Engineering, Technical University of Benmark, Denmark (2003)

Pierre, T., Colinart, T., Glouannec, P.: Measurement of thermal properties of biosourced building materials. Int. J. Thermophys. doi:10.1007/s10765-013-1477-0 (2013)

Pretot, S., Collet, F., Glouannec, P., Lang, V.: Variation des propriétés thermiques de bétons de chanvre en fonction de la formulation, Congrès français de thermique-Efficacité énergétique, Vannes 2009, Tome 2, pp. 865–870, Editions Société Française de Thermique, ISBN: 2-905267-67-2

Pretot, S., Collet, F.: Experimental study of hygrothermal behavior of a hemp concrete wall (without and with coating). In: 2nd International Conference on Building Energy and Environment, communication orale, Boulder, Colorado, 1er au 4 août 2012

Réglementation Thermique.: Règles Th-U—Fascicule 2 (2005)

Rode, C.: Moisture Buffering of Building Materials, Report BYG•DTU R-126, ISSN 1601-2917, ISBN 87-7877-195 (2005)

Sassoni, E., Manzi, S., Motori, A., Montecchi, M., Canti, M.: Novel sustainable hemp-based composites for application in building industry: physical, thermal and mechanical characterization. Energy Build. **77**, 219–226 (2014)

Stefanidou, M., Assael, M., Antoniadis, K., Matziaroglou, G.: Thermal conductivity of building materials employed in the preservation of traditional structures. Int. J. Thermophys. **31**, 844–851 (2010)

Stevulovaa, N., Kidalovaa, L., Cigasovaa, J., Junaka, J., Sicakovaa, A., Terpakovaa, E.: Lightweight composites containing hemp hurds. Proced. Eng. **65**, 69–74 (2013)

Suleiman, B.M., Larfeldt, J., Leckner, B., Gustavsson, M.: Thermal conductivity and diffusivity of wood. Wood Sci. Technol. **33**, 465–473 (1999)

Taoukil, D., El bouardi, A., Sick, F., Mimet, A., Ezbakhe, H., Ajzoul, T.: Moisture content influence on the thermal conductivity and diffusivity of wood–concrete composite. Constr. Build. Mater. **48**, 104–115 (2013)

Tran Le, A.D.: Etude des transferts hygrothermiques dans le béton de chanvre et leur application au bâtiment. Ph.D. thesis, Université de Reims Champagne- Ardenne; November (2010) [in French]

Tran Le, A.D., Maalouf, C., Mai, T.H., Wurtz, E., Collet, F.: Transient hygrothermal behavior of a hemp concrete building envelope. Energy Build. **42**, 1797–1806 (2010)

Walker, R., Pavía, S.: Moisture transfer and thermal properties of hemp–lime concretes. Constr. Build. Mater. **64**, 270–276 (2014)

Zaknoune, A.: Etude du comportement thermohydrique de matériaux « chanvre-chaux » lors de la phase de séchage – estimation par technique inverse des propriétés hydriques. Ph.D. thesis, Université de Bretagne Sud (December 2011) [in French]

Chapter 7
Bio-aggregate Based Building Materials Exposed to Fire

Christophe Lanos

Abstract This chapter reports the state of the art of several investigations on behaviour of bio-aggregate based building materials exposed to fire. Discrepancies between fire reaction and fire resistance is highlighted in this chapter. Various results of fire reaction test performed on bio based materials are presented. Bio-aggregates are often in Class F while concretes range in class B1. In the case of fire performances, limited tests are performed and reported in the literature. Such test must be performed on wall or part of building to give the more realistic overview of the behaviour of building structure exposed to standard fire. In some of presented case studies, render and plaster play a key role in the fire resistance. EI 90 fire resistance appeared to be accessible with conventional technologies.

Keywords Fire reaction · Fire resistance

7.1 Introduction

A distinction appeared between the behaviours of fire reaction of a material or a building product and of fire resistance of a building structure or element (wall, slab…).

The construction works must be designed and build in such a way, that in the event of an outbreak of fire:

– the generation and spread of fire and smoke within the works are limited,
– the spread of fire to neighbour construction works is limited,

C. Lanos (✉)
Laboratoire de Génie Civil et Génie Mécanique de Rennes équipe
Matériaux-Thermo-Rhéologie, Université de Rennes 1, Rennes, France
e-mail: christophe.lanos@univ-rennes1.fr

C. Lanos
IUT Génie Civil Construction Durable, 3 Rue Du Clos Courtel, BP 90422-35704
Rennes Cedex, France

S. Amziane and F. Collet (eds.), *Bio-aggregates Based Building Materials*,
RILEM State-of-the-Art Reports 23, DOI 10.1007/978-94-024-1031-0_7

- the load bearing resistance of the construction can be assumed for a specified period of time,
- the occupants can leave the works or can be rescued by other means,
- the safety of rescue teams is taken into consideration.

Construction products are tested in respect of their reaction to fire, which is important in determining how a fire is likely to start and/or how it will develop. These reaction properties are ease of ignition, spread of flame, evolution of smoke and toxic gases, and heat release rate of the burning material. Many test methods are applicable to evaluate these properties. The product can then be classified in terms of the relevant fire reaction. Test data is often also used in calculations of the spread of fire or for estimation of design fire ratings.

The fire resistance of building elements are evaluated to determine their behaviours when exposed to a particular temperature, normally representing a fire in an enclosed space (a room). Fire resistance is one of several properties of the structure/product, and thus is not simply a property of the specific materials used in the structure or product. Fire resistance is influenced by the properties of the assembled/combined structure or product, and is not a property of the materials themselves. Fire resistance testing can be performed for many different load bearing structures, such as walls and glazed structures, floor/ceiling structures and roofs, beams, pillars, doors, ceiling claddings, ducts, cable penetrations and fire dampers. Test can be performed with both vertical and horizontal exposure. The structure can then be classified into the appropriate fire resistance Class.

It can be important to highlight that safety in the event of fire is one of the six important requirements applicable to the finished building of the Construction Products Directive (89/106/EEC). The main purpose of this directive is to facilitate free movement of construction products throughout the EU. As a consequence, the classification of the fire properties of construction products is based on a EU standard defining fire reaction classes associated to the definition of harmonized test methods and classification rules.

According to EU standards a wide range of Fire Resistance classes of building elements and structures are defined in respect of their fire separation performance, load bearing capacity and smoke tightness. Tests methods are proposed by specific standards depending on the type of the studied structure element.

Other available data concerning fire behaviour of bio-aggregate based building materials complete the one linked to the fire reaction and resistance classification. Many rustic tests of fire direct exposure of formulated material are mentioned in references. The test protocols is even undefined and the evaluation of fire performances is subjective. In addition, the case of some realistic fire test performed on a real scale building or room have to be mentioned.

7.2 Fire Reaction

7.2.1 European Class of Fire Reaction

European standards are used as reference to structure the discussion of this state of the art. However, the reported data can be easily transposed regarding other standards. The reaction to fire classification of products is performed according to the terminology of Euroclass system [(EN 13501-1 2007)].

The Euroclass system for construction products affects mainly surface covering materials, insulation materials, floor coverings, pipe insulation materials and cables. The product groups are all treated in a similar manner. Fire reaction classes are divided into seven main classes: A1, A2, B, C, D, E and F. Additional criteria are taken into account: smoke release (s) and droplet formation (d) with 3 levels (1 for low release to 3 for large release).

Euroclass for insulation materials are:

- A1 (non-combustible material)
- A2-s1, d0 (limited combustibility material)
- B-s1, d0 (Class I surface lining)
- C-s2, d0 (Class II surface lining)
- D-s2, d0 (Class III surface lining)

Euroclass for floor coverings are:

- A1fl (non-combustible floor covering material)
- Cfl-s1 (Class G floor covering for exit routes)
- Dfl-s1 (Class G floor covering for meeting halls and similar)

The conditions for fulfilling the requirements of a particular class can be complicated [(EN 13501-1 2007)]. The main fire reactions associated to each class are described in Table 7.1.

The additional classifications for smoke production corresponds to:

- s3: No limitation of smoke production required,
- s2: The total smoke production as well as the ratio of increase in smoke production are limited,
- s1: More stringent criteria than s2 are satisfied.

The additional classifications for flaming droplets/particles: corresponds to:

- d2: If no performance is declared,
- d1: If no flaming droplets/particles persisting longer than a given time allowed (typically 10 s),
- d0: If no flaming droplets/particles occur within 600 s when tested in accordance with EN 13823.

Table 7.1 Main characteristics of fire reaction Euroclass for building products excluding floorings acc. to (EN 13501-1 2007)

Euro-class	Contribution to fire/aspired safety level
A1	Class A1 products will not contribute in any stage of the fire including the fully developed fire. For that reason they are assumed to be capable of satisfying automatically all requirements of all lower classes
A2	Satisfying the same criteria as Class B for the SBI-test according to EN 13823. In addition, under conditions of a fully developed fire these products will not significantly contribute to the fire load and fire growth
B	As Class C but satisfying more stringent requirements
C	As Class D but satisfying more stringent requirements. Additionally under the thermal attack by a single burning item they have limited lateral spread of flame
D	Products satisfying criteria for Class E and capable of resisting, for a longer period, a small flame attack without substantial flame spread. In addition, they are also capable of undergoing thermal attack by a single burning item with sufficiently delayed and limited heat release
E	Products capable of resisting, for a short period, a small flame attack without substantial flame spread
F	Products for which no reaction to fire performances are determined or which cannot be classified in one of the classes A1, A2, B, C, D, E

7.2.2 Tests Methods

Testing for the Euroclass system is performed in accordance with test methods, defined in harmonised European standards. Criteria leading to the classification are describe in standards [(NF EN 13 501-1, 2007)+A1 2009]. The test methods are summarized in this section. More details of test protocols and classification criteria can be found in references listed in this chapter. The choice of test depends on the expected classification.

7.2.2.1 Non Combustibility Furnace [Test According to (ISO 1182 2010)] and Calorimeter [Test According to (ISO 1716 2010)]

Such tests are used for Euroclass A1 and A2 (non-combustible materials). The masse release during heating of sample or superior calorific value are estimated during tests.

7.2.2.2 Single Burning Item, Fire Technical Testing of Building Products [Test Method (EN 13823 2013)]

Building products, except for flooring materials, are exposed to thermal heat from a gas burner with a heat release rate of 30 kW placed horizontally in a room corner. The sample (1 × 1.5 m) form the vertical wall upper the burner. The mounting of

Fig. 7.1 Example of FBI test realized on straw wall coated with earth rendering [illustration (cd2e)]

the sample should be as close to reality as possible. The test time is 21 min. The combustion gases are collected through a hood where heat release rate and smoke production are measured. In the test method, the products heat release rate (kW), total heat release (MJ) and smoke production rate (m^2/s) are measured. Flame spread and burning droplets/particles are observed visually. This test is usable according to Euroclass A1, A2, B, C, or D according to the European System (Fig. 7.1).

7.2.2.3 Cone Calorimeter [Test According to (ISO 5660 2015)]

This test is able to produce results leading to the same classification than obtained with SBI tests.

7.2.2.4 Reaction to Fire Tests for Floorings (ISO 9239 2010)

The flooring surface test specimen (230 × 1050 mm) is placed in a horizontal position below a gas-fired radiant panel inclined at 30° where it is exposed to a defined heat flux. A pilot flame is applied to the hotter end of the specimen. Following ignition, any flame front which develops is noted and a record is made of the progression of the flame front horizontally along the length of the specimen in terms of the time it takes to spread to defined distances. Smoke production during the test is recorded as light transmission in the exhaust stack. All types of flooring can be tested with this method, including wood floor, plastic floor, rubber floor, linoleum floor etc. This test is usable for floorings according to Euroclass A2fl, Bfl, Cfl or Dfl.

7.2.2.5 Reaction to Fire Tests (ISO 11925-2 2010)

This direct fire test is quite equivalent to the previous one but the specimen is in vertical position (surface test specimen 90 × 250 mm) exposed to direct flam for 15 or 30 s. Droplets formation and velocity of fire propagation are analysed. This test is usable for materials according to Euroclass B, C, D or E (Fig. 7.2).

7.2.3 Euroclass of Bio-agreggate Based Building Materials and Products

Data results gathered for various bio-aggregate based buildings materials are listed below. It appears that Euroclass of the various tested materials correspond to Class B in the case of straw but classes E or F in the most of cases.

Fig. 7.2 Example straw sample exposed to burner flam [illustration (cd2e)]

Euroclass of bio-aggregate and bio-aggregate based materials plus others:

Bio-aggregates or bio-fibers		
Sheep wool	D s3, d0	(GMI)[a]
Duck feathers	E or F	(GMI)
Metisse	E s1, d0	(GMI)
Cellulose wadding	B s2, d0	(GMI)
Flax	E	(GMI)
Hemp fibers	E	(SPTech 2009)
Hemp shiv	E	(GMI)
Hemp shiv with bitume	B	(GMI)
Black expanded cork	B	(GMI)
Wood fiber soft/hard panel	E	(GMI)
Fibragglo	B	(GMI)
Straw	E	(cd2e)
Straw bale	B	(GMI)

[a]Guide des Matériaux Isolants

Bio-aggregate based materials		
Straw wall with earth render	B s1, d0	(cd2e)
Straw panel	C	(stramit)
Hemp concrete block	F	(GMI)
Hemp concrete	F	(GMI)
Hemp concrete (floor 600 kg/m^3)	B_{fl} s1	(CSTB 2013a)
Hemp concrete (wall 600 kg/m^3)	B s1, d0	(CSTB 2013b)
Hemp concrete (wall 330 kg/m^3)	B s1, d0	(LNE 2014b)
Hemp concrete (roof 220 kg/m^3)	B s1, d0	(LNE 2014a)
Hemp lime render (750 kg/m^3)	A2 s1, d0	(CSTB 2014a)
Hemp lime render (935 kg/m^3)	A2 s1, d0	(LNE 2013)
Hemp concrete with lime render	A2 s1, d0	(CenC)

The listed data are often proposed without indications of samples characteristics as density, thickness, type of coating, thickness of render.

The Euroclass B for straw bale is questioning. Such performances is possibly obtained with a layer of render (earth or lime based coating).

Use of Euroclass E product as hemp, flax and straw in building impose interposition of fire proofing layer. A mineral render or plaster can be used as fire proofing layer. Another way is to increase the fire performances of the product using treatment. As example, coating of bio-based aggregate with a solution of phosphate salt or bore salt act as flame retardant. Introduction of mineral coating around the bio-based aggregate is an efficient method too (lime, cement paste, silica coating…).

In all cases, the density of treated aggregate increased and the product formulation must be adjusted to satisfy to the targeted density, thermal conductivity and mechanical performances.

The example of fireproofing increase of hempwool was carried out by (Glé, 2013). Initial Euroclass of selected hempwool was Class E for density of 323 kg/m^3. After treatment and adjustment of formulation necessary to master the acoustical and thermal properties, the obtained Euroclass was equivalent to Class B s3, d0 with density of 331 kg/m^3.

7.3 Fire Resistance

7.3.1 Fire Resistance Classes

The fire resistance is evaluated in fire labs using large scale furnaces. The structure element is exposed to a thermal loading corresponding to a theoretical fire. Different thermal solicitations are defined in standards. Building elements and structures are to be tested and classified in respect of their fire separation performance and smoke tightness according to a system that indicates the properties by a letter—e.g. R, E or I—and an index that indicates the time for which the property is maintained, e.g. RE60. According to standard (EN 13501-2 2007), a wide range of Fire Resistance classes E, EI, REI is possible. The criteria defining the fire resistance duration are presented in Table 7.2.

The test protocols are proposed in various standards following the type or elements of the tested structure and the fire configuration (horizontal, vertical). The Fire Resistance of sandwich panels or multilayer systems relies on a complex mix of factors including: joint design, installation format, thickness of the sandwich panel, density of the core insulation. The fire resistance must be evaluated for a

Table 7.2 Fire resistances classes and associated criteria acc. to (EN 13501-2 2007)

Fire resistance classes EN13501-2	Range of classes	Fire safety level	Failing criteria
EI (integrity, insulation)	EI 15 up to EI 240	Standard	Limited temperature at unexposed metal sheet Average < 140 °C Each part < 180 °C
REI (load-bearing capacity, integrity, insulation)	REI 15 up to REI 240	Premium	As for EI classes Average < 140 °C Each part < 180 °C –with additional loads
E (integrity)	E 15 up to E 240	None	Time till cracks/openings or sustained flaming –ignition of cotton pad

wide range of the possible values of each parameter. This requires more than one test to adjust a model.

7.3.2 *Examples of Fire Tests Performed on Bio-aggregate Based Products*

Few tests on bio-aggregate based materials are performed. Unfortunately access to the test data remains confidential and the understanding of the products behaviour is penalized. The reported tests results are realized on wall made with straw ball and hemp concrete.

7.3.2.1 Straw Wall with Renders: Ecological Building Network (USA) (Intertek 2007a)

This test was performed by Intertek Testing Service (TX USA, 2006) in accordance with the Fire Tests of Building Construction and Materials ASTM E 119-05a. The fire curve is quite similar to ISO R834 fire used in EU. The 10 ft × 10 ft wall assembly was constructed with rectangular wheat straw bales completely filling the test frame. The gaps at the intersections of the stacked bales were stuffed with a mud and straw mixture that was prepared using locally available dirt plus a small amount of chopped straw mixed with enough water. Each side of the wall was covered with galvanized self-furred stucco reinforcing mesh (KEYMESH). The cement/stucco was applied in two layers, each nominally ½″ thick. The mix consisted of 1 part lime, 3 parts Portland cement, 10 parts sand, and water to a workable consistency. The stucco was applied 36 days prior the fire test. The average moisture content of the straw before the fire test was 18.4%.

The test wall, contained in a non-loadbearing frame assembly, was placed in front of the vertical wall furnace for a period of 2 h. After 15 min fire exposure, steam/smoke issuing from small cracks on the unexposed side. At 20 min, popping noises coming from the exposed side. Small cracks have formed in the exposed stucco at 30 min, with light flaming. At 1 h, cracks and flames increase on the exposed side. The test is stopped at 2 h and the sample is removed for hose stream test. The exposed render layer falls down and the straw burn (Fig. 7.3). External render is still on place and no large cracks was observed (no passage of flame, of gases hot enough to ignite cotton waste, or of the passage of water from the hose stream). Transmission of heat through the wall during the fire exposure did not raise the average temperature on the unexposed surface more than 121 °C, nor any individual temperature more than 162 °C. Figures 7.3 and 7.4 show the state of the sample during fire test and the temperatures of thermocouples placed on cold face, respectively. The fire resistance referring to (EN 13501-2 2007) of the tested solution is probably equivalent to EI 120.

Fig. 7.3 Image of the 2 h fire resistance test performed on straw wall (Intertek 2007a)

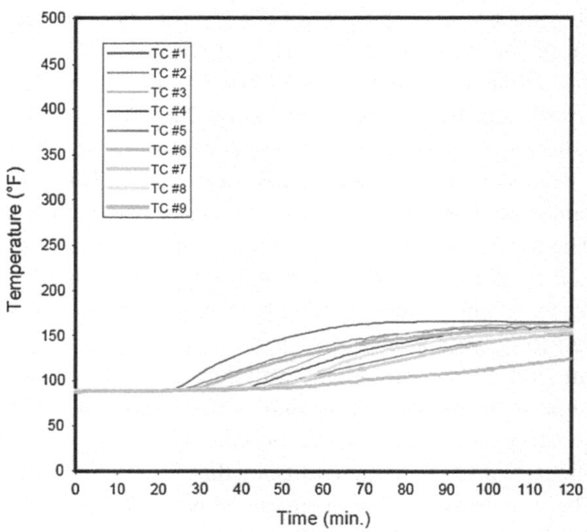

Fig. 7.4 Temperature of thermocouples placed on cold face of straw wall during two hours fire resistance test (Intertek 2007a)

The same test realized without reinforcement of the render was performed by the same authors. The results was equivalent to EI 60 due to large cracks on unexposed render (Intertek 2007b).

7.3.2.2 Straw Wall with Timber Frame Structure: "Maison de Montholier"(France) (CEBTP 2004)

This test was performed by CEBTP (France 2004). Two tests are performed using real fire (without furnace): combustion of heptane.

A test is performed on 1.80 × 2.00 × 0.39 m roof element realized with a timber frame structure filled with straw bale. A plywood panel (27 mm) was applied on the internal surface. The roof element was placed at 45° and exposed to the fire (Fig. 7.5). Thermocouple were placed in the sample. The fire test duration was about 20 min. The temperature at the interface straw plywood reached 230 °C at the end of the test. After such fire exposure, the straw was not altered.

Roof element fire test (combustion of heptane)

Straw wall : before fire test

straw wall during fire test

Fig. 7.5 Fire tests performed on roof and wall elements. "Maison de Montholier" (France) (CEBTP 2004)

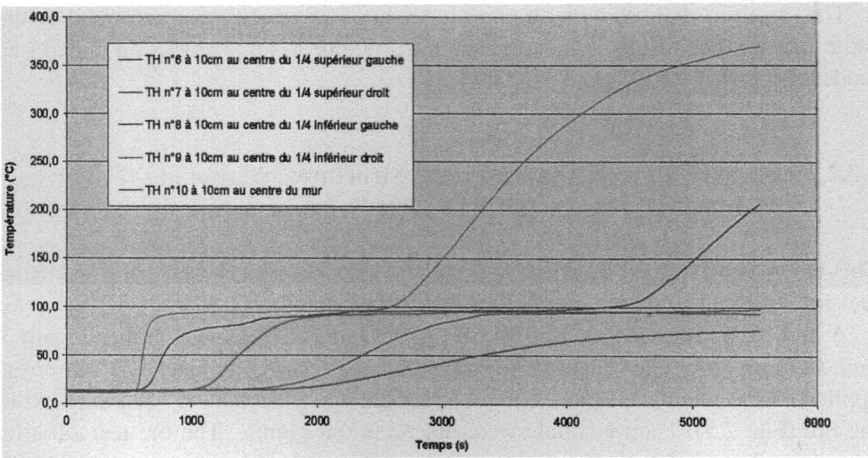

Fig. 7.6 Temperatures of thermocouples placed at 10 cm of the cold face. "Maison de Montholier" (France) (CEBTP 2004)

A second test was performed on $1.85 \times 1.90 \times 0.44$ m^3 wall element realized with a timber frame structure filled with straw bale. An internal lime coating was applied on the internal side (20 mm) and a hemp lime mortar (20 mm with two layers) was applied on the external side. The heptane tank was placed at the bottom of the wall (Fig. 7.5). The fire test duration was 85 min without loose of element stability. The exposed render lost a first layer after 45 min but the second layer reminded on place. The temperatures of thermocouples placed into the straw at 10 cm of the cold face are dispersed (Fig. 7.6). The straw combustion was locally started after 30 min and smouldering fires were suspected.

7.3.2.3 Straw Ball Wall: Performed by Pavus (Czech) (Pavus 2011)

Fire Resistance test of load bearing straw bale wall has been carried out by (Pavus 2011). The wall was made by 7 layers of straw bales of the size $50 \times 42 \times 50$ cm^3. The bales were compressed with the timber during the construction and at the end with the load which simulates the construction load. Basic volume weight of the bales was 77 kg/m^3 and after the compression it was 93 kg/m^3. The surface was covered with steel mesh and plastered with clay plaster from interior and lime plaster from exterior. The fire resistance of this structure is EI 120. After 146 min the wall collapsed due to limits in load bearing capacity.

7.3.2.4 Facade Element Made with Wood and Straw (France) (CSTB 2009)

This test was performed by CSTB (2009) in france for the "wood and construction association". The test is realized on a facade element made with a timber frame structure filled with straw ball. A plywood panel was applied on the internal and external surface. The facade element corresponds to the junction of wall and floor between two levels of the building. The detail of window frame was reproduced. The fire started in the down floor and the fire propagation to the second floor was analysed.

A real fire was obtained with 600 kg of wood placed on the down floor. The obtained temperature follows the fire test curve ISO R834 during the 30 min of the test. Upper floor windows glass brake at 3 min and plywood was carbonated on upper floor wall at 10 min. At 20 min the upper floor wall was burnt (Fig. 7.7).

After fire extinguishment at 30 min, the state of the wood slab and wall were analysed. In the down floor wall, the straw was carbonated on 10 cm. The tightness at the junction of floor and facade was not affected.

Fig. 7.7 Facade element made with wood and straw, test performed by CSTB (France) (CSTB 2009)

The tested building solution was then validated for a fire resistance of 30 min and was approved for buildings with three levels.

7.3.2.5 Wall Made with Hemp Concrete Blocks: BCB (France) (CSTB 2005)

This test was performed by CSTB (2005) on a wall made with block of hemp concrete. The blocks "chanvribloc" (30 cm of thickness) were coated with a render (20.2 mm thickness) on the exposed side (Fig. 7.8). The density of hemp concrete blocks is 486 kg/m^3. The blocks were glued with a lime mortar applied in layer of 5 mm thickness. The wall was placed above a gas furnace. The furnace temperature follows the fire test curve ISO R834 during 111 min.

After 14 min of fire exposure, the render fallen down. Some smoke emissions were noted on all the wall surface until the end of the test. The colour of joins between blocks changed with eat. After 90 min, hollows formed in the joins. At 102 min, a join was perforated (direct vision of the fire) and the blocks turned black

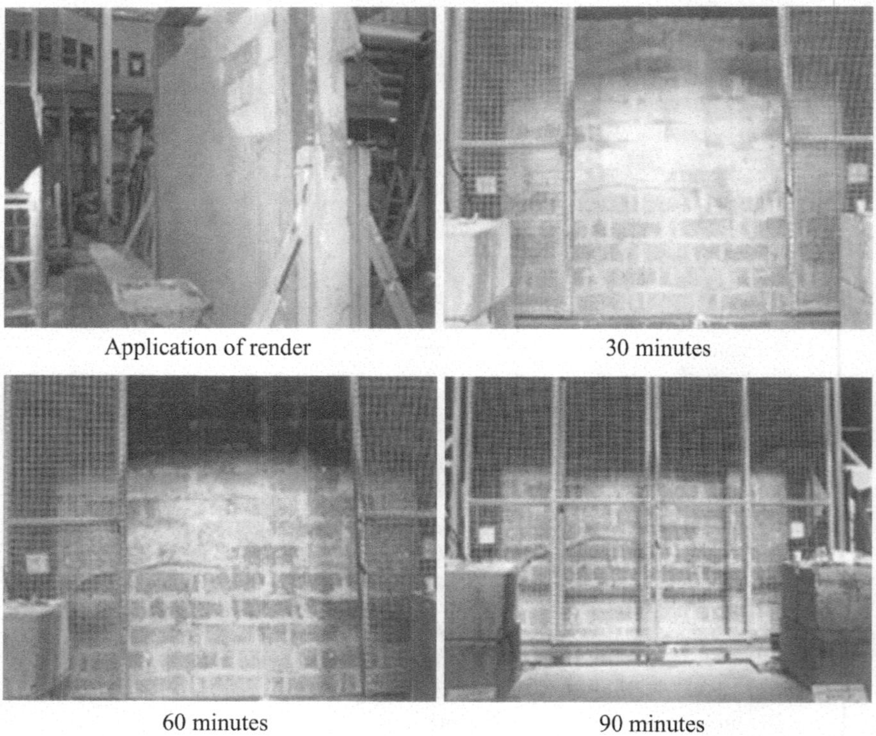

Application of render 30 minutes

60 minutes 90 minutes

Fig. 7.8 Wall realized with hemp concrete blocks. Fire test performed by CSTB (France) (CSTB 2005)

Fig. 7.9 Detail of a join between blocks at 102 min. Fire test performed by CSTB (France) (CSTB 2005)

colour (Fig. 7.9). At 105 min the fire proofing is broken. An increase of temperature higher than 180 °C was quoted and the test is stopped at 111 min. After cooling, the entire of the wall fallen down.

7.3.3 *Fire Resistance of Bio-aggregate Based Products*

The bio-aggregate based products are currently non loadbearing. As a consequence, fire resistance is only based on EI classes. As previously mentioned, limited studies were reported. Some additional references can be quoted:

–Hemp concrete wall (>30 cm thickness)	90 min	(CenC)
–Straw wall + earth render (1 single face)	90 min	(IBMB 2014)
–Straw panel (stramit)	EI30	(stramit)

Due to the high cost and the complexity of the test, fire resistance test are not sufficiently performed on bio-aggregate based products to propose a model. The behaviour of such material ranges between the behaviour of mineral binder based product (cellular concrete, gypsum) and the behaviour of combustible materials as timber frame.

In the case of straw wall, the persistence of the render during fire exposure appears as a key parameter. Such a conclusion is not appropriate in the case of hemp concrete wall.

7.4 Real Scale Fire Tests

Some other real scale fire tests are performed to evaluate the efficiency of building solutions. Such tests are realized with real fire. The tested room or building can be partially instrumented. The test duration and temperature level are not systematically controlled. Interest of such tests is to test the building solution under increase and decrease of temperature, including effect of fireman intervention. The test can be interesting to really understand the building behaviour under fire: loose of stability, smoke production, temperature level. Such test are inappropriate to predict the expected Euroclass or to predict the fire resistance of the tested product. An example of Straw house exposed to bushfire is mentioned in (Csiro 2012). In such fire exposure various heat rates are tested reproducing external fire.

7.5 Other Quaint Matter Fire Test

Many fire test are presented to prove the fire performances of some new bio-aggregate based products. These tests usually consist in the exposition of a sample of product directly to the flame of a burner (floweringelbow 2012; Allin 2013).

The low temperature level and the delay of temperature increase of the un-exposed sample face is shown as a demonstration of the good performances of tested material against fire. But, in such tests, the heat flux in the sample is not controlled and nothing regarding smoke release or droplet formation was analyzed.

Such test are also inappropriate to predict the expected Euroclass or to predict the fire resistance of the tested product.

References

Allin, S.: Building with hemp 2013 (fire test video on https://www.youtube.com/watch?v=FeW6kuZgPY4)

Association Création Développement Eco-Entreprises: cd2e website: www.cd2e.com

Association «Construire en chanvre» CenC website: www.construire-en-chanvre.fr

CEBTP A. Grelat, Using sustainable materials as walling for individual housing with wood structure—Final report extracts Volume 2—Laboratory experiments, Site instrumentation, Update—Convention CEBTP ADEME-ITFFB July 2004

CSTB BCB Rapport d'essais RS05-048 Résistance au feu d'un élément de construction: cloison en chanvribloc avec enduit (2005)

CSTB Association Bois et Construction Rapport d'essais 26021044 Comportement au feu d'un élément de façade, 2009 (video of fire test on http://www.videoprofil.fr/ISSY/pagefeu.html)

CSTB CESA Rapport d'essais RA13-0023 Réaction d'un matériau au feu: Béton de chanvre – application sol (2013a)

CSTB CESA Rapport d'essais RA13-0024 Réaction d'un matériau au feu: Béton de chanvre – application mur (2013b)

EN 13501-1: Fire classification of construction products and building elements—Part 1: classification using test data from reaction to fire tests (2007)
EN 13501-2: Fire classification of construction products and building elements—Part 2: classification using data from fire resistance tests, excluding ventilation services (2007)
Guide des matériaux isolants, website: http://www.cg43.fr/sites/cg43/IMG/pdf/guide_des_materiaux__isolants.pdf
Glé, P.: Acoustique des Matériaux du Bâtiment à base de Fibres et Particules Végétales - Outils de Caractérisation, Modélisation et Optimisation (ref traitement au feu these entpe acoustique) Ph. D. thesis, ENTPE, France (2013)
How durable is straw bale with clay plaster? Website http://www.floweringelbow.org/2012/workshop/destruction-testing-straw-bale-construction/
Institut fur Baustoffe Massivbau und brandschutz TU Braunschweig, Allgemaines bauaufsichtliches prufzeugnis P-3048/817/08-MPA BS, December 2014
Intertek Testing Service Fire Tests of Building Construction and Materials 2hr fire resistance test of a non loadbearing wheat straw ball wall for Ecological Building Network Project No. 3098054A July 31, 2006 Revised: July 9, 2007 (2007a)
Intertek Testing Service Fire Tests of Building Construction and Materials 1hr fire resistance test of a non loadbearing straw ball wall for Ecological Building Network Project No. 3098054B July 31, 2006 Revised: July 9, 2007 (2007b)
ISO 1182: Reaction to fire tests for products—Non-combustibility test (2010)
ISO 1716—Reaction to fire tests for products—Determination of the gross heat of combustion (calorific value) (2010)
ISO 9239-1: Reaction to fire tests for floorings—Part 1: determination of the burning behaviour using a radiant heat source (2010)
ISO 11925-2: Reaction to fire tests—Ignitability of products subjected to direct impingement of flame—Part 2: single-flame source test (2010)
ISO 5660-1: Reaction-to-fire tests—Heat release, smoke production and mass loss rate—Part 1: heat release rate (cone calorimeter method) and smoke production rate (dynamic measurement) (2015)
LNE BCB Rapport d'essais P113132 DE/4 Réaction d'un matériau au feu: enduit hygrothermique (2013)
LNE BCB Rapport d'essais P113132 DE/11 Réaction d'un matériau au feu: Béton de chanvre—application toiture (2014a)
LNE BCB Rapport d'essais P113132 DE/5 Réaction d'un matériau au feu: Béton de chanvre—application mur (2014b)
Pavus a.s. protokol o zkousce pozarni odolnosti Pr-11-2.097, Nosná stěna z balíků slámy, September 2011 (final report is available at http://kps.fsv.cvut.cz/index.php?lmut and fire test video on https://www.youtube.com/watch?feature=player_detailpage&v=sF8s2ULM8Eg)
SP Technical Research Institute of Sweden, Reaction-to-fire classification report ref 913013A, 1-10-2009
Staw house by Joost Csiro Bushfire—Simuilation and fire safety test—Mogo testing facility New South Wales AU, 15 February 2012 in (http://csironewsblog.com/2012/02/16/straw-house-1-fire-0/)
Stramit International website: http://stramitinternational.com/
Test method EN 13823 2013—SBI, fire technical testing of building products (2013)

Chapter 8
Durability of Bio-based Concretes

Sandrine Marceau and Guillaume Delannoy

Abstract Used for several decades for building insulation, concretes containing plant aggregates have thermal, acoustic and hygrothermal properties that greatly improve the comfort of homes (Amziane and Arnaud 2013). Nevertheless, during their life time, they are submitted to a hydrothermal environment (humidity and temperature variations) that can change these functional properties and/or induce the development of microorganisms on their surface. The objective of this chapter is to present the state of the art on the evolution of the properties of these vegetal concretes after different types of aging in laboratory.

Keywords Durability · Aging · Environment · Temperature · Humidity · Fungal growth · Carbonation

8.1 Introduction

The assessment of the durability of buildings is an activity for sustainable development, which results in requirements of the long term-performance of the structures in the whole life of structure. This is defined by the overall behaviour of the functions required by the product. This performance is studied for the product lifetime, from its commissioning to its failure.

Different terms can be used to characterize the lifetime of the construction materials (Talon 2006):

- The durability is the ability to perform a function until a limit state is reached.
- Aging refers to the functional changes decreasing the ability of a product to perform a function.
- The evolution of the properties refers to positive or negative changes of the characteristics of the material according to its expected performance.

S. Marceau (✉) · G. Delannoy
Ifsttar, Mast/Cpdm, Université Paris-Est, 77447 Marne-La-Vallee Cedex 2, France
e-mail: sandrine.marceau@ifsttar.fr

S. Amziane and F. Collet (eds.), *Bio-aggregates Based Building Materials*,
RILEM State-of-the-Art Reports 23, DOI 10.1007/978-94-024-1031-0_8

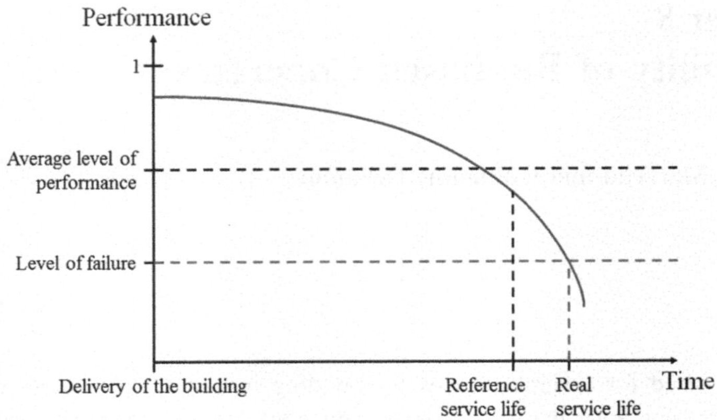

Fig. 8.1 Concepts of performance and service life (Talon 2006)

- The effective life of a material is the time between commissioning and the moment when its performance level becomes lower than a failure threshold (Fig. 8.1).

The evolution of the properties of construction materials in a building is related to several parameters: the initial properties of the material, its environment and its conditions of use. Based on the previously described properties of vegetal concretes, several factors can influence their performances: concerning environmental and use conditions, relative humidity, temperature, or exposure to liquid water can play a role. The alkaline character of the binder and the presence of ions may also be responsible of properties modifications. Finally, because of the nature of the plant aggregates, microorganisms' growth is possible. It also depends on the environment in which the material is used.

These factors play a role in modifying microstructural properties of materials. Their thermal, acoustic and mechanical performance, highly dependent on their microstructure, can also be modified over time.

A common method to study the durability of building materials is to achieve accelerated aging tests in laboratories, according to the specific properties of materials and their environment of use. The comparison of the characteristics of materials after these accelerated aging and in natural conditions is then used to model the evolution of their performance and determine their lifetime. In the case of vegetal concretes, few studies exist and they are based on previous works performed on concrete durability. These studies aim to analyse the impact of humidity, of immersion in water, immersion and drying, wetting and drying or freeze/thaw cycles, accelerated carbonation on material properties. Protocols have also been developed to observe the potential development of microorganisms on the surface of the bio-based building materials.

In this chapter the different accelerated aging protocols used in laboratory to assess the durability of bio-based construction materials is firstly described. The behaviour of plant-based concrete submitted to these tests is then presented. To better understand these performances, the state-of-the art is expanded to several investigations of aging of vegetal fibres and composites containing vegetal fibres embedded in a mineral matrix.

Table 8.1 Different types of aging tests used for vegetal concretes

Reference	Material	Description	Duration
Arizzi et al. (2016)	Hemp concrete	Dynamic: 3 Climatic simulations: variation of temperature, humidity, rainfall, influence of salts	12 days
Walker et al. (2014)	Hemp-lime concretes	Freeze/thaw cycles	10 cycles between −15 and 20 °C
		Exposition to salts (NaCl) during 12 h and drying during 12 h	20 °C during two weeks and 40 °C during the next two weeks
		Biological aging	Inoculation and storage at 30 °C and 80% HR during 7 months
Hellebois 2013, Hellebois et al. (2013)	Hemp concrete	Wetting and drying cycles	30 °C, 40 and 90% RH
		Immersion and drying cycles	20 °C, 2 cycles, 41 days, Drying at 40 °C
		Biological aging	3 months at 30 °C and 98% RH
Abdellaoui (2014), Marceau et al. (2015)	Hemp concrete	Wetting and drying cycles	30 °C, 40 and 98% RH
		Biological aging	3 months at 30 °C and 98% RH
Sonebi et al. (2015), Castel et al. (2016)	Hemp concrete	Full Immersion in water and drying cycles	Immersion at 20 °C, drying at 50 °C during 48 h
Magniont et al. (2012)	Hemp concrete	Storage at 25 °C and RH > 95%	Until 2.5 years
Bessette et al. (2015)	Precast hemp concrete	Storage in inside climate	90 days
		Storage in external climate	One year
Le Bayon et al. (2015)	Different bio-based construction materials	Development of a mould test method	

8.2 Accelerated Aging Protocols for Bio-based Construction Materials

Various types of aging protocols were used to accelerate the evolution of the properties of plant-based concretes in laboratory: environmental aging, based on temperature and/or humidity variations in static or dynamic mode and biological aging. These experiments were performed essentially on hemp concretes. The main parameters of these studies are reported on Table 8.1.

8.2.1 Environmental Aging

The protocols used to study the durability of bio-based concretes subjected to environmental constraints are varied: cyclic variations of humidity at constant temperature (Hellebois et al. 2013; Marceau et al. 2015), simultaneous variations of temperature and humidity (Arrizi et al. 2015; Arrizi et al. 2016), freeze/thaw cycles (Walker et al. 2014), immersion and drying cycles (Hellebois et al. 2013; Sonebi et al. 2015; Castel et al. 2016) or storage in static conditions of temperature and humidity, accelerated carbonation (Chabannes et al. 2015). Moreover, the properties of the plant-based concretes measured before and after these aging protocols are also very different. Therefore, the comparison of the results of the different studies is very difficult.

On the other hand, when the conditions of aging are cyclic, the duration of the test is often short, up to three months. This aging time is probably too short to highlight significant variations in the properties of the materials by keeping aging mechanisms similar to those that occur naturally in a building.

8.2.2 Biological Aging

Plants, either in the form of aggregates or fibres, naturally contain microorganisms. Their proliferation can affect the indoor air quality in buildings where bio-based construction materials are used. Moreover, it may also modify the intrinsic properties of the materials. It is therefore useful to identify the necessary conditions for their growth and their impact on the performances of plant concretes.

8.2.2.1 Description of the Microorganisms

Different kinds of microorganisms are able to grow on the surface of materials: moulds and bacteria (CSHPF 2006; Dehoux and Dehoux 1997). Bacteria may form colonies with agglutinated cells remaining in an aqueous gel (biofilm).

Moulds have to draw into their environment the water and organic and inorganic substances necessary for their development. Their vegetative system consists of filaments or hyphae which constitute a network called mycelium. Fungi are propagated by spores formed from the mycelium. Spores are kinds of microscopic seeds which are dispersed by air currents, runoff water or by sticking on objects. Under favourable conditions of temperature and moisture, spores can germinate and create mycelium again that can, in turn, resporulate and recontaminate.

Fungi need humidity to develop and the presence of different genera depends on the water activity A_w in the materials: for low water activity ($A_w < 0.8$), *Penicillium* and *Aspergillus* will colonize the substrate. *Clostrium* will appear when the humidity increases ($0.8 < A_w < 0.9$) and for wet surfaces ($A_w > 0.9$), *Strachybotrys* can be observed (Gueguen et al. 2015; Pasanen et al. 1992). The nature of the support may also influence the installation of adapted species.

Their presence is not always dangerous for inhabitants. However, they may present risks to human health: allergies, infections…

8.2.2.2 Mechanisms of Biodegradation

Colonization of materials by moulds generally induces their biodegradation and results from two mechanisms:

- A physical action, linked to the development of hyphae in the material, which can lead to its breakage,
- A chemical action, due to the production of various metabolites, which act by assimilation or dissimilation.

During the assimilation process, the constituents of the material are used as nutrients after their reduction by different extracellular enzymes that facilitate penetration of the hyphae in the material. The dissimilation processes are related to the production of organic acids and pigments. Organic acids (such as gluconic, citric, oxalic acids…) are produced in varying amounts during the metabolic activity and can react with the substrate. In addition to the direct action of these acids, their production promotes the growth of acidophilic fungal species that can continue the degradation of the support.

8.2.2.3 Methods for Determination of Fungal Resistance of Construction Products

Several standards of various origins exist to determine the performance of construction products against a fungal contamination. The assessment procedures proposed include three recurring phases:

- Product contamination, by spraying a liquid suspension, or by deposition of microorganisms from a liquid inoculum.

- Incubation of the contaminated material in static conditions: samples are generally incubated between 25 and 32 °C, with relative humidity close to saturation for periods during several weeks.
- Evaluation of microbial growth: the evaluation techniques consist in determining, quantitatively, by measuring the fungal biomass, or semi-quantitatively, the level of development of fungi. Less common techniques rely on mass variation or modification of various physical properties.

8.2.2.4 Proposal of a Fungal Resistance Test Development Tailored to Bio-based Insulation Materials

A laboratory test method has been developed in order to assess the resistance of bio-based insulation materials against moulds (Le Bayon et al. 2015). According to this study, the parameters having the highest impact on fungal growth are humidity, the composition of bio-based material and the additives it contains.

The parameters of the mould test take into account the climatic conditions of insulation materials used in buildings. The test is performed at 85% RH and 26 °C. The samples are inoculated with a fungal solution containing three fungal strains (*Aspergillus Niger, Penicillium Brevicompactum, Cladosporium Sphaerospermum*). The mould growth is visually assessed and quantified by measuring the number of cultivable fungal units.

8.3 Aging of Bio-based Concretes

The evaluation of the lifetime of construction materials requires to characterize the evolution of their properties by subjecting them to accelerated aging protocols in laboratory. The results obtained are then compared to those measured after aging under natural exposure conditions in a building.

In this section, the results published about the impact of natural aging are firstly presented. Then, the results concerning different types of laboratory aging tests are then detailed.

8.3.1 Natural Aging of Bio-based Concretes

There are limited results published regarding the evolution of the properties of bio-based concretes under natural conditions. Three investigations on hemp concretes were reported.

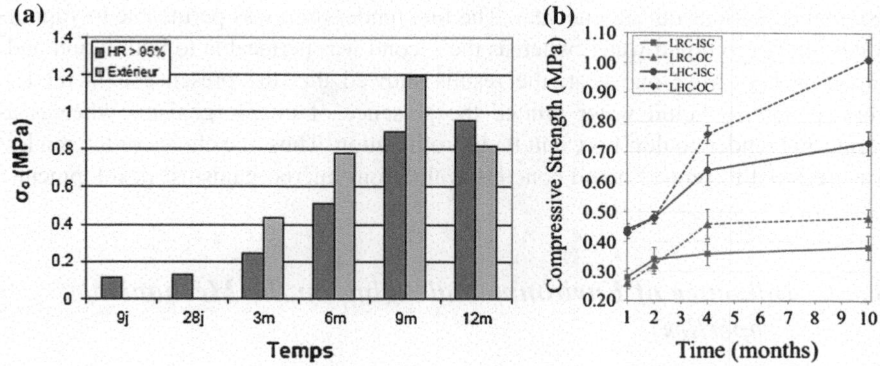

Fig. 8.2 Evolution of the compression strength **a** for hemp concretes stored in controlled atmosphere and outdoor conditions (Magniont 2010) and **b** for hemp concretes (LHC) and rice husk concretes (LRC) stored outdoor (*OC* outdoor conditions) and in indoor standard conditions (ISC) (Chabannes et al. 2015)

In the first study, (Magniont 2010) the compression strength of hemp concretes stored until one year in different environments: controlled atmosphere (20 °C and relative humidity above 95%) and outdoor conditions without protection were measured (Fig. 8.2a).

Figure 8.2 indicates that the compression strength increases up to 12 months for samples stored in controlled conditions. For samples stored in outdoor conditions, it increases until 9 months, and after it drops at 12 months. This reduction of compressive strength could be attributed to the experimental dispersion of the results or to the high water content of the specimens at the time of the test. This result can also result from long-term degradation mechanisms under conditions of outdoor exposure, like leaching phenomena of the mineral binder.

The same type of study has been performed on hemp concretes (LRC) and rice husk concretes (LHC) (Chabannes et al. 2015): samples were stored outdoor (OC: outdoor conditions) and in indoor standard conditions (ISC) 24 h after mixing. The evolution of the compressive strength of the concretes is reported on Fig. 8.2b. It can be observed that the compressive strength of both concretes increased as a function of the time, but this increase is limited after four months for rice husk concretes. These evolutions of compressive strength are in agreement with the previous study (Magniont 2010).

Another result is the difference between the two curing conditions: The strength enhancement was higher for specimens exposed outdoors whether the type of concrete. This was explained by favourable conditions to the carbonation process in the outdoor conditions. Indeed, in this environment, the relative humidity varied from 45 to 75% and these conditions were beneficial for CO_2 diffusion and dissolution.

The third investigation (Bessette et al. 2015) focussed on the development of mould on hemp concrete walls coated with two different renders and exposed to external conditions during one year. The first render used was permeable to vapour water but not to liquid water, whereas the second was permeable to both liquid and vapour water. After one year, the results showed that the presence of a render impermeable to liquid water limited the presence of microorganisms whereas a permeable render couldn't prevent their proliferation. Thus, the choice of the render was essential to protect hemp concrete walls from microorganisms' development.

8.3.2 Influence of Environmental Aging on the Mechanical Properties

Several studies have focussed in laboratory on the influence of different environments on the properties of bio-based concretes.

8.3.2.1 Static Conditions

In order to study the long-term performances of hemp concretes, Magniont (Magniont et al. 2012) stored hemp concretes in a humid room with relative humidity higher than 95% during one year. The evolution of the compressive strength of hemp concrete and of the pure binder paste is presented on Fig. 8.3.

It can be observed that the evolutions of the compressive strength of the pure binder paste and of the concrete are decoupled. The maximum compressive strength is constant after about 50 days for the binder paste, whereas it increases continuously until one year for the hemp concrete. This decoupling may be explained by a setting delay due to the presence of vegetal aggregates (Sedan 2007). However, other studies mentioned delays of about several minutes to several hours and therefore, this statement cannot be used to support the shift of several months observed here.

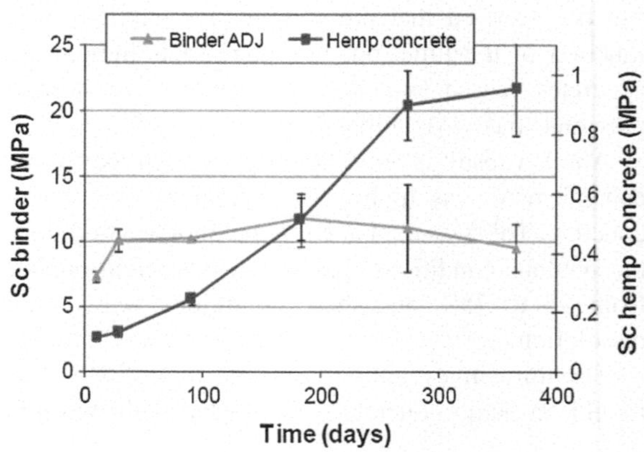

Fig. 8.3 Evolution of compressive strength of the pure binder paste and hemp concrete with time (Magniont et al. 2012)

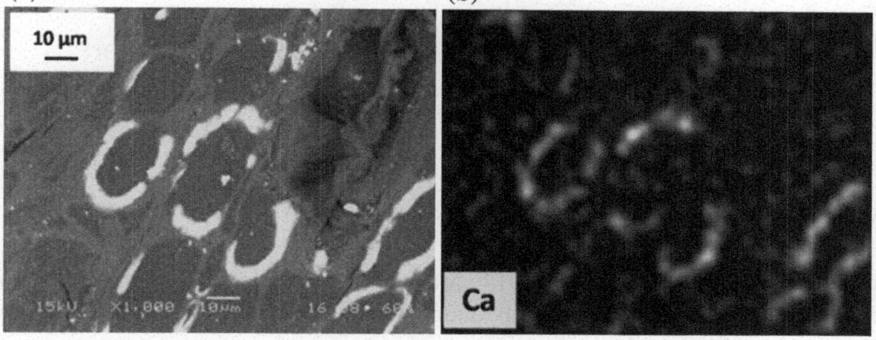

Fig. 8.4 SEM observations and CA EDS mapping of cross sections of hemp shiv extracted from a 2.5 years concrete (Magniont et al. 2012)

Another phenomenon could explain the increase of the strength of the hemp concrete: the mineralization of plant aggregate. This was demonstrated by SEM on hemp shiv extracted from a 2.5 years concrete (Fig. 8.4). Mineral products are visible in the walls of the pores of the hemp shiv. During the storage in a wet chamber, calcic phases of the binder dissolved and calcium species could diffuse into the pores of hemp shiv, where they could re-precipitate as calcite. This mineralization of the vegetal aggregate could explain the enhancement of the compressive strength of concretes and the stiffening of hemp aggregates.

8.3.2.2 Dynamic Conditions

More complex aging protocols have also been set up to simulate actual use conditions of bio-based concretes used in buildings.

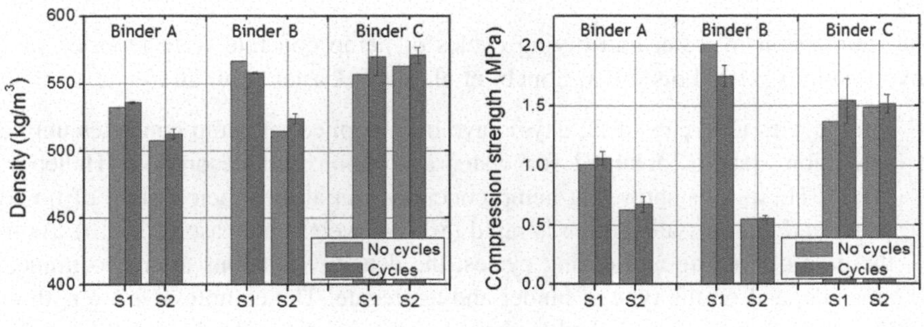

Fig. 8.5 Variation of the density and of the compressive strength of hemp concretes (Hellebois et al. 2013)

Table 8.2 Variation of the thermal conductivity of two hemp concretes before and after 75 days of wetting and drying cycles (Abdellaoui 2014; Marceau et al. 2015)

Thermal conductivity (W/(m.K))	PNC—shiv A	PNC—shiv B
100 days	0.102 ± 0.006	0.105 ± 0.005
After aging	0.103 ± 0.003	0.112 ± 0.008

Wetting and Drying Cycles

In several studies (Abdellaoui 2014; Marceau et al. 2015), the aging of hemp concretes has been simulated in laboratory with humidity cycles. The temperature was fixed at 30 °C for both investigations. It corresponds to the temperature range where fungal growth is possible. The humidity was varied from 40% to 90 and 98%. The duration of the cycles in these investigations varied from one to two weeks.

The variation of density and of compressive strength was small after the wetting and drying cycles (Fig. 8.5). In this study (Hellebois et al. 2013), three different binders and two type of shiv have been used. It can be observed that the density and the compressive strength of the concretes depended on both the type of binder and shiv. The variation of these properties after wetting and drying cycles (duration: 45 days) was also different depending on the composition of the material. The variation of compressive strength was linked to the variation of density of the concretes after the cycles. Without any further experiment, no explanation of these results can be obtained.

In the second study (Abdellaoui 2014; Marceau et al. 2015), a slight increase of the density of the two hemp concretes was observed after 75 days of cycles, corresponding to a weak reduction of the porosity of the material. However, none of these studies have shown significant variations of the insulation properties of the hemp concretes (Table 8.2): no evolution of the thermal conductivity and of the acoustic behaviour has been observed.

Full Immersion in Water and Drying Cycles

Full immersion in water and drying cycles of hemp concrete were reported in 3 investigations (Hellebois 2013; Sonebi et al. 2015; Castel et al. 2016):

- Long cycles (2 cycles of 22 days) have been applied on hemp concretes until a saturation state is obtained for water absorption and desorption (Hellebois 2013). These tests show that hemp concretes can absorb their weight of liquid water and that this absorption is rapid (80% of weight increase in 100 h). As in the case of wetting and drying cycles, the density variations after two immersions depend on the type of binder and aggregate. The compressive strength of the concretes was then linked to their density variations. In the case of prompt natural cement, a hydraulic binder, the increase of density and compression

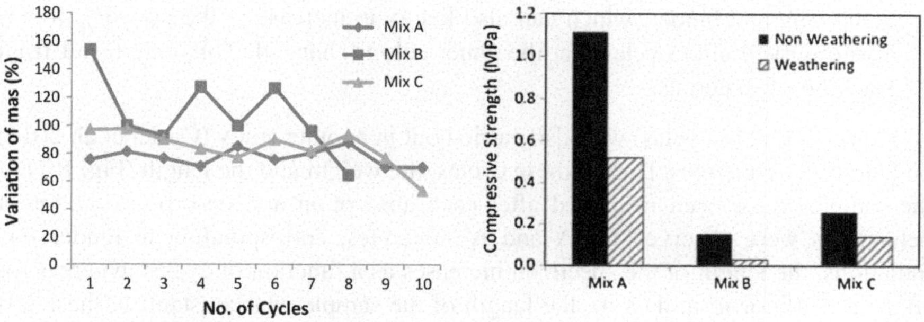

Fig. 8.6 Variation of weight and of compressive strength after 10 cycles of full immersion and drying (Sonebi et al. 2015)

strength may suggest that hydration of the binder continued during the immersions. This result is not observed when the binder contains lime. Leaching of the binder was also observed, especially for the first immersion, wherein the pH of the solution increases up to 12 in a few hours.

– Sonebi et al. (2015) submitted hemp concrete specimens to 10 cycles of full immersion in water and drying in oven. The samples are successively immersed in a water bath at 20 °C during 48 h and then placed in a ventilated oven to dry at 40 °C for 48 h. Figure 8.6 presents the variation of mass after 10 cycles. The masses of hemp concrete are reduced after the cycles. Depending on the mix, the reduction of compressive strength varies from 53 to 81% (Fig. 8.6). The weathering affected significantly the compressive strength after only 10 cycles. This reduction after immersion and drying cycles can be attributed to the softening of the hemp concrete and the weakening of the interface zone between

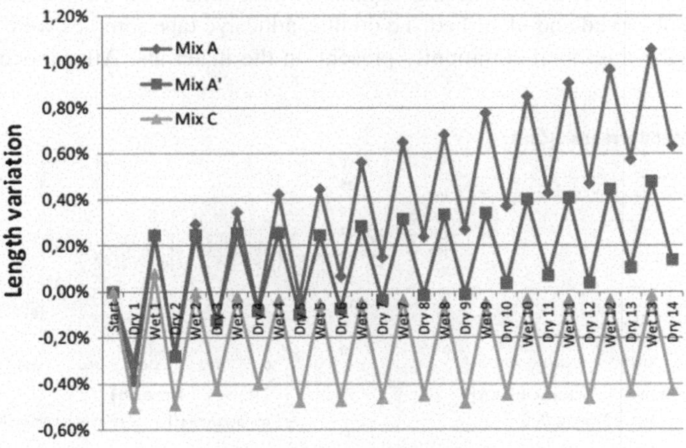

Fig. 8.7 Length variation of the samples during immersion and drying cycles (Castel et al. 2016)

hemp shiv and binder, which can also led to an increase in the porosity. It was also observed after cycles that the water colour changed. This may result from leaching of materials.

Short cycles (13 cycles) was also carried out in another study (Castel et al. 2016) on hemp shiv concretes. During these cycles, the weight and the length (Fig. 8.7) of the samples have been measured after each absorption and desorption. Different behaviours were observed for A and A' concretes, corresponding to render formulations, the length of the specimen increases as a function of cycles, whereas for concrete C (floor formulation), the length of the sample was constant. In these last samples, the porosity was high enough to compensate the swelling of the vegetal aggregates during the water immersion. After these cycles, the compressive strength decreased for all the concretes, whatever their composition.

Climatic Simulations

Real climatic conditions have been simulated to study the long-term durability of hemp concretes (Arizzi et al. 2015, 2016). Three types of climate were selected: Mediterranean, tropical and semi-arid climates (Fig. 8.8).

The mean values of temperature, relative humidity and rainfall for one month are applied during one day in the accelerated aging test. Thus, the total duration of the aging in laboratory is 12 days, corresponding to one year. The effect of airborne salt in coastal areas has also been taken into account, by soaking a part of the samples in a NaCl solution before the aging test.

During the cycles, the weight of the samples remained constant, except for rainfall events that cause a massive increase proportional to the spraying duration and to the amount of sprayed water.

At the end of each weathering simulation, the microbial colonisation of the samples is observed and identified. To do this, adhesive tape samples were collected to analyse the microbial community present on the materials. After inoculation on

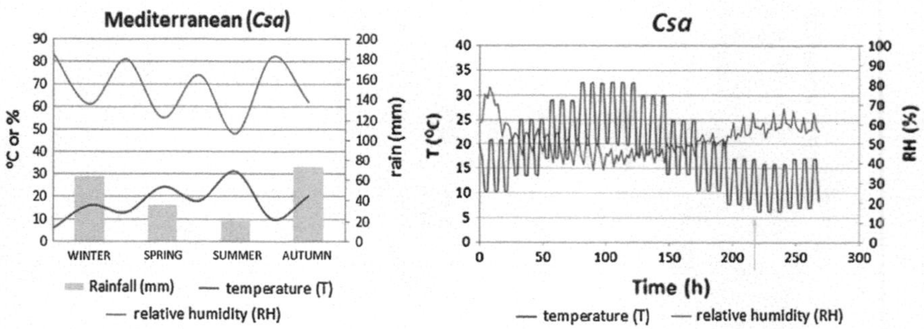

Fig. 8.8 Average conditions of temperature, relative humidity and rainfall recorded in a whole year in the Mediterranean zone and corresponding test conditions of temperature and relative humidity (Arizzi et al. 2016)

Petri plates and incubation at 28 °C during one week, the different colonies were separated and identified. In both control and test samples, different common bacteria have been detected. Some fungi were also isolated.

Due to the higher temperature and relative humidity in the tropical climate, microbial colonisation was more intense in these conditions. The colonisation was less important for the arid climate. However, no predominance of one type of microorganism over another has been observed in this study and most of the detected microorganisms were often present under a large range of environmental conditions.

The presence of a biofilm of bacteria in the surface of specimen was also visible by ESEM. It could have an impact on the properties of the materials by closing the porosity and decreasing the water permeability of the concretes.

The characterisation of the sample mineralogy after the different weathering tests has shown the presence of vaterite at the surface of the samples. This component can be produced by bacilli, that may precipitate carbonates, and by fungi, which have the capacity to transform and precipitate minerals. The presence of vaterite was higher after the Mediterranean and the tropical simulations, where the biological activity was the more important.

Accelerated Carbonation Test

The carbonation reaction is a natural reaction taking place between the CO_2 of the surrounding air and hydrated lime or portlandite $Ca(OH)_2$ (Cizer et al. 2012). It proceeds from the surface to the core of the concretes by diffusion of gaseous CO_2 in the open pores. The overall carbonation reaction is:

$$Ca(OH)_2(s) + CO_2(g) \rightarrow CaCO_3(s) + H_2O(aq) + 74\ kJ/mol$$

The reaction is very slow, due to the low concentration of CO_2 in the atmosphere. It depends on the nature of the binder phase, to the pore network and to the relative humidity. The formation of $CaCO_3$ induces an increase of the bulk density of the material and a decrease of its porosity and its alkalinity. Thus, these variation of microstructure can have an influence on the functional properties of the materials, such as mechanical, thermal or acoustical behaviour.

During 10 months of aging in outdoor conditions, Chabannes (Chabannes et al. 2015) showed that this environment enhanced the compressive strength of bio-based concretes by increasing the carbonation of the binder (Fig. 8.2b).

In the same study, the influence of accelerated carbonation curing (ACC) during one month is also investigated: after 40 days of drying at 20 °C and 50% RH, the concrete specimens were exposed to CO_2 curing. The conditions in the enclosure are fixed to 20 °C and 65% RH. CO_2 was injected with an initial concentration of 50% v/v and regular injections are performed when the CO_2 is entirely consumed by the carbonation reaction. The compressive strength of the concretes was

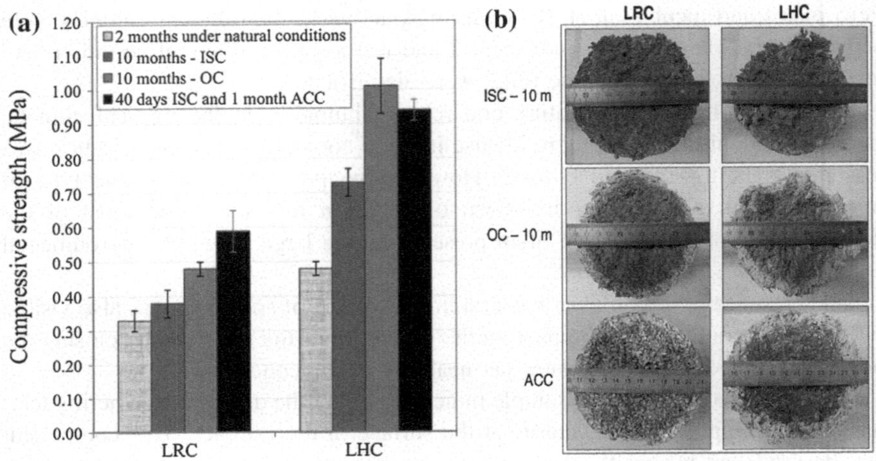

Fig. 8.9 **a** Compressive strength of hemp concretes (LHC) and rice husk concretes (LRC) after accelerated carbonation curing (ACC) compared to natural conditions (*ISC* indoor standard conditions, *OC* outdoor conditions), **b** comparison of the cross sectional views of the samples after spraying with phenol-phthalein (Chabannes et al. 2015)

measured after the initial curing and after one month of accelerated carbonation curing (Fig. 8.9a).

These results showed that the compressive strength after ACC is almost equivalent to that observed after 10 months of outdoor exposure. It was doubled if it is compared to that measured after two months of natural exposure.

The carbonation profiles of theses samples (Fig. 8.9b) show that:

- For sampled stored in indoor standard conditions (ISC), the binder was uncarbonated,
- A carbonation front was visible for samples stored 10 months in outdoor conditions (OC), varying from 0.8 to 1.5 cm depending on the plant aggregate,
- After one month of ACC, the carbonation front was almost the same as that observed after 10 months of outdoor exposure. Moreover, the core of the hemp concretes appeared more carbonated and was coloured in pale pink.

These results are confirmed by the measurements of the $CaCO_3$ contents in the core of the samples. Moreover, this work showed that the slow C_2S hydration is linked to the carbonation of the hydrated lime of the binder. Indeed the water released by the carbonation reaction enables the C_2S hydration.

Other Aging Tests

Other aging tests, based on those performed on civil engineering concretes, have also been applied to hemp concretes, such as freeze/thaw cycles and salt exposure (Walker et al. 2014).

Freeze/thaw cycles didn't create any cracks in the materials, and no visible variations of the microstructure can be observed by SEM. Similarly, sodium chloride exposure during one month does not modify the compressive strength of the samples. This can be explained by the high ductility of the plant aggregates that can accommodate expansive salt crystallisation pressures.

8.3.3 Microbial Aging

The development of microorganisms created a real obstacle to wide spreading of bio-based materials in buildings. Due to the hydrophilicity of the materials and the presence of vegetal aggregates, the microorganisms find conditions very favourable to their proliferation.

In the last section, results regarding the microorganisms' growth after weathering tests were presented. It has reported that their presence has, among other things, an impact on the mineralogical composition of the concretes with the formation of vaterite (Arizzi et al. 2016).

Before the definition of a fungal resistance test developed for bio-based construction materials (Part 8.2.2.3), several studies have been performed in order to analyse specifically this type of aging on hemp concretes. Most of these studies consist in storing hemp shiv or hemp concretes in conditions favourable to the growth of bacteria and fungi (temperature between 22 and 35 °C, high relative

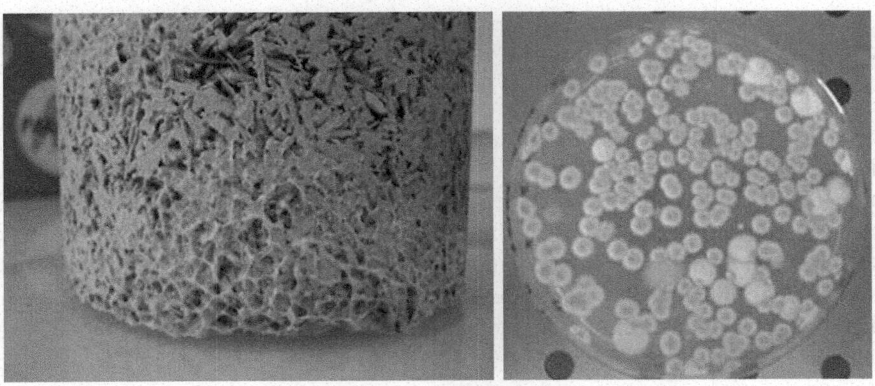

Fig. 8.10 Presence of fungi after conservation of hemp concrete at 30 °C and 98% RH (Hellebois et al. 2013)

Table 8.3 Influence of the material pH on fungal growth (Abdellaoui 2014)

Age of the concrete	14 days		120 days	
Type of shiv	A	B	A	B
Surface pH	10.5	10.4	8.7	9.2
Presence of moulds?	No	No	Yes	Yes

humidity). This can be done with microorganisms naturally present in the material or after inoculation with a stain solution (Marceau et al. 2015).

After 7 months of storage in a humidity chamber at 30 °C and 80% RH, for hemp concretes inoculated with a culture of microorganisms, no microbial development has been observed (Walker et al. 2014).

The natural microbial growth has also been studied on different hemp concretes after 3 months of storage at 30 °C and a relative humidity higher than 95% (Hellebois et al. 2013). Fungal growth is visible on the concretes, whatever their composition (Fig. 8.10).

These fungi were collected, cultivated and used to inoculate new samples of hemp concretes (Abdellaoui 2014; Marceau et al. 2015). At the time of their inoculation, the concretes were aged of 14 and 120 days. They were then incubated in environmental conditions favourable to microorganisms' development (30 °C, 98% RH) and the presence of moulds was observed 100 days after (Table 8.3). The surface pH of the concretes has also been measured at the same time.

In Table 8.3, it can be observed that the presence of moulds is only visible for samples inoculated 120 days after their manufacturing and the surface pH of the material is about 9. For samples inoculated 14 days after their manufacturing, no fungal growth is observed and the pH surface is around 10.5. The pH value is reduced, due to the natural carbonation reaction of the binder. This lower value of pH allowed the microorganisms to grow at the surface of the hemp concretes. The pH value seems to be one of the factors to take into account to assess the risk of mould growth in plan concretes.

8.3.4 Conclusion

Because of their recent development, few studies have focused on the assessment of the durability of hemp concretes. The protocols used in laboratory and the measured properties were very different. Therefore, it is still difficult to identify precisely the main degradation mechanisms of these materials. However, many others studies have focused on the durability of composites containing a mineral matrix reinforced with vegetal fibres, used for their mechanical properties. The main results of these investigations, in connection with the bio-based concretes, are presented in the next section.

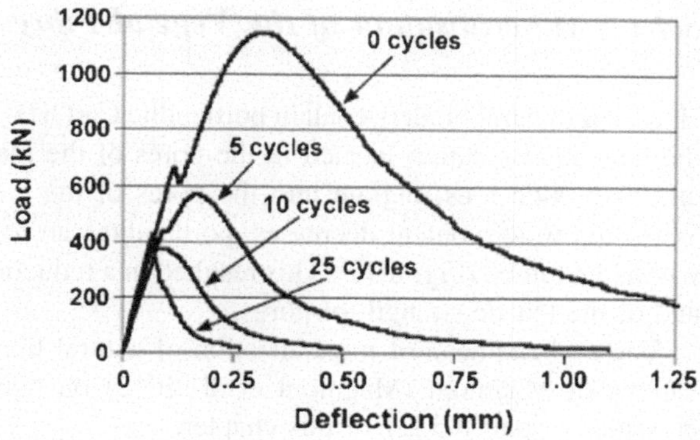

Fig. 8.11 Typical load-deflection curves of kraft pulp fibre-cement composites exposed to wetting and drying cycles (Mohr et al. 2005)

8.4 Aging of Natural Fibres-Cement Composites

Composites containing natural fibres reinforcements in a mineral matrix are promising structural materials for construction. Several studies gave results about the evolution of their mechanical properties when they are exposed to different kinds of environments, especially humidity.

The aging protocols are generally immersion and drying cycles and the mechanical properties of the composites are measured as a function of time (Melo Filho et al. 2013; Mohr et al. 2005, 2006; Ramakrishna and Sundararajan 2005; Sivaraja et al. 2010; MacVicar et al. 1999; Juarez et al. 2007; Ferreira et al. 2015; Toledo Filho et al. 2009). A reduction of mechanical strength of the composites, accompanied by a decrease of their ductility was observed in the early aging cycles (Fig. 8.11). Physicochemical and microstructural analyses were performed to understand these results. Two explanations are proposed to understand these results: the mineralisation of the fibres and their degradation in the alkaline binder.

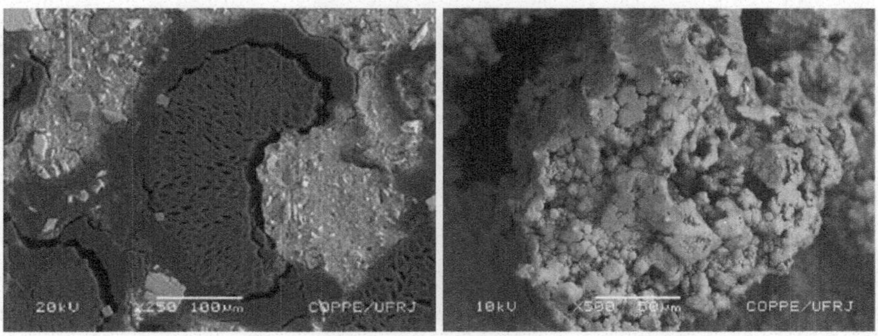

Fig. 8.12 Initial structural fibre geometry and mineralized fibre inside the composite (Toledo Filho et al. 2009)

8.4.1 Mineralisation of the Vegetal Fibres

Hydrated mineral binders contain portlandite $Ca(OH)_2$. This compound is soluble in the interstitial solution located in the pores of the material. By diffusion, calcium and hydroxide ions migrate into the pores of the elementary fibres of the plant. When the water content decreases, portlandite can re-precipitate on the fibres and within the lumen (Fig. 8.12). This resulted in a reduction of the porosity of the plant and of the tensile strength of fibres.

These phenomena of mineralization of natural fibres are consistent with observations of Magniont (Magniont et al. 2012) on hemp shiv in a mineral matrix presented in part 8.3.2.1 of this chapter.

8.4.2 Degradation Mechanisms of Vegetal Fibres

The degradation of natural fibres in an alkaline medium is described by Wei (Wei and Meyer 2014, 2015) in four steps (Fig. 8.13). Lignin plays the role of glue in the cell wall and protects the fibre against microbial and chemical degradation of polysaccharides, whereas hemicelluloses bind the cellulose micro-fibrils. These two components are amorphous and sensitive to the alkaline environment of the cement matrix. Therefore, the first step of the mechanism corresponds to the degradation of lignin and of a part of hemicellulose, leading to the exposure of holocellulose. Then, the degradation of the hemicellulose induces the decreasing of the integrity and of the stability of the vegetal cell walls. The third step is the destruction of the intra-molecular hydrogen bonding, leading to the dispersion of the cellulose

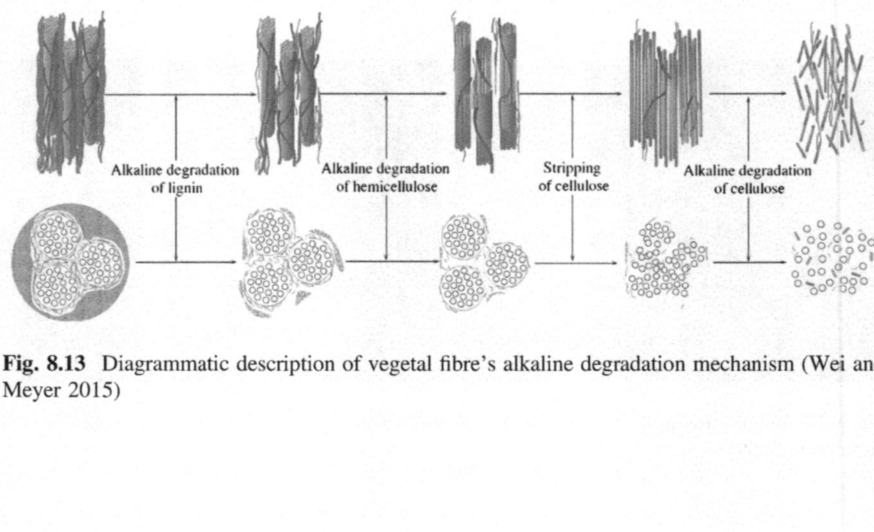

Fig. 8.13 Diagrammatic description of vegetal fibre's alkaline degradation mechanism (Wei and Meyer 2015)

micro-fibrils. Then, the amorphous regions of cellulose are hydrolysed, inducing the complete degradation of cellulose micro-fibrils.

During this degradation process, the hydration products of the binder, such as C-S-H and soluble portlandite, can diffuse in the cell wall, leading to a faster mineralization and embrittlement of natural fibres.

After the first step of the alkaline degradation, Pejic (Pejic et al. 2008) observed that the degradation of lignin and hemicellulose hemp fibre led to a reduction of the water vapour permeability.

8.5 Concluding Remarks

Because of their relatively recent development, few studies have focused on the durability of plant concrete until now. The aging protocols applied in laboratory and the analysed properties are diverse, making the comparison of obtained results very complex. In addition, the aging times used are rather short, and significant variations of the material properties are perhaps not visible for these aging durations. The protocols implemented until now are focused on similar factors that could influence the material properties: temperature, relative humidity and potential microbiological growth. These can lead to physicochemical and microstructural modifications of the materials and therefore impact the performances of materials in a building.

The understanding of aging mechanisms of plant concretes and the prediction of their lifetime requires comprehensive studies, including multidisciplinary and multi-scale analysis, which take into account the variability of the formulations and of the final properties of the materials. Data on the evolution of their performances in real conditions are also needed. Finally, the construction systems in which plant concretes are integrated into a building, such as the bearing structure and the interior and exterior plasters, have also to be taken into account.

Available data on composites containing vegetal fibres in a mineral matrix, used for their structural properties, allow to understanding the chemical interactions between plants components and a cement matrix. They can be transposed to the plant aggregates used in the bio-based insulation.

References

Abdellaoui, L.: Etude de la durabilité du béton de chanvre, Mémoire de fin d'études, Polytech'Clermont-Ferrand (2014)

Amziane, S., Arnaud, L. (eds.) Bio-aggregate-based Building Materials, Applications to Hemp Concrete. ISTE Ltd and Wiley (2013)

Arizzi, A., Viles, H., Martin-Sanchez, I., Cultrone, G.: Predicting the long-term durability of hemp-lime renders in inland and coastal areas using Mediterranean. Tropical and Semi-arid climatic simulations, Science of the Total Environment **542**, 757–770 (2016)

Arrizi, A., Viles, H.: The "naturalime" project: new insights into the study of hemp-based mortars for new construction and heritage conservation. In: First International Conference on Bio-Based Building Materials, Clermont-Ferrand, France (2015)

Bessette, L., Trémerie, B., Béjat, T., Piot, A., Jay, A., Barnes Davin, L.: Study of the developement of mould on prompt naturel cement-based hemp concrete. In: First International Conference on Bio-Based Building Materials, Clermont-Ferrand, France (2015)

Castel, Y., Amziane, S., Sonebi, M.: Durabilité du béton de chanvre: résistance aux cycles d'immersion hydrique et séchage. 1ère Conférence EuroMaghrébine des BioComposites, Marrakech (2016)

Chabannes, M., Garcia-Diaz, E., Clerc, L., Bénézet, J.-C.: Studying the hardening and mechanical performances of rice husk and hemp-based building materials cured under natural and accelerated carbonation. Constr. Build. Mater. **94**, 105–115 (2015)

Cizer, Ö., Van Balen, K., Elsen, J., Van Gemert, D.: Real time investigation of reaction rate and mineral phase modifications of lime carbonation. Constr. Build. Mater. **32**, 741–751 (2012)

Conseil supérieur d'hygiène publique en France: Groupe de travail «Moisissures dans l'habitat». Diagnostic, effet sur la santé respiratoire, conduites à tenir, Contaminations fongiques en milieux intérieurs (2006)

Dehoux, S., Dehoux, P.: Habitat Qualité Santé en mains—Le guide de l'habitat sain, 286 (1997)

Ferreira, S.R., Silva, F.A., Lima, P.R.L., Toledo Filho, R.D.: Effect of wetting and drying cycles on the interface of natural fibers with a Portland cement based matrix. In: First International Conference On Bio-Based Building Materials, Clermont-Ferrand, France (2015)

Guéguen, M., Moscardelli, S., Van Schoors, L., Nour, I., Marceau, S.: Study of the microbial development impact on bio-based materials. In: First International Conference on Bio-Based Building Materials, Clermont-Ferrand, France (2015)

Hellebois, P.: Etude de l'évolution des propriétés de bétons de chanvre pendant leur vieillissement, Mémoire de fin d'études, Polytech'Clermont-Ferrand (2013)

Hellebois, P., Marceau, S., Guéguen, M., Amziane, S.: Influence of wetting and drying cycles on the properties of hemp concrete. In: 1st International RILEM Conference on Rheology and Processing of Construction Materials, Paris (2013)

Juarez, C., Duran, A., Valdez, P., Fajardo, G.: Performance of "Agave Lecheguilla" natural fiber in portland cement composites exposed to severe environment conditions. Build. Environ. **24**, 1536–1541 (2007)

Le Bayon, I., Draghi, M., Gabille, M., Prégnac, M., Lamoulie, J., Jequel, M., Roger, M., Kutnik, M.: Development of a laboratory test method to assess the resistance of bio-based insulation materials against moulds. In: First International Conference on bio-based building materials, Clermont-Ferrand, France (2015)

MacVicar, R., Matuana, L.M., Balatinecz, J.J.: Ageing mechanisms in cellulose fiber reinforced cement composites. Cement Concr. Compos. **21**, 189–196 (1999)

Magniont, C.: Contribution à la formulation et à la caractéri-sation d'un écomatériau de construction à base d'agroressources. Ph.D. Thesis, Université de Toulouse (2010)

Magniont, C., Escadeillas, G., Coutand, M., Oms-Multon, C.: Use of plant aggregates in building ecomaterials. Eur. J. Environ. Civil Eng. **16**, 17–33 (2012)

Marceau, S., Glé, P., Guéguen, M., Gourlay, E., Moscardelli, S., Nour, I., Amziane, S., Abdellaoui, L.: Assessment of the durability of bio-based building materials. In: First International Conference on bio-based building materials, Clermont-Ferrand, France (2015)

Melo Filho, J.A., de Andrade Silva, F., Toledo Filho, R.D.: Degradation kinetics and aging mechanisms on sisal fiber cement composite systems. Cement Concr. Compos. **40**, 30–39 (2013)

Mohr, B.J., Nanko, H., Kurtis, K.E.: Durability of thermomechanical pulp fiber-cement composites to wet/dry cycling. Cem. Concr. Res. **35**, 1646–1649 (2005)

Mohr, B.J., Biernacki, J.J., Kurtis, K.E.: Microstructural and chemical effects of wet/dry cycling on pulp fiber-cement composites. Cem. Concr. Res. **36**, 1240–1251 (2006)

Pasanen, A.-L., Juutinen, T., Jantunen, M.J., Kalliokoski, P.: occurrence and moisture requirements of microbial growth in building materials. Int Biodeterior. Biodegradation **30**, 272–283 (1992)

Pejic, B.M., Kostic, M.M., Skundric, P.D., Praskalo, J.Z.: The effects of hemicelluloses and lignin removal on water uptake behavior of hemp fibers. Bioresour. Technol. **99**, 7152–7159 (2008)

Ramakrishna, G., Sundararajan, T.: Studies on the durability of natural fibres and the effect of corroded fibres on the strength of mortar. Cement Concr. Compos. **27**, 575–582 (2005)

Sedan, D.: Etude des interactions physico-chimiques aux interfaces fibres de chanvre/ciment. Influence sur les propriétés mécaniques du composite. Ph.D. Thesis, Université de Limoges (2007)

Sonebi, M., Wana, M., Amziane, S., Khatib, J.: Investigation of the mechanical performance and weathering of hemp concrete. In: First International Conference on bio-based building materials, Clermont-Ferrand, France (2015)

Sivaraja, M., Kandasamy, Velmani, N., Pillai, M.S.: Study on durability of natural fibre concrete composites using mechanical strength and microstructural properties. Bull. Mater. Sci. **33**, 719–729 (2010)

Talon, A.: Evaluation des scénarii de dégradation des produits de construction. Ph.D. Thesis, Université Blaise Pascal, Clermont-Ferrand (2006)

Toledo Filho, R.D., de Andrade Silva, F., Fairbairn, E.M.R., de Almeida Melo Filho, J.: Durability of compression molded sisal fiber reinforced mortar laminates. Constr. Build. Mater. **23**, 2409–2420 (2009)

Walker, R., Pavia, S., Mitchell, R.: Mechanical properties and durability of hemp-lime concretes. Constr. Build. Mater. **61**, 340–348 (2014)

Wei, J., Meyer, C.: Degradation rate of natural fiber in cement composites exposed to various accelerated ageing environment conditions. Corros. Sci. **88**, 118–132 (2014)

Wei, J., Meyer, C.: Degradation mechanisms of natural fiber in the matrix of cement composites. Cem. Concr. Res. **73**, 1–16 (2015)

Chapter 9
Effect of Testing Variables (Method of Production)

Sara Pavía

Abstract Production parameters affect hydration/carbonation and density of hemp concrete which consequently determine the strength and the hygric and thermal properties of the concrete. This chapter investigates the effect of production parameters including curing conditions (65% vs. >95% RH), time of demoulding and specimen geometry (cylinder vs. cube) on the concrete's strength which relates to density and therefore to thermal and hygric properties. It studies hydration in the concrete's microstructure and measures the compressive strength development at intervals between 1 day and 1 month. Moulding time and curing conditions influence drying and therefore may impact binder hydration and consequently strength evolution. Specimen geometry may affect drying and can also determine how strain builds up in the concrete and thus when failure occurs. The chapter concludes that curing hemp-lime concrete with hydraulic content (50%CL90:50% CEMII) at high RH (>95%) lowers compressive strength (65.4% drop at 10 weeks). It is unclear why this happens, as the presence of water vapour during curing at high RH should enhance hydration and consequently increase strength. It was also found that delaying specimen demoulding increases compressive strength of the CL90: CEMII concrete (22.9% increase at 10 weeks), probably due to the presence of moisture for longer enhancing hydration. The specimen geometry does not significantly impact the ultimate compressive strength of hemp-lime concrete however, it affects behaviour in compression. Initially, cylinders and cubes deform on load application up to a similar yield point. However, following this yield point, the cylinders fracture showing a more brittle behaviour while the cubes keep crushing to finally experience an additional stiffness produced by mechanical bridges being formed between opposing cell walls.

Keywords Production method · Curing · Compressive strength · Microstructure

S. Pavía (✉)
Department of Civil Engineering, Museum Building, Trinity College,
Dublin 2, Ireland
e-mail: pavias@tcd.ie

S. Amziane and F. Collet (eds.), *Bio-aggregates Based Building Materials*,
RILEM State-of-the-Art Reports 23, DOI 10.1007/978-94-024-1031-0_9

9.1 Introduction

A significant amount of research has been undertaken on lime hemp concrete in the last decade however, testing variables are not yet standardised. This chapter provides an overview on the effect that some testing variables have on the compressive strength of the concrete with a view to standardise tests methods and procedures.

Some of the most outstanding features of hemp concrete such as its a high thermal capacity and low thermal conductivity are closely related to production parameters as these determine properties such as density therefore, it is interesting to investigate how production variables affect the properties of the concrete.

An attempt was made at comparing results from previous authors to conclude on the effect of production variables in hemp concrete properties however, it was found that the results were hardly ever comparable due to the amount of different testing variables used by the different authors. Therefore, a testing program was set out to investigate the effect of some production parameters, that impact hemp concrete properties including curing conditions (65% vs. >95% RH), time of demoulding, specimen geometry (cylinder vs. cube) and re-hydration by re-immersion. Concretes made with two different binders, a 50% CL90s:50% CEMII binder and a 100% NHL3.5, were produced and the impact of the above variables on compressive strength studied.

An overview of the impact of some production parameters on concrete properties

Variables that impact compressive strength. A significant number of variables can affect the strength of hemp concrete including compaction, drying and hydration/carbonation. Moulding time and curing conditions influence drying and therefore may impact binder hydration/carbonation and consequently strength evolution over time. Specimen geometry can affect drying and also determines how strain builds up in the concrete and thus when failure occurs.

In addition, competition for mixing water between the binder and hemp particles (resulting in the binder not fully hydrating) can undermine strength.

High-humidity curing, long periods before demoulding and small surface area of specimens can result in longer drying of mixing water affecting hydration and compressive strength.

The effect of the binder on compressive strength has yielded varying opinions. Hirst et al. (2010) found that the concrete strength does not increase with the strength of the binder. Nevertheless, higher compressive strengths are usually obtained for cement-rich binders (Murphy et al. 2010; De Bruijn et al. 2009). Nguyen (2010) found that, at 90 days, lime binders reached higher compressive strengths than commercial binders that are typically more hydraulic. Walker and Pavía (2014a, b) observed that strength development of hemp concrete was a function of the binder's hydraulic strength up to 6 months but, at 1 year, all concretes displayed a similar compressive strength (0.32–0.41 MPa). The authors however noticed that the most hydraulic binders did not fully hydrate, and later reintroduction of water increased hydration significantly enhanced strength.

The curing environment (temperature and humidity) impact drying and speed of carbonation/hydration which lead to hardening, consequently affecting strength development and ultimate strength.

Following manufacture, lime-hemp concrete is placed in a constant environment of temperature and relative humidity (RH) until the time of testing.

Increasing temperature usually enhances carbonation and hydration rates (speeding strength development) while low temperatures can impede hardening and the development of strength. High temperatures may speed up drying however, hemp concrete is typically cured at ambient temperature (approximately 20 °C).

Hemp concrete is typically cured at relative humidity ranging from 50 to 65% although a wide variation has been reported including 20 ± 1 °C and 60 ± 5% RH (Hirst et al. 2010); 20 °C and 60% RH (Evrard 2006); 22–26 °C and 30–60% RH (Colinart et al. 2012) and 20 °C and 75% RH (Nguyen et al. 2009); 20 °C and humidity saturation (Nozahic et al. 2012). Arnaud and Gourlay (2012) determined 20 °C and 50% RH as the optimum curing conditions for the evolution of strength of four hydraulic binders at 28 days. Higher and lower RHs (30, 75 and 90%) were found to reduce mechanical strength.

Retention of samples in their moulds during curing affects the early moisture content and the drying rate of the concrete, variables which determine hydration/carbonation of the binder which impact early strength development. Early strength is important as it relates to building speed and the ability of the concrete to initially stand its own weight when it is heaviest (at highest moisture content).

The effect of the time of demoulding has not been yet investigated. Demoulding has widely ranged, with some authors demoulding immediately after manufacture (Walker 2013) and others keeping the concrete in the moulds for the full duration of curing with only 1 or 2 faces exposed (Arnaud and Gourlay 2012). Demoulding after 1–6 days is a popular option used by several authors: 1 day (Nguyen et al. 2009); 1 day followed by sealing until 5 days (Colinart et al. 2012); 2 days (Nozahic et al. 2012) and 6 days (Hirst et al. 2010). As a result of the high moisture permeability of the concrete, fast drying can occur when the specimens are quickly removed from their moulds and left to cure unwrapped at 60% RH. As aforementioned, fast drying reduces available water and halts and retards hydration of binders with quick set however it can be beneficial for binders with low hydraulic content.

Specimen geometry may affect ultimate strength. Previous authors have found that using hydraulic limes, half prisms are on average 37% stronger than cubes, and that this may be due to the ratio of length to height which determines how strains build up in the specimen (Patterson and Pavía 2012). In PC mortar and concrete, cubes are reported to be stronger than cylinders. A factor of 1.2 is used to convert cylinder to cube strength for normal strength concrete. However, this factor becomes smaller as strength increases so that, for high-strength concrete, the influence of shape is much less significant (Yi et al. 2006 citing Gonnerman 1925; Gyengo 1938; Murdock and Kesler 1957). In PC composites, it has also been found that strength is an inverse function of the specimen size for cubic and prismatic samples whereas for larger cylinders the effect of size on strength is almost negligible (Yi et al. 2006; Del Viso et al. 2008).

In hemp-lime concrete, specimen geometry widely varies in research—cylinders, blocks and small scale walls of varying scales have been investigated. Some examples include 50 mm cubes (Elfordy et al. 2008; Collet et al. 2013); 40 * 40 * 160 mm prisms (Nozahic et al. 2012); 300 * 300 * 160 mm blocks (Colinart et al. 2012); 100 mm (d) * 50 mm (h) cylinders (Collet et al. 2013), 190 mm (d) * 35 mm (h) cylinders (Evrard 2006); 100 mm cubes (Walker and Pavía 2014); 160 mm (d) * 320 mm (h) cylinders (Arnaud and Gourlay 2012); 150 mm (d) * 300 mm (h) cylinders (Hirst et al. 2010). Collet (2004) notes that 50 mm cubes are representative of the material however, cylinders are typically more common for compression testing although 100 mm cubes (Elfordy et al. 2008; Walker and Pavía 2014) have also been used.

Glouannec et al. (2011) compared different geometry specimens and observed similar compressive strengths although their behaviour on load application differed. Tall specimens (height > width) showed a clear fracture plain, and their maximum compressive strength was followed by a decrease in stress. In contrast, stout specimens (height < width) crushed continuously.

9.2 Materials and Methods

As aforementioned, in order to investigate the effect of some production parameters, standard tests were performed with two binders: a 50%CL90s/50% CEMII binder and a 100% NHL3.5. Testing parameters that impact the material properties including the effect of curing conditions (65% vs. >95% RH), time of demoulding, specimen geometry (cylinder vs. cube) and re-hydration on compressive strength were studied.

9.2.1 Manufacture (Mixing, Curing and Compaction)

The mix proportions were constant at 1:2:3.1 (hemp:binder:water). The binder was dry mixed by hand to ensure it was homogenous. The binder and ¾ of the water were placed in a drum cement mixer and mixed for 2.5 min to form a slurry. The hemp was then gradually added along with the remaining water. The mixer was stopped half way through mixing to break up any clumps formed in the material. The total mixing time was 7 min.

Control mixes and further specimens were produced according to the parameters in Table 9.1. For the preparation of control mixes, the concrete was weighed and put into 100 mm cubic moulds in a single layer. It was lightly compressed by hand as it was put into the mould. The control samples were immediately removed from

Table 9.1 Summary of variables in concrete specimens tested

Testing variables	Wet weight (g)	Size/geometry	Curing conditions	Time in mould (days)
Control	680	100 mm^3	20 ± 2 °C 60 ± 5% RH	0
Curing humidity (95% RH)	680	100 mm^3	20 ± 2 °C >95% RH	0
Time of demoulding	680	100 mm^3	20 ± 2 °C 60 ± 5% RH	70
Specimen shape (cylinders)	1070 (to maintain a wet density of 680 kg/m^3)	Cylinder d = 100 h = 200 mm	20 ± 2 °C 60 ± 5% RH	70

Four specimens of each binder (CL90:CEMII and NHL3.5) were fabricated with each of the 4 testing variables in the table. Hemp:binder:water = 1:2:3.1. All dried for 24 h prior to testing

their moulds and transferred to a curing room at 20 ± 2 °C and 60 ± 5% RH until testing.

The specimens retained in moulds were coated with three layers of oil to facilitate removal. The samples had an initial wet density of 680 kg/m^3. Following drying, the control concrete achieved an approximate density of 400 and 379 kg/m^3 for the CL90:CEMII and NHL3.5 binders respectively.

Four specimens of each binder were fabricated with each of the testing variables in Table 9.1 and tested at 1 day and 1, 2, 4 and 10 weeks.

In addition, five cubes of each mix were immersed for two days in a water bath after one month of curing to induce re-hydration, these cubes were then removed and left to cure at normal temperature and humidity and tested at 10 weeks.

9.2.2 *Compressive Strength*

Compressive strength was measured using a Zwick testing apparatus. The cubes were removed from the curing room at 1 day and 1, 2, 4 and 10 weeks and oven dried at 50 °C for 24 h prior to testing. Oven drying was necessary to measure early changes in strength development as samples up to 2 weeks are so wet (the CL90: CEMII and NHL3.5 concretes retain approximately 51 and 64% of the original mixing water respectively at 2 weeks), that they only compress under load application and consequently small changes in strength development are not discernible. Oven drying enhanced the sample's strength and allowed early changes in strength to be measured. Concrete strength arises from a combination of drying and binder carbonation/hydration. The contribution of drying is considered to remain the same for all samples dried and therefore the impact of binder hydration/carbonation can be seen in the increasing compressive strength.

No standards currently apply to hemp concrete thus the testing procedures of EN 459-2[1] and EN 196-1[2] were used to guide the test. The cubes did not break but continuously deformed in a plastic manner. Failure was considered as the point at which the stress/strain curve departs from linear behaviour. Student's t-Tests were carried out to determine if the results were statistically significant ($P < 0.05$).

9.2.3 *Microstructure*

Investigating the microstructure of the concrete should reveal the presence of carbonates and hydrates that contribute to strength. The microstructure of the binder and the surface of the hemp aggregate were investigated using a Tescan MIRA Field Emission Scanning Electron Microscope (SEM). The samples were freshly fractured and covered with a gold coating in an 'Emscope SC500' plasma coating unit. Individual hemp particles were extracted from fractured surfaces and mounted on pin stubs prior to coating. Samples were sealed in air tight conditions until the analysis was undertaken.

9.3 Results

9.3.1 *Compressive Strength and Microstructure*

Compressive strength increase (Fig. 9.1) is evident up to 28 days however, later, the increase was not statistically significant ($P > 0.05$). This indicates that both binders achieved most of their entire compressive strength in the first month.

SEM analysis evidenced hydrates responsible for strength development at 1 day, increasing significantly by 1 month (Figs. 9.2, 9.3, 9.4 and 9.5). Hydrates were present in both concretes however, they were substantially more prolific in the CL90:CEMII binder than in the NHL3.5 binder (Figs. 9.2, 9.3, 9.4 and 9.5).

The CL90:CEMII concrete achieved a considerably greater strength than the NHL3.5 concrete (Fig. 9.1) due to its greater hydraulic content—see Figs. 9.2, 9.3, 9.4 and 9.5.

Re-immersion in water in order to trigger re-hydration (for two days after the 1st month of curing) shows a decrease in compressive strength for both concretes but it is not statistically significant.

Curing at 95% RH reduces the compressive strength of CEM:CL90 (50/50) and has no significant effect on NHL3.5

[1]EN 459-2:2010 Building lime. Test methods.

[2]EN 196-1-2005 Methods of testing cement. Determination of strength.

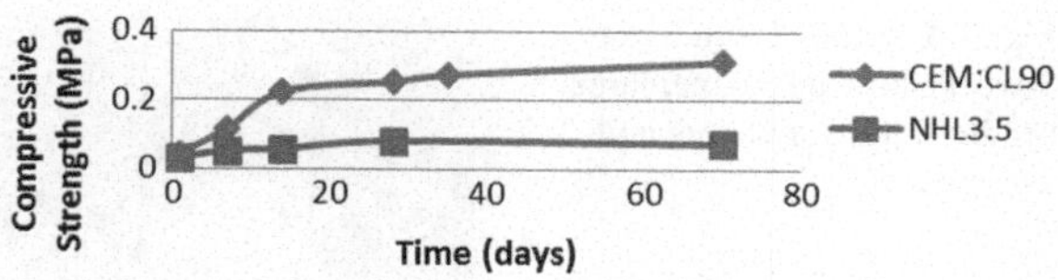

Fig. 9.1 Strength development of the concretes over time

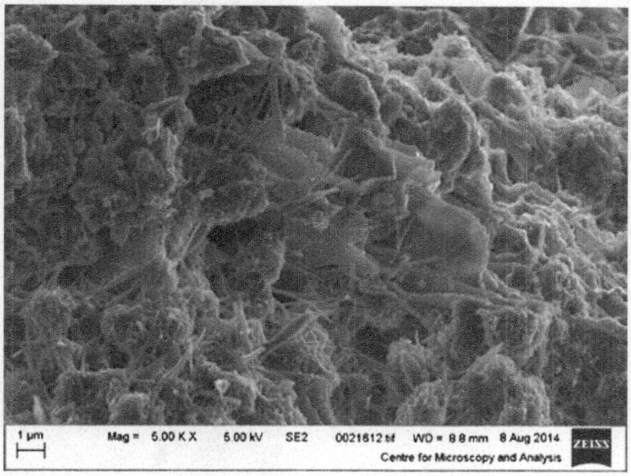

Fig. 9.2 SEM microstructure of the hydrated CL90:CEMII binder at 1 day showing significant, calcium silicate hydrate (CSH), ettringite and portlandite

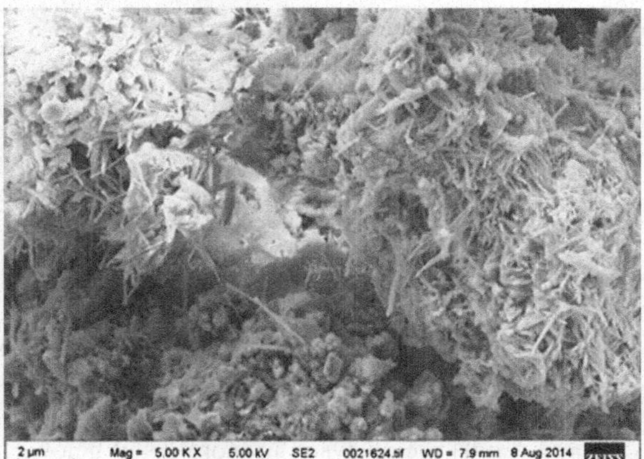

Fig. 9.3 SEM microstructure of the hydrated CL90:CEMII binder at 1 month with abundant CSH, ettringite, cubic hydrates and carbonates

There is no significant difference in the compressive strength of cubes and cylinders however the strain on load application is different in cubes and cylinders (different stress vs. stain curves). Some of the NHL3.5 cylinders broke during demoulding (binder was too weak and the results not valid).

Curing in moulds significantly increases compressive strength of CEM:CL90 binder but has no significant effect on the NHL3.5 concrete. Keeping the concrete in moulds during curing allows closer conditions to real curing inside the wall.

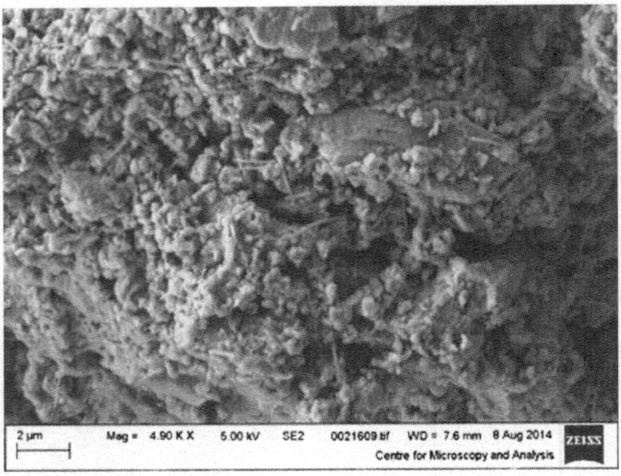

Fig. 9.4 NHL3.5 binder at 1 week showing scarce, large, needle-shaped hydrates and carbonates

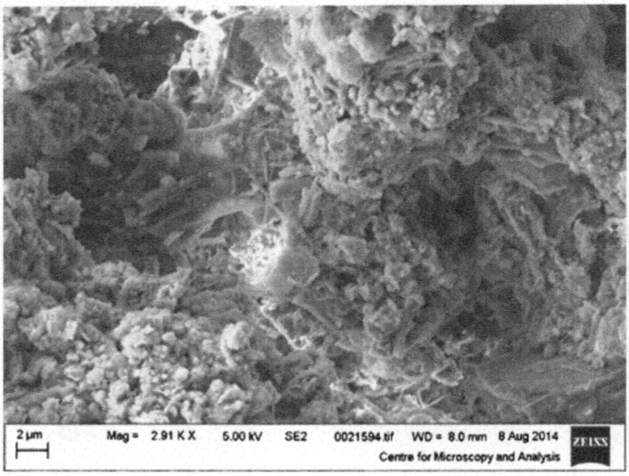

Fig. 9.5 NHL3.5 binder at 1 month including significant hydrates and carbonates

The excess moisture in the concretes kept in the moulds enhances hydration to a greater extent than the high moisture vapour in high (90%) RH curing environments.

These results are analysed in more detail below.

9.3.2 *Effect of Oven Drying on Compressive Strength*

As aforementioned, the concretes were oven dried to remove the high moisture content and be able to measure early strength changes. The compressive strength of the concrete is due to carbonation and hydration of the binder together with drying which increases stiffness at early ages. Strength contribution due to drying should be consistent for all samples independent of age or binder type.

The compressive strength of the NHL3.5 concrete at 1 day is 0.03 MPa and the SEM analysis evidenced some hydrates at this stage. This suggests that the contribution of drying towards compressive strength is less than 0.03 MPa.

The strength of CL90:CEMII concrete specimens that were oven dried for 24 h following 28 days of curing was compared with the strength of those not oven dried. The average compressive strength was 0.25 and 0.27 MPa for the 28 day (non-oven dried) and 28-day oven dried respectively. The difference in results is not statistically significant ($P > 0.05$). This suggests that oven drying at low temperatures, does not impact the strength performance of hemp concrete.

9.3.3 *Effect of Curing at High RH on Strength*

Curing the hemp-lime concrete at high RH (>95%) was found to reduce the compressive strength of the CL90:CEMII concrete at 10 weeks (Fig. 9.6). The NHL3.5 concrete shows the same trend, although the results are not statistically significant. The lack of statistical significance may be on account of the little strength of the NHL3.5 concrete (0.07 MPa) resulting in low measurement sensitivity of the testing equipment.

The findings are similar to those of Arnaud and Gourlay (2012) who observed that 90% RH during curing reduced mechanical strength of four hydraulic binders.

The CL90:CEMII binder shows significant hydrates at 1 month (Fig. 9.3). It is unclear why the presence of water vapour during curing at high RH does not contribute to binder hydration and consequently increase strength as it would in the case of mortars.

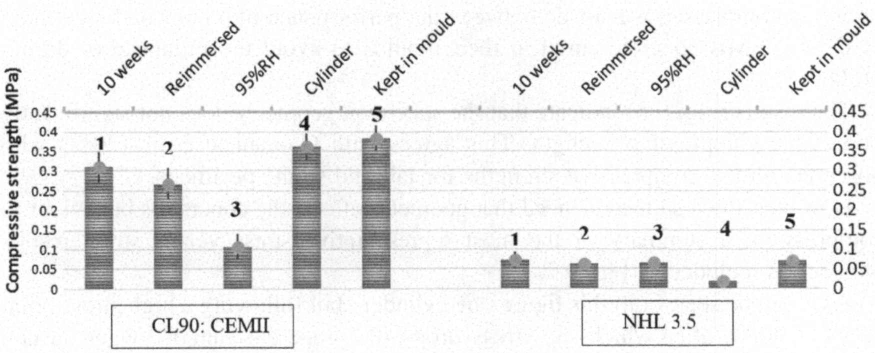

Fig. 9.6 Effect of testing variables on compressive strength of concretes after 10 weeks of curing. *1* Control specimes (100 mm cubes, cured at 65% RH, straight out of mould); *2* cubes reimmersed; *3* cubes cured at >95% RH; *4* cylindrical concrete specimens; *5* cubes retained in moulds

9.3.4 *Effect of Retention in Moulds During Curing*

Retention of the concrete in its mould will delay drying, as moisture is blocked from escaping through the mould and drying is largely restricted to the uncovered sides.

Retaining the hemp-lime concrete in its mould during curing was found to increase the compressive strength of the CL90:CEMII concrete at 10 weeks (Fig. 9.6). It is likely that the presence of moisture for longer periods (delayed from drying by the mould) facilitates the formation of additional hydrates which enhance strength.

The effect of retaining the NHL3.5 concrete in its mould during curing on strength is not statistically significant. This may be due to lack of measuring instrument sensitivity at low strength values. However, it is also likely that that the presence of moisture is less beneficial to the NHL3.5 binder on account of its lower hydraulic content and higher hydrated lime content.

9.3.5 *Effect of Specimen Geometry on Strength*

Specimen size and shape determine the surface area available for drying which may affect binder hydration/carbonation therefore compressive strength. For example, a 100 mm cube has a larger surface area than a cylinder with a 100 mm diameter and 100 mm height and will consequently dry faster. As aforementioned, geometry also determines the ratio of length to height which dictates how strains build up in the specimen and hence its strength.

In order to establish the effect of geometry on strength and exclude the effect of drying, a comparison was made between the performance of a cube and a cylinder of CL90:CEMII concrete cured in their moulds to avoid the influence of drying (Table 9.1).

The results (Fig. 9.6) indicate that the specimen geometry does not significantly impact the compressive strength. This agrees with Glouannec et al. (2011) who observed similar compressive strengths for tall and stout specimens.

However, the results evidenced that geometry affects the concrete's behaviour in compression; a summary of the most representative stress versus strain results obtained is included in Fig. 9.7.

As it can be seen from this figure, the cylinders fail following a high stress point or yield point, after which the stress drops (the concrete cannot longer sustain stress). In contrast, following a similar high stress point to that of the cylinders, the cubes continuously deform, showing a large plateau (region of increasing strain for small stress increase) followed by a raising branch where stress increases rapidly in relation to the strain. Elfordy (2008) attributed this to the irreversible compaction of the porous hemp shiv.

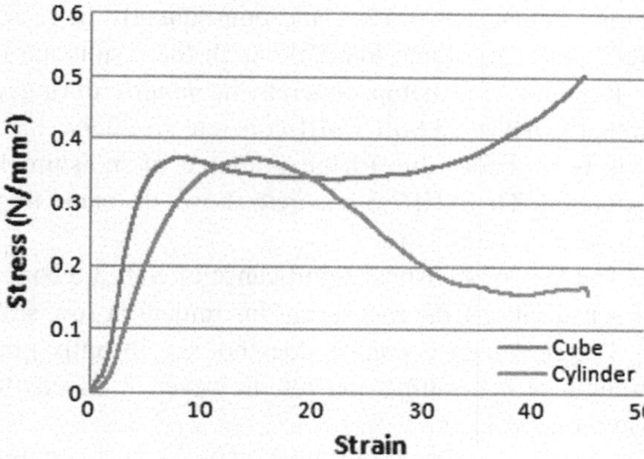

Fig. 9.7 Typical stress versus strain behaviour of hemp concrete varying with specimen geometry

As discussed in Walker and Pavía (2014), the final behaviour of the cubes (stress increase) is produced when most hemp cells have collapsed and, as the cells are further compressed, contact between opposing cell walls occurs, resulting in the formation of mechanical bridges which lead to an increase in the stiffness of the material. A similar behaviour has been observed in other cellular solids (Daxner 2010).

Therefore, both geometries typically show three stages in their stress-strain evolution: the first two (linear and plateau stages) are common. However the plateau stage is short for the cylinders and is followed by failure whereas the cubes show a long plateau (increased deformation) followed by a stress increase due to the additional stiffness produced by contact between opposing cell walls.

The variation specimens behaviour is similar to that reported by Glouannec et al. (2011) who observed that tall specimens (height > width) had a clear fracture plain and the maximum compressive strength was followed by a decrease in stress while stout specimens (height < width) crushed continuously. The cylindrical NHL3.5 binder samples crumbled during demoulding and the results are therefore disregarded.

9.4 Conclusion

Both concretes achieved most of their compressive strength in the first month. The CL90:CEMII concrete achieved a considerably higher strength than the NHL3.5 concrete probably due to its greater hydraulic content (hydrates were much more prolific in the CL90:CEMII binder).

Curing hemp-lime concrete with hydraulic content (50%CL90:50%CEMII) at high RH (>95%) lowers compressive strength (65.4% drop at 10 weeks). It is unclear why this happens, as the presence of water vapour during curing at high RH should enhance hydration and consequently increase strength. The NHL3.5 concrete shows the same trend although the results are not statistically significant.

Retaining the hemp concrete in *moulds* during curing increases compressive strength of the CL90:CEMII concrete (a 22.9% increase measured at 10 weeks). This is probably due to the presence of moisture for longer enhancing hydrate formation. The NHL3.5 concrete shows the same trend although the results are not statistically significant.

The lack of statistical significance of NHL3.5 concrete results may be due to lack of sensitivity of the measuring instrument at low strength values.

The specimen *geometry* does not significantly impact the ultimate compressive strength of hemp-lime concrete however, it affects the concrete behaviour under a compressive load.

Initially, cylinders and cubes deform on load application up to a similar yield point. Following this yield point, the cylinders fracture whereas the cubes keep crushing to finally experience an additional stiffness produced by mechanical bridges between opposing cell walls.

Drying hemp concrete at 50 °C, allows to monitor early strength development and does not impact strength performance. Strength contribution by drying in hemp concrete is small (e.g. less than 0.03 MPa in NHL3.5 concrete).

Acknowledgements All testing was carried out in the Department of Civil Engineering, Trinity College Dublin. The authors thank Dr. Kevin Ryan and Michael Grimes for their assistance with testing. The authors also wish to thank the Traditional Lime Company and Clogrennane Lime for the provision of materials. The authors thank Dr. Heath Bagshaw (Centre for Microscopy and Analysis) for his assistance with the SEM. This research follows up from a former research project funded by the Environmental Protection Agency and it is largely based upon experimental work published in: 'Effect of testing variables on the hydration and compressive strength of lime hemp concrete' S. Pavía, R. Walker, J. McGinn. First Int. Conf. on Bio-based Building Materials June 22nd–24th 2015 Clermont-Ferrand, France.

References

Arnaud, L., Gourlay, E.: Experimental study of parameters influencing mechanical properties of hemp concretes. Constr. Build. Mater. **28**(1), 50–56 (2012)

Colinart, T., Glouannec, P., Chauvelon, P.: Influence of the setting process and the formulation on the drying of hemp concrete. Constr. Build. Mater. **30**(9), 372–380 (2012)

Collet, F.: Caractérisation hydrique et thermique de matériaux de génie Civil à faible impacts environnementaux. Ph.D. thesis, Institut National des Sciences Appliquées de Rennes, France (2004)

Collet, F., Chamoin, J., Pretot, S., Lanos, C.: Comparison of the hygric behaviour of three hemp concretes. Energy Build. **62**, 294–303 (2013)

Daxner, T.: Finite Element Modelling of Foams. CISM International Centre for Mechanical Sciences, Italy (2010)

De Bruijn, P.B., Jeppssona, K.H., Sandin, K., Nilssona, C.: Mechanical properties of lime–hemp concrete containing shives and fibres. Biosyst. Eng. **103**(4), 474–479 (2009)

Del Viso, J.R., Carmona, J.R., Ruiz, G.: Shape and size effects on the compressive strength of high-strength concrete. Cem. Concr. Res. **38**(3), 386–395 (2008)

Elfordy, S., Lucas, F., Tancret, F., Scudeller, Y., et al.: Mechanical and thermal properties of lime and hemp concrete ("hempcrete") manufactured by a projection process. Constr. Build. Mater. **22**(10), 2116–2123 (2008)

Evrard, A.: Sorption behaviour of Lime-Hemp Concrete and its relation to indoor comfort and energy demand. In: The 23rd Conference on Passive and Low Energy Architecture, Geneva, Switzerland, 6–8 Sept 2006

Glouannec, P., Collet, F., Lanos, P., Mounanga, C. et al.: Proprietes physiques de betons de chanvre. Mater. Tech. **99**:657–665 (2011)

Hirst, E., Walker, P., Paine, K., Yates, T.: Characterisation of low density hemp-lime composite building materials under compression loading. In: Second International Conference on Sustainable Construction Materials and Technologies, Italy (2010)

Murphy, F., Pavia, S., Walker, R.: An assessment of some physical properties of lime hemp concrete. In: NiNuallain, N.A., Walsh, D., West, R., Cannon, E. et al. (eds.) Bridge Infrastructure Concrete Research Ireland, University of Cork, pp. 431–438 (2010)

Nguyen, T.: Contribution a l'etude de la formulation et du procede de fabrication d'elements de construction enbeton de chanvre. Ph.D. thesis, Universite de Bretagne Sud, France (2010)

Nguyen, T.T., Amziane, S., Picandet, V., Baley, C.: Influence of compactness and hemp hurd characteristics on the mechanical properties of lime and hemp concrete. Eur. J. Environ. Civil Eng. (EJECE) 1039–1050 (2009)

Nozahic, V., Amziane, S., Torrent, G., Saïdi, K., DeBaynast, H.: Design of green concrete made of plant-derived aggregates and a pumice–lime binder. Cement Concr Biocompos **34**:231–241 (2012)

Patterson, R., Pavía, S.: Influence of loading rate and specimen geometry on lime mortar strength. In: Caprani, C., O'Connor, A. (eds.) Bridge Infrastructure Concrete Research Ireland (BCRI), Dublin, pp. 361–365 (2012)

Walker, R.: A study of the properties of lime-hemp concrete with pozzolans. Ph.D. thesis, Trinity College Dublin, Ireland (2013)

Walker, R., Pavía, S.: Mechanical properties and durability of hemp-lime concretes. Constr. Build. Mater. **61**, 240–348 (2014)

Walker, R., Pavía, S: Impact of hydration on the properties of hemp-lime concrete. Civil Engineering Research in Ireland, CERI, pp. 211–216 (2014b)

Yi, S.T., Yang, E.I., Choic, J.C.: Effect of specimen sizes, specimen shapes, and placement directions on compressive strength of concrete. Nucl. Eng. Des. **236**(2), 115–127 (2006)

Appendix
Technical Committee Report—Rilem TC 236 BBM—Bio based Building Materials—Round Robin test for hemp shiv CHARACTERISATION

AUTHORS OF THE TC REPORT

Sofiane Amziane[(1)], Florence Collet[(2)], Mike Lawrence[(3)], Camille Magniont[(4)], Vincent Picandet[(5)]

1: Chair of RILEM TC-BBM 236, Institut Pascal, Clermont Université, France, sofiane.amziane@univ-bpclermont.fr
2: LGCGM, Université Rennes 1, Rennes, France, florence.collet@univ-rennes1.fr
3: BRE Centre for Innovative Construction Materials, University of Bath, Bath, UK m.lawrence@bath.ac.uk
4: Université de Toulouse, UPS, INSA, LMDC (Laboratoire Matériaux et Durabilité des Constructions), camille.magniont@insa-toulouse.fr
5: Université de Bretagne Sud, EA 4250, LIMatB, F-56100 Lorient, France, 18 vincent.picandet@univ-ubs.fr

TC 236 BBM Members

Chairman:
Sofiane AMZIANE, Institut Pascal, Clermont Université, France.

Secretary:
Florence COLLET, Université de Rennes 1, LGCGM, France.

TC Members:

Laurent ARNAUD, Ecole Nationale Supérieure d'Arts et Métiers, ENSAM Cluny, France.
Laëtitia BESSETTE, Centre Technique Louis VICAT, L'Isle d'Abeau, France.
Paulien DE BRUIJN, Lund University, Faculty of Engineering (LTH), Dept of Building Materials, Lund, Sweden.
Samuel DUBOIS, Université de Liège (Ulg), Gembloux Agro-Bio Tech, Belgique.
Gilles ESCADEILLAS, Université de Toulouse, UPS, INSA, LMDC, France.

S. Amziane and F. Collet (eds.), *Bio-aggregates Based Building Materials*,
RILEM State-of-the-Art Reports 23, DOI 10.1007/978-94-024-1031-0

Harald GARRECHT, Universität Stuttgart, Institut für Werkstoffe im Bauwesen (IWB), Stuttgart, Deutschland.
Etienne GOURLAY, CETE de l'Est - Laboratoire Régional de Strasbourg, France.
André KLATT, Universität Stuttgart, Institut für Werkstoffe im Bauwesen (IWB), Stuttgart, Deutschland.
Christophe LANOS, Université de Rennes 1, LGCGM, France.
Mike LAWRENCE, University of Bath, BRE Centre for Innovative Construction Materials, UK.
Camille MAGNIONT, Université de Toulouse, UPS, INSA, LMDC, France
Sandrine MARCEAU, Université Paris-Est, IFSTTAR, France.
Sara PAVÍA, Department of Civil Engineering, Trinity College, Ireland.
Ulrike PETER, Lhoist, Lhoist Recherche et Développement S.A., Belgium.
Vincent PICANDET, IRDL, Université de Bretagne Sud, France
Mohammed SONEBI, School of Planning, Architecture and Civil Engineering, Queen's University Belfast, UK.
Peter WALKER, University of Bath, BRE Centre for Innovative Construction Materials, UK.

Other contributors:

Thibaut COLINART, IRDL, Université de Bretagne Sud, France.
Guillaume DELANNOY, Université Paris-Est, IFSTTAR, France.
Yunhong JIANG, University of Bath, BRE Centre for Innovative Construction Materials, UK.
César NIYIGENA, Institut Pascal, Clermont Université, France.
Vincent NOZAHIC, Institut Pascal, Clermont Université, France.
Sylvie PRETOT, Université de Rennes 1, LGCGM, France.
Pierre TRONET, IRDL, Université de Bretagne Sud, France.

Table of Contents

(continued)

(continued)

ROUND ROBIN TEST FOR HEMP SHIV CHARACTERISATION

PART 1: EVALUATION OF INITIAL WATER CONTENT AND WATER ABSORPTION

Authored by: Sofiane Amziane[(1)], Florence Collet[(2)], Mike Lawrence[(3)], Camille Magniont[(4)], Vincent Picandet[(5)]

1: Chair of RILEM TC-BBM 236, Institut Pascal, Clermont Université, France, sofiane.amziane@univ-bpclermont.fr
2: LGCGM, Université Rennes 1, Rennes, France, florence.collet@univ-rennes1.fr
3: BRE Centre for Innovative Construction Materials, University of Bath, Bath, UK m.lawrence@bath.ac.uk
4: Université de Toulouse, UPS, INSA, LMDC (Laboratoire Matériaux et Durabilité des Constructions), camille.magniont@insa-toulouse.fr
5: Université Bretagne Sud, IRDL - CNRS FRE 3744, F-56100 Lorient, France, vincent.picandet@univ-ubs.fr

With the contribution of:

6: Laurent Arnaud, Art et Métiers, ENSAM Cluny, laurent.arnaud@ensam.eu
7: Sylvie Prétot, LGCGM, Université de Rennes 1, France, sylvie.pretot@univ-rennes1.fr
8: Christophe Lanos, LGCGM, Université de Rennes 1, France, christophe.lanos@univ-rennes1.fr
9: Thibaut Collinart and Pierre Tronet, Université Bretagne Sud, IRDL - CNRS FRE 3744, F-56100 Lorient, France, thibaut.collinart@univ-ubs.fr
10: Vincent Nozahic, Institut Pascal, Clermont Université, France, vincent.nozahic@hotmail.fr
11: Etienne Gourlay, CETE de l'Est - Laboratoire Régional de Strasbourg, etienne.gourlay@developpement-durable.gouv.fr
12: Sandrine Marceau, Paris-Est, IFSTTAR, MAST, Marne-La-Vallée, France, sandrine.marceau@ifsttar.fr

S. Amziane and F. Collet (eds.), *Bio-aggregates Based Building Materials*,
RILEM State-of-the-Art Reports 23, DOI 10.1007/978-94-024-1031-0

13: Gilles Escadillas, Université de Toulouse, UPS, INSA, LMDC (Laboratoire Matériaux et Durabilité des Constructions), Gilles.Escadillas@u
14: S. Dubois, Université de Liège (Ulg), Gembloux Agro-Bio Tech, Belgique, s.dubois@student.ulg.ac.be
15: Laetitia Bessette, Centre Technique Louis VICAT, L'Isle d'Abeau, France, L'Isle d'Abeau, l.bessette@vicat.fr
16: Paulien Debruijn, Lund University, Faculty of Engineering (LTH), Dept of Building Materials, Lund, Sweden, pbdebruijn@gmail.com
17: Ulrike Peter, Lhoist, Lhoist Recherche et Développement S.A., Belgium, ulrike.peter@lhoist.com
18: Sara Pavia, Dept of Civil Engineering, Trinity College Dublin, Ireland, PAVIAS@tcd.ie

Abstract

The paper presents the experience of a working group within the RILEM Technical Committee 236-BBM 'Bio-aggregate based building Materials'. The work of the Technical Committee (TC) will be to study construction materials made from plant particles. These materials are obtained from the processing of hemp, flax, miscanthus, pine, maize, sunflower, bamboo and others. The first round robin test of the TC-BBM was carried out to compare the protocols in use by the different laboratories (labs) to measure initial water content, bulk density, water absorption, particle grading and thermal conductivity. The aim is to define a characterisation protocol derived from those used by the different labs. This first round robin test was carried out on one variety of hemp shiv. Nine laboratories from European universities and research centers were involved (Table 1). The test results of 7 laboratories constitute a set of statistically representative data in order to propose recommendations to characterise hemp shiv after analysing the different methodologies in use in these labs.

Table 1 Participating Labs

Letter	*City*	*Labs*
A	***Bath (UK)***	***BRE Centre for Innovative Construction Materials/University of Bath***
B	***Clermont Ferrand (France)***	***Institut Pascal***
C	***Lorient (France)***	***LIMatB/Université de Bretagne Sud***
D	***Lyon (France)***	***DGCB/ENTPE***
E	***Paris (France)***	***IFSTTAR***
F	***Rennes (France)***	***LGCGM/Rennes 1***
G	***Toulouse (France)***	***LMDC/Université de Toulouse/UPS/INSA***
I	***Combloux (Belgium)***	***Combloux-Agro ressource – Université de Liège***

1. INTRODUCTION

This study focus on bio-based aggregate coming from the stem of plants cultivated either for their fibers (hemp, flax, etc.) or for their seeds (oleaginous flax, sunflower, etc.). Owing to the structure of the stem of the plant they are made from, such aggregates are generally malleable, elongated and highly porous with a low apparent density. They are very different from the mineral aggregates typically used in concretes, for which there are standardised tools and techniques for characterisation. Amongst these, hemp shiv (the woody core of the stem of the hemp plant) is probably the most widely used in alternative or eco-friendly building materials in Europe and is also representative of most of the aggregate coming from the stem of an annual crop. This is usually mixed with a lime-based binder and the resultant 'bio-concrete' is known as 'hemp-lime'.

This kind of aggregate is a co-product of hemp industry that is renewable and produced in an annual cycle while the price of mineral aggregates is steadily increasing as resources become less readily available. The characterisation of these aggregates, however, which is crucial to a proper understanding of the quality of the materials in which they are incorporated, requires adaptations to be made to the techniques usually employed for mineral aggregates, or the devising of new characterisation procedures.

The first round robin test of the Rilem TC-BBM was carried out to compare the protocols in use by the different labs. The aim was to define a test method to measure initial water content and water absorption of bio-aggregates. This first round robin test was carried out with one variety of hemp shiv (the woody core of the plant stalk chopped into lengths of a few centimeters) coming from the same production of a processing factory located in France. Seven labs conducted the measurements (Table 2).

Table 2 ***Description of the interlaboratory test***

Characteristics	Participating Labs
Initial water content	B, F, G
Water absorption	A, B, C, D, E, F, G

2. MATERIAL

One variety of hemp shiv was selected for this inter-laboratory test. It comes from the same processing factory where the bast fibers are stripped off (de-cortication), leaving the shiv behind. This shiv was provided by LCDA producer under the commercial name "KANABAT" (Tables 3 and 4). This shiv is in line with the French national recommendation provided by "Construire en Chanvre" association (Table 5) [1].

Table 3 Physical properties of shiv as supplied by the producer

Characteristic	Measurement
Density	100 to 110 kg/m^3 (depending on ambient relative humidity) Loosely packed, not compressed
Water absorption	198% (NFV 19 002)
Water absorption of mineral elements	24 meq per 100 g of raw material
Calorific value	3804 cal/g (NF M 07-030 12/9G)
Thermal Conductivity (10 °C in a dry state)	0.0486 W/m.K (NF EN12667)

Table 4 ***Chemical composition of the shiv*** as supplied by the producer*

Water	9 to 13%
Dry material	85 to 90% of which
Total organic material	97.5% on a dry basis of which:
Net cellulose:	52%
Lignin:	18%
Hemicellulose:	9%
Minerals:	Calcium: 1% on a dry basis Magnesium: 0.03% on a dry basis Phosphorus: 9 mg/100 g Potassium: 0.8% on a dry basis Total Nitrogen: 0.4 to 1% on a dry basis Total carbon: 496 g/kg on a dry basis C/N: 87 Ash: 2% pH in suspension at 10%: 6.7

*These data are supplied as indicative, certain values can vary depending on the year, the variety, the cultivation location etc.

Table 5 French national recommendations for the shiv characteristics

Characteristics	Recommendation
Apparent density of particles in kg/m^3	Quoted apparent density +/−15%
Maximal length (L_{dmax} in %)	Quoted L_{dmax} +/−10%
Initial moisture content	<19%
Dust content	<2% passing a 0.25 mm sieve
Colour	% of particles not conforming <5%

2.1 Microscopical description

Specimens of hemp shiv were examined at the University of Bath, both longitudinally with the capillaries and transversely across them, using a JEOL 6840 scanning electron microscope in the secondary electron mode at 15 kV voltage and a working distance of 10 mm. Figure 1 shows a scanning electron microscope view at x250 magnification of a transverse section.

The surface of the shiv has been torn by the sawing process used in the preparation of the thin sections, which makes this image slightly more difficult to interpret. However the two porosities are still visible. In the centre of the image is a xylem capillary, and it can be seen that there are much smaller pores visible on the cell wall. Figure 2 shows this capillary at x1000 magnification. It can be seen that the cell wall is penetrated by numbers of pores with diameters of between 1 μm and 5 μm. In fact these pores are known as 'pits' and consist of a thin membrane which allows water vapour to pass, but not liquid water. This is the mechanism by which the parenchyma are fed with nutrients and moisture from the xylem.

We thus have three distinct macroporosities in hemp shiv – the xylem at ~50 μm, the parenchyma at ~20 μm and the pits at ~1 μm to 5 μm. Given that the parenchyma are closed cells accessible to each other and to the xylem via the pits, this creates an interesting scenario which can be used to explain the latent heat effects believed to occur in hemp-lime. Figure 3 shows a schematic of water vapour flow through hemp shiv. In this schematic, the water vapour passes along the large

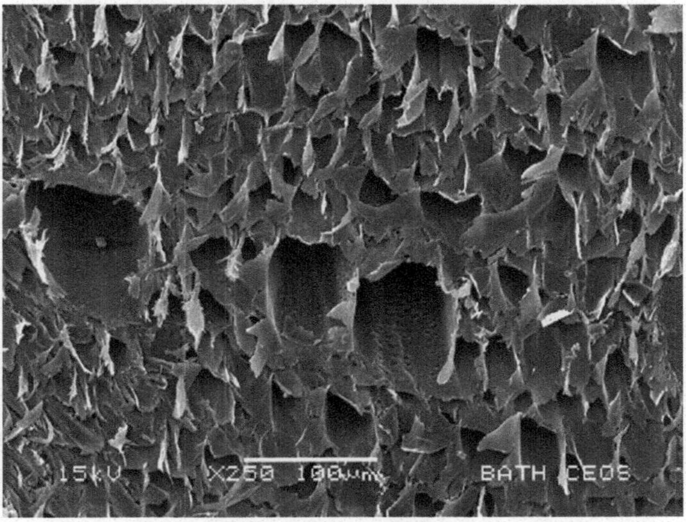

Figure 1 Scanning Electron Microscope view of hemp shiv (x250)

Figure 2 Scanning Electron Microscope view of hemp shiv (x1000)

Figure 3 Schematic of water vapour flow through hemp shiv

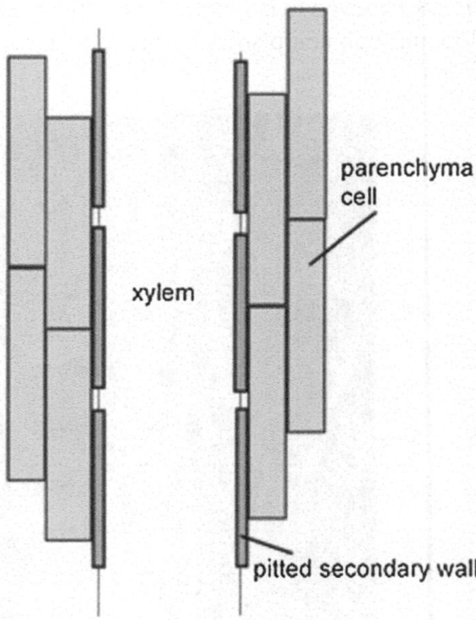

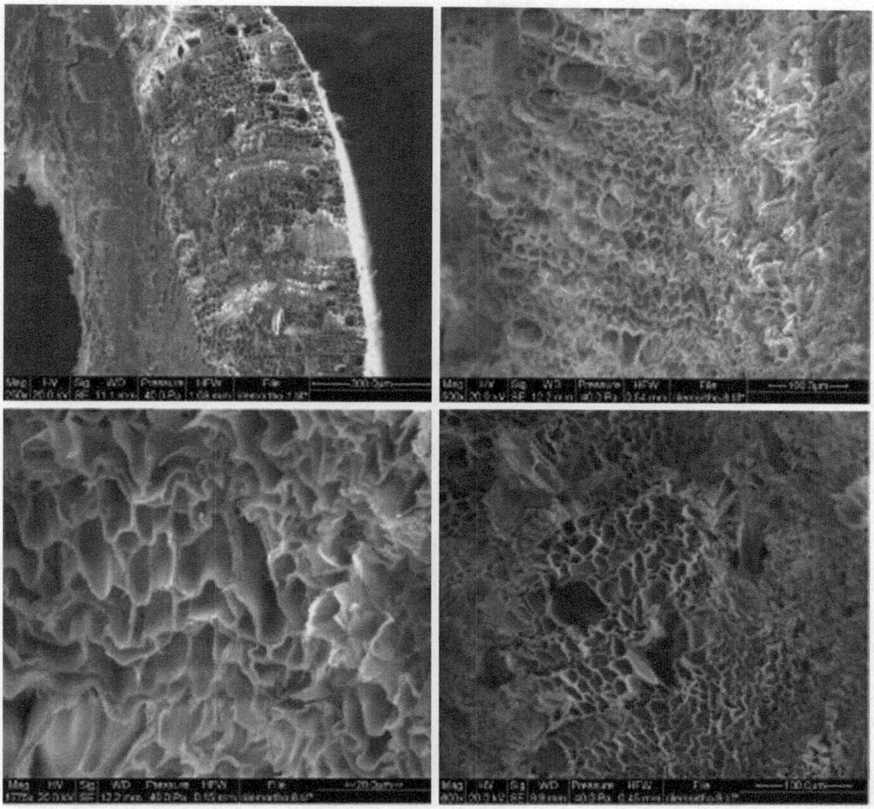

Figure 4 Transverse section of hemp shiv (Clockwise from top L 250x; 500x; 600x; 1775x) (IFSTTAR)

xylem capillaries (diameter ~50 μm). In order to access the parenchyma pores (~20 μm in diameter) the vapour has to pass through the 1 μm to 5 μm pits. In order for this to happen, there must be a partial pressure differential between the vapour pressure in the xylem and that in the parenchyma.

At IFSTTAR similar images have been taken using a Philips XL SEM in secondary electron mode at 20 kV and a working distance of 10 mm.

The mean diameter of the pores was measured at 10-30 μm (Figure 4).

Similar images were taken of longitudinal sections (Figure 5) showing lengths of the pores of between 20 μm and 80 μm.

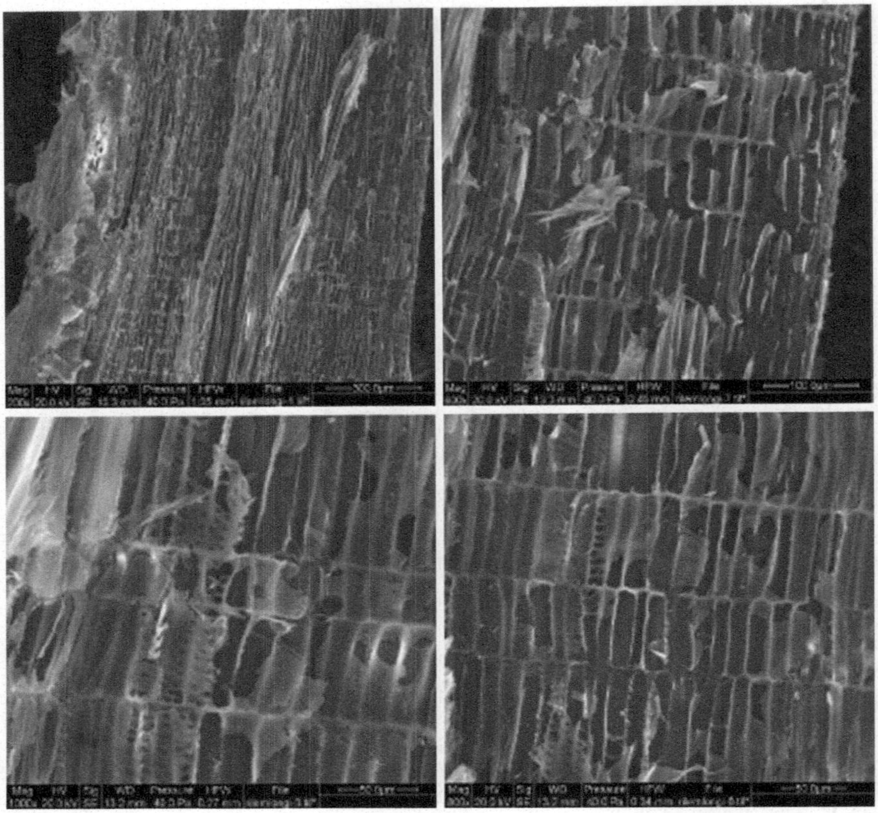

Figure 5 Longitudinal section of hemp shiv (Clockwise from top L 250x; 500x; 600x; 1775x) (IFSTTAR)

3 INITIAL WATER CONTENT

As with other lignocellulosic materials, hemp shiv is known to present a high hygroscopic behaviour. The water content of hemp shiv will then be largely influenced by relative humidity of air at the time of bagging or mixing. This property needs to be assessed as it can strongly influence the mix proportioning of hemp-lime and consequently the properties of hardened hemp-lime.

3.1 Methods

The initial water content of hemp shiv is calculated from the following relation:

$$W_0 = \frac{M_0 - M_D}{M_D} \times 100 \qquad \text{(eq.1)}$$

Where W_0 is the initial water content [%], M_0 is the initial mass of the sample [g] and M_D is the mass of dry sample [g].

Experimental protocols applied by labs B, F and G differ one from the others in the size of sample and the drying methods described in Table 6.

Table 6 Experimental protocols to measure the dry mass of the sample

Labs	Size of the sample and drying method
B	– Size of the sample around 120 g – Drying of the plant aggregates in an oven at 60 °C for 48 h
F	– Size of the samples: between 150 and 650 g – Drying of the plant aggregates: Method 1: in an oven at 60 °C with silica gel until the change in mass of the sample was less than 0.1% over 24 hours Method 2: under vacuum with silica gel (evaporation by all the surfaces of the sample) until the change in mass of the sample was less than 0.1% over 24 hours Method 3: under vacuum with silica gel (evaporation by the upper surface only) until the change in mass of the sample was less than 0.1% over 24 hours
G	– 3 different sizes of samples: 5, 10 and 250 g. – Drying of the plant aggregates in an oven at 50 °C until the change in mass of the sample was less than 0.1% over 24 hours.

3.2 Results

Figure 6 shows a synthesis of the results obtained by the three participating labs.

We can see on Figure 7 that the initial water contents measured by the different labs are similar. The average value from all these data is 10.9% with a standard deviation of 0.6%.

Results do not reveal any significant effect due to the size of sample.

Figure 7 shows a comparison between the five different drying methods used. No significant differences can be identified. Including the three different drying methods applied by Lab F results in comparable values of initial water content. Nevertheless, comparing the kinetics of the three drying methods, Lab F found that the drying method in the oven at 60 °C allows equilibrium to be achieved faster than the methods using silica gel under vacuum (7 days as against 1 month).

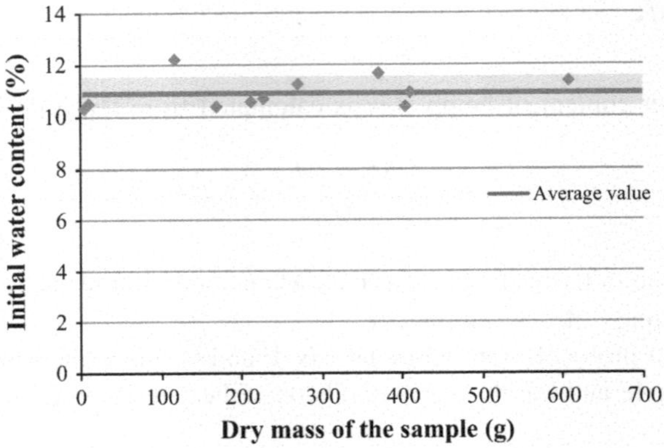

Figure 6 Synthesis of initial water content from the three participating labs

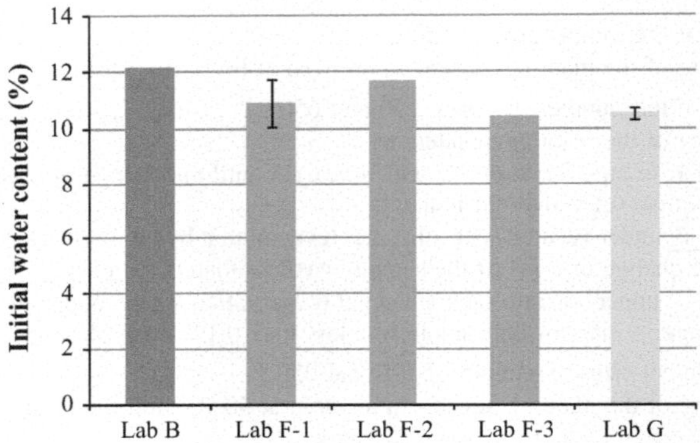

Figure 7 Initial water content versus drying method

Comparing the results of Lab F-1 (drying in an oven at 60 °C) and Lab G (drying in an oven at 50 °C) we can conclude that a drying temperature between 50 °C and 60 °C is suitable for the measurement of initial water content of plant aggregates.

3.3 Concluding remarks on initial water content

The results from all the labs are to each other. The average value of initial water content is 10.8% with a standard deviation of 0.6%. To reduce the discrepancy of the results and propose a fast and easy to process testing method, we propose the adoption of the following protocol:

1. Weigh an initial mass of aggregates m_0 (g) ($m_0 > 50$ g).
2. Dry the material at a temperature between 50 °C and 60 °C until constant mass is reached (change in mass of the sample less than 0.1% over 24 hours).
3. Weigh the dry mass of aggregates m_D (g).

   Calculate the initial water content W_0(%):

$$W_0 = \frac{M_0 - M_D}{M_D} \times 100 \qquad \text{(eq.2)}$$

4. Repeat the test 3 times (with 3 different samples of shiv)

4 WATER ABSORPTION

The hygroscopic behavior of lignocellulosic plants is largely due to their hydrophilicity. Their complex architecture is marked by a multi-scale porosity in order to conduct the necessary fluids for their development (sap and water). Even after cutting and processing, this porosity continues to play its role and is therefore the main way of absorption of water following the Laplace laws. This absorption occurs mainly by conducting vessels or tracheids before the water spreads to the rest of the cells by diffusion through the cell walls and punctuation.

4.1 Description of the experimental methods

Table 7 Experimental protocol to measure the water absorption

Labs	Method
A	Three tests are presented by Lab A following this protocol: Approximately 50 g shiv placed in a metal wire sieve Second sieve placed on top of shiv Two sieves sealed together with waterproof tape Approximate thickness of shiv layer is 5-10 mm Assembly is fully submerged into water When assembly is removed, it is shaken thoroughly to remove all free water This is critical, as considerable amounts of free water can be held between the particles Tests 2 and 3 were shaken more thoroughly than test 1
B	The test is based on the following protocol: Fill with water and drain the container (Step 1 of Figure 16); Weigh 20 g of dry hemp shiv (60 °C, 48 h) in the humid container (Step 2 of Figure 16);

(continued)

Table 7 (continued)

Labs	Method
	Cover the container with a sieve and fill it with water through the sieve (Step 3 of Figure 16); After a determined immersion time, drain the container from its water through the sieve. User must wait for the last drop to fall! (Step 4 of Figure 8) Weigh the wet particles and calculate the water absorption ratio W(t) (Step 5 of Figure 16).
C	The test is based on the following protocol: – For each measurement, one sample (100 g) is soaked continuously till measurement and then dried until constant weight in an oven at 60 °C – For each sample, at each time interval, two measurements are conducted (except after one hour).
D	– Weigh approximatively 50 g of hemp shiv sieved under 80 μm and dried 48 hours in an oven at 80 °C. – Immerse the shivs in a bucket containing about 5 liters of water. – At time t, pour the content of the bucket into another bucket on which is disposed a 80 μm sieve. We get wet shiv on one side and water on the other. – Weighing the bucket of water before and after immersion of the shivs gives access to the amount of water absorbed during the time interval t.
E	– Drying of the particles in an oven for 48 hours at 40 °C, Three specimen of particles were tested for each drying temperature – g of particles are immersed in water at 20 °C – After each measurement, the particles are quickly spread on an absorbent tissue and weighed
F	– Samples of dry hemp shiv (around 18 g) are placed in a basket – They are immersed and weighed after 30 minutes, 4 days and 5 days of immersion – The water content is then calculated
G	The initial state of the plant aggregates was obtained by drying in an oven at 50 °C until the change in mass of the sample was less than 0.1% over 24 hours. Samples constituted of 5 g of particles were then immersed in water and the gain in mass was measured after 5, 10, 15, 30, 60 minutes and 24 hours. Before each weighing, plant aggregates were strongly drained and promptly superficially dried with absorbent paper, this method eliminated part of the water adsorbed at the surface of particles or located between particles. The results presented are mean values of measurements taken on 5 different samples.

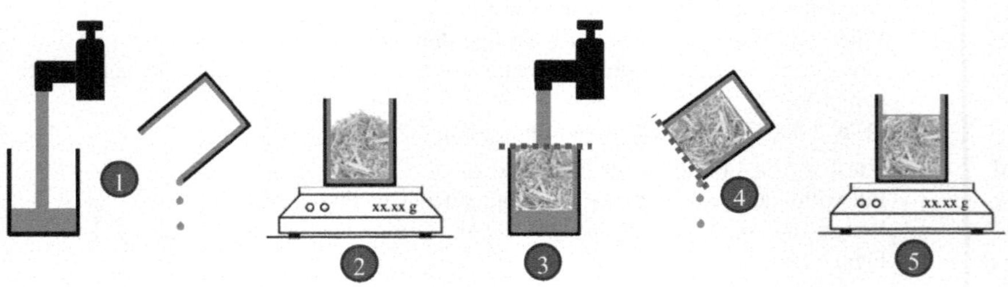

Figure 8 Protocol of Lab B

4.2 Results

The water absorption capacity of these aggregates was determined gravimetrically by applying the following expression:

$$W(t) = \frac{M(t) - M_0}{M_0} \times 100 \quad \text{(eq.3)}$$

where W(t) [%] is the water absorption ratio at time t [g], M(t) the soaked bio-aggregate mass at time t [g], and M_0 is the initial oven-dried aggregate mass [g].

W_{MAX} [%] was measured after soaking for 48 hours (see Table 7).

High porosity and internal structuring of shiv are responsible for a high water absorption and retention capacity (Figures 9 and 10).

Two distinct absorption phases can be observed for aggregate soaking (Figure 9).

A rapid mass increase is observed during the first minute. This is due to fast particle internal porosity filling by free water, mainly driven by xylem sap canal capillarity forces [2, 13]. Aggregates surface covering with water, which cannot be removed, is also responsible for a part of this rapid mass increase.

The second step is related to diffusion and water bonding through plant cell wall openings such as pits (diameter between 20-40 nanometers) [3]. As it is a diffusion

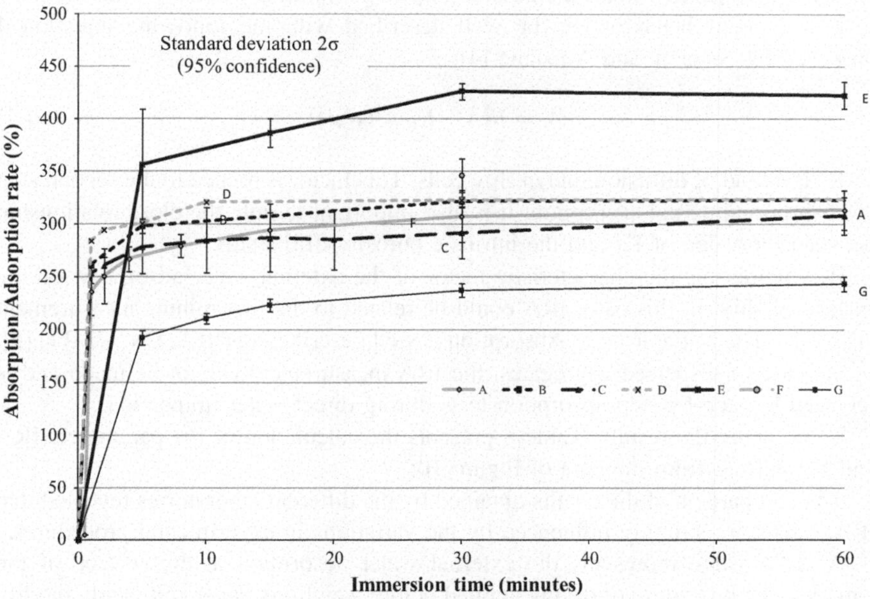

Figure 9 Absorption diagram

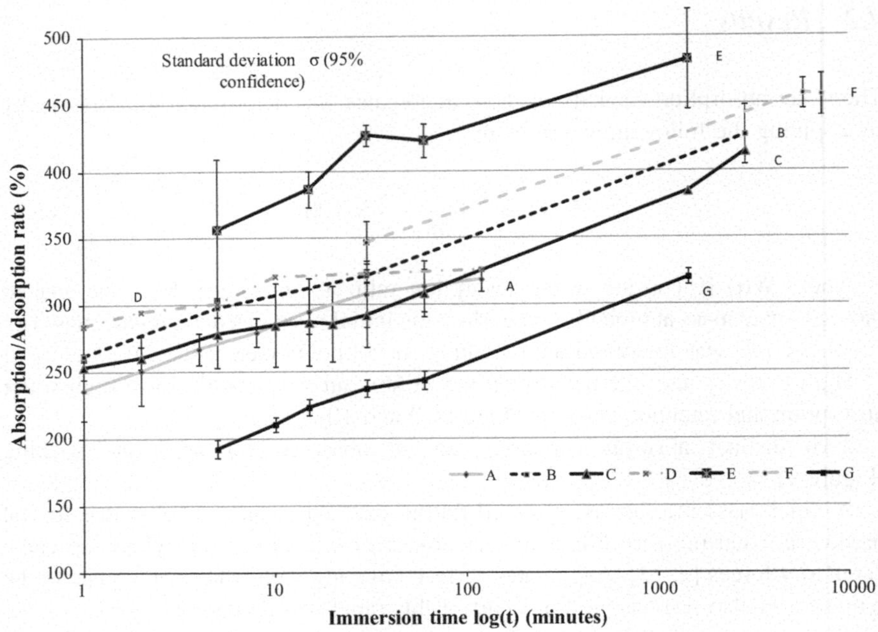

Figure 10 Logarithmic diagram of absorption

process, it depends on the aggregate particle size distribution. After 30 minutes, around 300% percent mass increase compared to initial dry mass is measured.

This type of behavior can be well described with the following equation 4 proposed by Nozahic and Amziane [4]:

$$W = IRA + K_1 \times Log(t) \qquad \text{(eq.4)}$$

K_1 is a kind of diffusion rate in shiv cells. This factor is related to the second step of absorption/internal adsorption. It is also important to note the close relationship between the value of K_1 and the intrinsic porosity of the shiv.

IRA represents the characteristic factor of the external water adsorption on the surface of shiv. In this case, IRA could be related to the first minute measurement (Table 8). The Initial Rate of Absorption is well described in (IRA, EN 772-11) [5]. In the case of dispersed aggregates, the IRA measurement cannot be made and is replaced by step-by-step absorption tests during direct water immersion.

To compare the results, Table 8 presents the calculation of the parameters IRA and K_1 starting from the data of Figure 10:

The comparison of the results obtained by the different laboratories revealed that IRA values are strongly influenced by the variations in experimental procedures.

As IRA value represents the external water adsorption at the surface of the particles, the procedure of drying applied before weighting can significantly modify this parameter. The quantity of water adsorbed at the surface of the particles or

Table 8 Comparison of IRA and K_1

Lab	$W = IRA + K_1 \times Log(t)$ eq. 5			
	IRA	K_1	R^2	W_{max} at 48 h
A	248.2	34.9	0.98	
B	260.7	47.2	0.993	426,8 ± 22,3
C	239.4	45.7	0.953	
D	288.8	21.1	0.862	
E	333.4	49.3	0.93	482,8 ± 36,7
F	277.1	47.2	0.998	
G	159.9	50.5	0.992	
Average	258.2 ± 106.7	42,3 ± 21,3		
Partial average*	262,8 ± 40,6 (A/B/C/D/F)	48,0 ± 3,8 (B/C/E/F/G)		

located between the particles is, for example, largely reduced when absorbent paper is used as showed by results obtained in lab G.

It can be seen in Figure 10 that the intra-particular absorption of water follows a logarithmic law as for diffusive phenomena within porous materials. The results of Round Robin Test revealed that in spite of the variations in experimental procedures, the determination of K_1 parameter, representative of this intra-particular diffusion, is quite repeatable.

4.3 Concluding remarks

With respect to the external adsorption (IRA parameter), the data obtained by the labs E and G differ widely from the average while other labs results are very close. This could be explained by differences in drying procedure before particle weighing. The key method of reducing the discrepancy of the results would be to standardise the drying stage. We recommend using only one protocol developed from an analysis of the initial tests with the addition of a repeatable drying stage, using common and widely available equipment:

1. Dry shiv at 60 °C until the change in mass of the sample is less than 0.1% over 24 hours.
2. Weigh 50 g of dry hemp shiv at 20 °C (M_0(g)) in a water permeable bag (plastic or metallic bag with a maximum hole size of 1 mm^2).
3. Immerse the bag for 1 min, 15 min, 4 h, and 48 h into the water.
4. After the specified time, drain the bag using a salad spinner for 1 min.
5. Weigh the bag M(t) (g) and calculate the water absorption ratio W(t) (%) at 5 min, 6 h, 24 h and 48 h (eq. 6):

$$W(t) = \frac{M(t) - M_0}{M_0} \times 100 \qquad \text{(eq.6)}$$

6. Repeat the test 3 times (with 3 different samples of shiv)

SUMMARY OF FINDINGS

To obtain the principal physical characterisation of plant aggregate as hemp shiv, the literature shows that several protocols are in use. Some of them are directly derived from the methodology to describe the properties of mineral aggregates or other materials. Hemp shiv exhibits some very specific properties in terms of geometry, water absorption, bulk density etc. which are very different from normal mineral aggregates. As a result, the objective of the inter-laboratory tests presented in this paper is to analyse the data obtained from several protocols in order to propose recommendations which are able to reduce the discrepancies resulting from the limitations and lack of adaptation of the protocols to the specificity of hemp shiv.

At this stage two characteristics are investigated. The discussion of the results allows the proposal of the following principal recommendations:

Dry the material at 60 °C until constant mass is reached (variation less than 0.1% between two readings over a 24 hour period). Sampling quality is a key point to consider before any determination od properties.

1. Initial Water Content: For each sample, a minimal mass of 50 g of hemp shiv is advised. The drying method in an oven between 50 °C and 60 °C is recommended for a fast equilibrium. Dry state will be considered to be reached when the change in mass of the sample is less than 0.1% over 24 hours.
2. Water Absorption: 50 g of hemp shiv are necessary. After immersion in a water bath, the drying procedure before particles weighing was identified as a critical stage. In order to facilitate and standardise this step, we recommend the use of a salad spinner to drain excess water from the bag over a one minute period.

References

[1] Construire en chanvre: Règles professionnelles d'exécution, Fédération Française du Bâtiment, Collection recherche, développement métier, 2009.

[2] Nozahic V, Amziane S, Torrent G, Saïdi K, De Baynast H. Design of green concrete made of plant-derived aggregates and a pumice-lime binder. Cement and Concrete Composites. 2012;34(2):231-41.

[3] Banks WB. Water uptake by scots pine sapwood, and its restriction by the use of water repellents. Wood Science and Technology. 1973;7(4):271-84.
[4] Nozahic, V., Amziane, S., Influence of sunflower aggregates surface treatments on physical properties and adhesion with a mineral binder, (2012) Composites Part A: Applied Science and Manufacturing, Volume 43, Issue 11, November 2012, Pages 1837-1849.
[5] Groot C, Larbi J. The influence of water flow (reversal) on bond strength development in young masonry. Heron. 1999;44(2):63-77.

ROUND ROBIN TEST FOR HEMP SHIV CHARACTERISATION

PART II: BULK DENSITY AND PARTICLE SIZE DISTRIBUTION

Authored by: Sofiane Amziane[(1)], Florence Collet[(2)], Mike Lawrence[(3)], Camille Magniont[(4)], Vincent Picandet[(5)]

1: Chair of RILEM TC-BBM 236, Institut Pascal, Clermont Université, France, sofiane.amziane@univ-bpclermont.fr
2: LGCGM, Université Rennes 1, Rennes, France, florence.collet@univ-rennes1.fr
3: BRE Centre for Innovative Construction Materials, University of Bath, Bath, UK m.lawrence@bath.ac.uk
4: Université de Toulouse, UPS, INSA, LMDC (Laboratoire Matériaux et Durabilité des Constructions), camille.magniont@insa-toulouse.fr
5: Université Bretagne Sud, IRDL - CNRS FRE 3744, F-56100 Lorient, France, vincent.picandet@univ-ubs.fr

With the contribution of:

6: Laurent Arnaud, Art et Métiers, ENSAM Cluny, laurent.arnaud@ensam.eu
7: Sylvie Prétot, LGCGM, Université de Rennes 1, France, sylvie.pretot@univ-rennes1.fr
8: Christophe Lanos, LGCGM, Université de Rennes 1, France, christophe.lanos@univ-rennes1.fr
9: Thibaut Collinart and Pierre Tronet, Université Bretagne Sud, IRDL - CNRS FRE 3744, F-56100 Lorient, France, thibaut.collinart@univ-ubs.fr, pierre.tronet@univ-ubs.fr
10: Vincent Nozahic, Institut Pascal, Clermont Université, France, vincent.nozahic@hotmail.fr
11: Etienne Gourlay, CETE de l'Est - Laboratoire Régional de Strasbourg, etienne.gourlay@developpement-durable.gouv.fr
12: Sandrine Marceau, Paris-Est, IFSTTAR, MAST, Marne-La-Vallée, France, sandrine.marceau@ifsttar.fr

S. Amziane and F. Collet (eds.), *Bio-aggregates Based Building Materials*,
RILEM State-of-the-Art Reports 23, DOI 10.1007/978-94-024-1031-0

13: Gilles Escadillas, Université de Toulouse, UPS, INSA, LMDC (Laboratoire Matériaux et Durabilité des Constructions), Gilles.Escadillas@u
14: S. Dubois, Université de Liège (Ulg), Gembloux Agro-Bio Tech, Belgique, s.dubois@student.ulg.ac.be
15: Laetitia Bessette, Centre Technique Louis VICAT, L'Isle d'Abeau, France, L'Isle d'Abeau, l.bessette@vicat.fr
16: Paulien Debruijn, Lund University, Faculty of Engineering (LTH), Dept of Building Materials, Lund, Sweden, pbdebruijn@gmail.com
17: Ulrike Peter, Lhoist, Lhoist Recherche et Développement S.A., Belgium, ulrike.peter@lhoist.com
18: Sara Pavia, Dept of Civil Engineering, Trinity College Dublin, Ireland, PAVIAS@tcd.ie

Abstract

The paper presents the experience of a working group within the RILEM Technical Committee 236-BBM 'Bio-aggregate based building Materials'. The work of the Technical Committee (TC) will be to study construction materials made from plant particles. These materials are obtained from the processing of hemp, flax, miscanthus, pine, maize, sunflower, bamboo and others. The second part of the first round robin test of the TC-BBM was carried out to compare the protocols in use by the different laboratories (labs) to bulk density and particle grading. The aim is to define a characterisation protocol derived from those used by the different labs. This first round robin test was carried out on one variety of hemp shiv. Nine laboratories from European universities and research centers were involved (Table 1). The test results of 7 laboratories constitute a set of statistically representative data in order to propose recommendations to characterise hemp shiv after analysing the different methodologies in use in these labs. This paper presents the bulk density and particle size distribution (PSD) measurements.

Table 1 Participating Labs

Letter	*City*	*Labs*
A	***Bath (UK)***	***BRE Centre for Innovative Construction Materials/University of Bath***
B	***Clermont Ferrand (France)***	***Institut Pascal***
C	***Lorient (France)***	***LIMatB/Université de Bretagne Sud***
D	***Lyon (France)***	***DGCB/ENTPE***
E	***Paris (France)***	***IFSTTAR***
F	***Rennes (France)***	***LGCGM/Rennes 1***
G	***Toulouse (France)***	***LMDC/Université de Toulouse/UPS/INSA***
I	***Combloux (Belgium)***	***Combloux-Agro ressource – Université de Liège***

5 INTRODUCTION

This study focus on bio-based aggregate coming from the stem of plants cultivated either for their fibers (hemp, flax, etc.) or for their seeds (oleaginous flax, sunflower, etc.). Owing to the structure of the stem of the plant they are made from, such aggregates are generally malleable, elongated and highly porous with a low apparent density. They are very different from the mineral aggregates typically used in concretes, for which there are standardised tools and techniques for characterisation. Amongst these, hemp shiv (the woody core of the stem of the hemp plant) is probably the most widely used in alternative or eco-friendly building materials in Europe and is also representative of most of the aggregate coming from the stem of an annual crop. This is usually mixed with a lime-based binder and the resultant 'bio-concrete' is known as 'hemp-lime'.

This kind of aggregate is a co-product of hemp industry that is renewable and produced in an annual cycle while the price of mineral aggregates is steadily increasing as resources become less readily available. The characterisation of these aggregates, however, which is crucial to a proper understanding of the quality of the materials in which they are incorporated, requires adaptations to be made to the techniques usually employed for mineral aggregates, or the devising of new characterisation procedures.

The first round robin test of the Rilem TC-BBM was carried out to compare the protocols in use by the different labs. The aim was to define a test method to measure bulk density, particle size distribution, water absorption and thermal conductivity of bio-aggregates. This first round robin test was carried out with one variety of hemp shiv (the woody core of the plant stalk chopped into lengths of a few centimeters) coming from the same production of a processing factory located in France. The physical properties and more description of this hemp shiv is given in the Part I of the RTT [1]. Seven labs conducted the measurements (Table 2). This part presents the bulk density and particle size distribution (PSD) measurements and analysis.

Table 2 ***Description of the interlaboratory test***

Characteristics	Participating Labs
Bulk Density	B, C, D, E, F
Particle Size Distribution	A, B, C, D, E, G, I

6 MATERIAL

One variety of hemp shiv was selected for this inter-laboratory test. It comes from the same processing factory where the bast fibers are stripped off (de-cortication), leaving the shiv behind. This shiv was provided by LCDA producer under the commercial name "KANABAT" (Tables 3 and 4). This shiv is in line with the French national recommendation provided by "Construire en Chanvre" association (Table 5) [2].

Table 3 Physical properties of shiv as supplied by the producer

Characteristic	Measurement
Density	100 to 110 kg/m^3 (depending on ambient relative humidity) Loosely packed, not compressed
Water absorption	198% (NFV 19 002)
Water absorption of mineral elements	24 meq per 100 g of raw material
Calorific value	3804 cal/g (NF M 07-030 12/9G)
Thermal Conductivity (10°C in a dry state)	0.0486 W/m.K (NF EN12667)

Table 4 ***Chemical composition of the shiv*** as supplied by the producer*

Water	9 to 13%
Dry material	85 to 90% of which
Total organic material	97.5% on a dry basis of which:
Net cellulose:	52%
Lignin:	18%
Hemicellulose:	9%
Minerals:	Calcium: 1% on a dry basis Magnesium: 0.03% on a dry basis Phosphorus: 9 mg/100g Potassium: 0.8% on a dry basis Total Nitrogen: 0.4 to 1% on a dry basis Total carbon: 496 g/kg on a dry basis C/N: 87 Ash: 2% pH in suspension at 10%: 6.7

*These data are supplied as indicative, certain values can vary depending on the year, the variety, the cultivation location etc.

Table 5 French national recommendations for the shiv characteristics

Characteristics	Recommendation
Apparent density of particles in kg/m^3	Quoted apparent density +/−15%
Maximal length (L_{dmax} in %)	Quoted L_{dmax} +/−10%
Initial moisture content	<19%
Dust content	<2% passing a 0.25 mm sieve
Colour	% of particles not conforming <5%

7 BULK DENSITY

The bulk density of hemp shiv is linked to the porosity of the particles and to the inter-particular porosity. In this study, bulk density of hemp shiv is measured in a cylindrical mould with a loose packing and without compaction. Four labs measured it with one mould and at one state (dry state or ambient relative humidity). The other lab studied the effect of the size of the mould on bulk density measurement at 23 °C 50%RH and then the effect of relative humidity on the bulk density of hemp shiv (see Table 6).

7.1 Methods

Table 6 Description of the methodology to measure the bulk density

Labs	
B	Bulk density is measured on hemp shiv at 23 °C and ambient relative humidity. The water content of hemp shiv is taken to be 12%. Measurements are made in a 10-litre container and are repeated nine times.
C	Bulk density is measured in a dry state. Hemp shiv are dried in oven at 60 °C for 72 hours. The measurements are made in a two litre cylinder and are repeated 10 times.
D	Bulk density is measured in a dry state. Hemp shiv are dried in oven at 80 °C for 72 hours. The measurements are made in a 10 litre cylinder and are repeated five times.
E	Bulk density is measured in a dry state. Hemp shiv are dried in oven at 40 °C for 48 hours. The measurements are made in a five litre cylinder and are repeated five times.
F	Bulk density is measured after stabilisation at 23 °C, 50%RH using 8 sizes of mould (Table 7) see Figure 1. Hemp shiv are unpacked and placed in a large pan to eliminate initial compaction. Moulds are filled with hemp shiv that are poured so as to cover the entire surface of the mould. The upper level is levelled. The effect of relative humidity on bulk density was then studied using hemp shiv stabilised in a dry state, at 23 °C and 33%RH, and at 23 °C and 80%RH. Each measurement was repeated five times.

Figure 1 Moulds used for the measurement of bulk density - Lab F

Table 7 *Size of the* moulds *used for the measurement of bulk density of hemp shiv*

Mould	1	2	3	4	5*	6	7	8
D (cm)	15.0	12.1	14.0	11.9	7.0	5.6	5.6	17.4
H (cm)	31.7	19.5	9.4	8.5	8.5	9.7	20.0	11.6

**Gentle slanting of lateral side*

7.2 Results

Figure 2 shows a synthesis of the results obtained by all the participating labs. For the lab F, only the values obtained at dry point and at 50%RH (at 23 °C) are given here.

The average value from all these data is 116 kg/m^3 with a standard deviation of 12.2 kg/m^3. One lab (E) gives a value much higher than the others labs. This may be due to the fact that hemp shiv was not sufficiently loosened to remove the initial packing compaction.

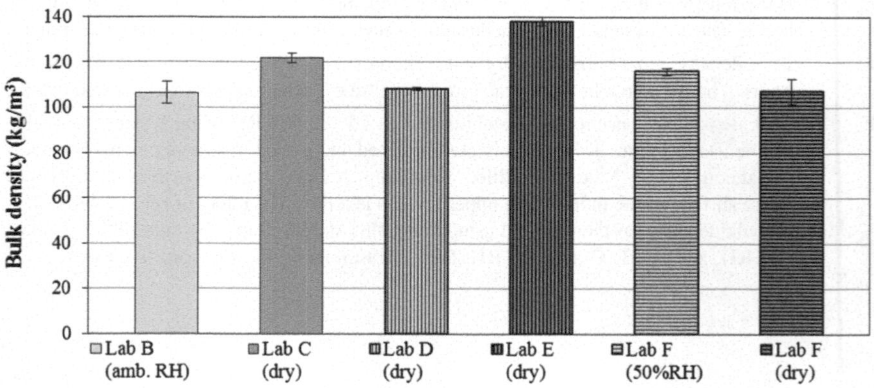

Figure 2 Synthesis of bulk density from all participant labs

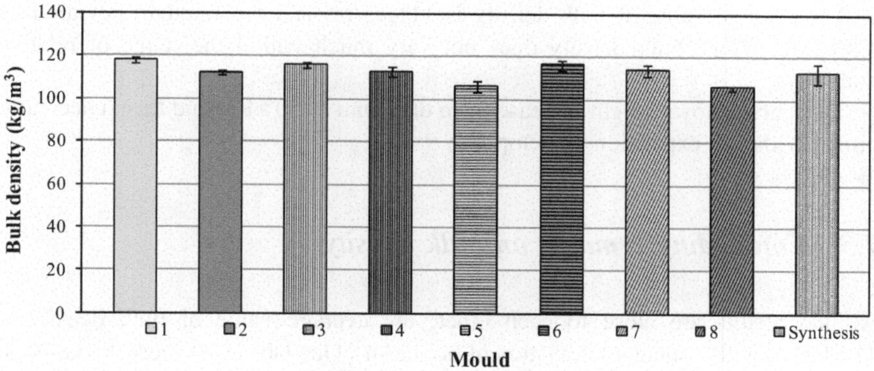

Figure 3 Variation of bulk density against mould (average value and standard deviation)

Ignoring lab E, the average value of bulk density is 112 kg/m^3 with a standard deviation of 6.7 kg/m^3. It can be seen that the results from the different participating labs are closely aligned.

Figure 3 shows the bulk densities measured against the mould type used. The average value of bulk density obtained with all the moulds is 112.7 kg/m^3 and the standard deviation is 4.6 kg/m^3.

Two moulds show a lower value of bulk density. For mould number 5, this is unlikely to be due to its size as mould number 6, with nearly the same size, gives similar results as the other moulds. This deviation may be due to a gentle slanting of its lateral side. For mould number 8, this may be due its large diameter. In this case, the top is more difficult to level than the tops of the other moulds.

The average value of bulk density obtained (excluding mould 5 and mould 8) is 114.9 kg/m^3 and the standard deviation is 2.2 kg/m^3. Thus, all moulds in the range of 5 to 15 centimeters in diameter and 8 to 32 centimeters in height give representative values of bulk density. For a larger value in diameter, the top of the sample is difficult to level.

Figure 4 shows the variation of bulk density of hemp shiv versus ambient relative humidity.

Figure 4 Variation of bulk density versus ambient relative humidity - Lab F (mould 3)

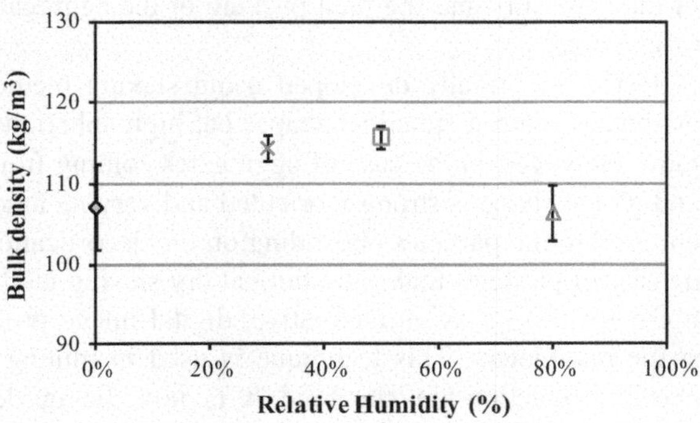

The average value of bulk density is 111 kg/m^3 and the standard deviation is 4.8 kg/m^3. Thus, bulk density does not vary much within the range of relative humidity.

This curve shows a light increase from dry point to 50%RH and then a decrease, probably due to expansion of hemp shiv.

7.3 Concluding remarks on bulk density

All the results are close to each other, the average value of bulk density is 112 kg/m^3 with a standard deviation of 6.7 kg/m^3. One lab shows more deviation, it is presumed to be due to insufficient loosening of the packed material. To reduce the discrepancy of the results, we recommend the adoption of the following protocol:

1. Dry the material at 60 °C until constant mass is reached (variation less than 0.1% between two readings at 24 hours).
2. Put hemp shiv in a sealed bag or a sealed bucket until equilibrated with room temperature.
3. Put the dried material in a glass cylinder 10 cm to 20 cm in diameter and at least twice the diameter in height. The quantity of the material has to be adjusted to be half the volume of the container.
4. Upend the glass cylinder ten times
5. Shake to obtain a horizontal surface
6. Use a cardboard disc and mark the level
7. Measure the volume with water
8. Calculate the bulk density
9. Repeat the test 3 times (with 3 different samples of shiv)

8 PARTICLE SIZE DISTRIBUTION

Particle Size Distribution (PSD), is commonly used to characterise mineral aggregates. It is a key factor in many casting process since it controls the flowability of the fresh mix and the final packing of the aggregates inside composite materials or concrete.

PSD was initially developed using sieving methods, when dealing with rigid aggregates with a spherical shape, i.e. high sphericity ratio and/or low elongation ratio. However, in the case of aggregates coming from plant stems, pore structures and global shape is strongly oriented and varying amounts of cortical fibers remain attached to the particles depending on the decortication process. Moreover, the low density of particles makes traditional dry sieving method using normalised aperture not relevant [3]. As an alternative, digital image processing can be used to assess shape parameters. This technique is used in mining industries to instantaneously check production on site, and it is now being developed in agricultural and biological engineering to control efficiency of plant milling operations, for instance in biomass-to-fuel process [4] [5], or to control pelletising of bulky biomass residues [6] and the associated dust management [7].

Starting from the traditional standardised sieving method, and using digital image processing method in addition, this study briefly describes the hemp shiv tested, and presents the applied methods of sieving and image analysis. The results have led to some PSD measurement recommendations and to focus on some relevant parameters.

8.1 Sieving method

Mechanical sieving is processed with normative square opening sieves. Metallic mesh, woven wire sieves, conforming to ISO 3310-1 in most of laboratories, or perforated plate square hole sieves conforming to ISO 3310-2 if the opening size is greater than 4 mm in some other laboratories.

The mechanical sieve shakers used induce horizontal circular motion overlying a vertical motion which is created by a tapping impulse. Due to the low density and elongated shape of the particles, the vibration duration should be longer than for mineral aggregates to obtain repeatable results. Drying of samples before testing may prevent the finest particles from sticking to the coarser ones.

The protocols used by each laboratory are summarised in Table 8.

Results are plotted in Figure 5. Since 50% approximately of particles are retained on the 2 mm sieves, the mass fraction passing this sieve is relevant. Laboratories do not use the same stack of sieves but 6 out of 7 use the sieve with 2 mm apertures. Extrapolated values have been deduced using a linear regression in logarithmic scale of size where intermediate values are missing and results are reported in Table 9.

Table 8 ***Experimental protocol used by different Laboratories***

Laboratory	A	B	C	D	E	G	I
Sample mass (g)	100	150	200	50	50	200	50
Sieve diameter (mm)	203	315	315	315	315	315	203
Vibration time (mins) ^{M}: Manual sieving	10	15	40	5^{M}	20	3 10 30*	10
Repetitions	5	3	3	10	5	3 3 3*	5
Drying temperature ^{Const}: until constant weight	105 °C^{Const}	None	60 °C^{Const}	None	40 °C 2 days	None	50 °C^{Const}

**Reported result*

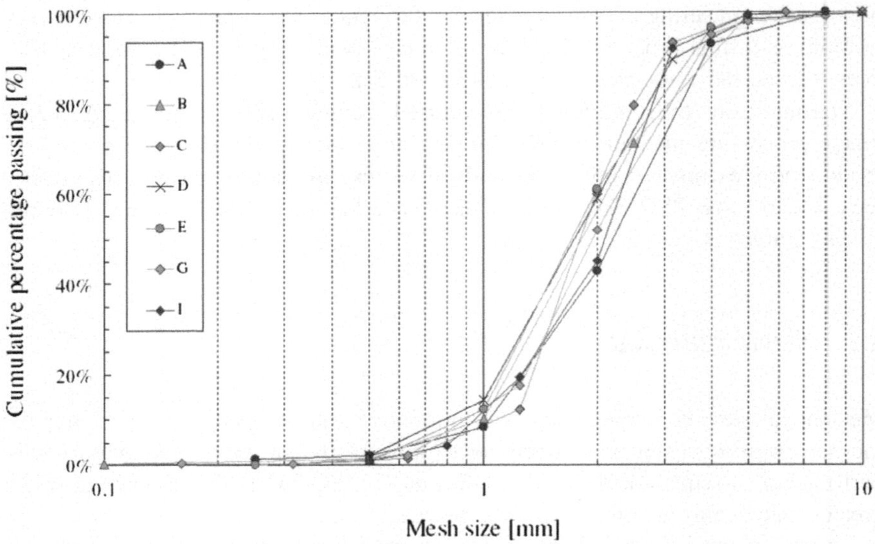

Figure 5 Cumulative mass distribution

Table 9 ***Mass fraction (%) passing through sieves with 0.5, 1, 2, 4 and 5 mm nominal opening size***

Nominal opening size	A	B	C	D	E	G	I	Mean value	Standard deviation
0.5 mm	2.0	1.2	1.1*	2.1	0.9	1.5*	0.8	**1.4**	0.5
1 mm	8.3	10.3	8.8*	14.4	12.3	12.5*	11.9*	**11.2**	2.2
2 mm	42.8	35.5*	60.3	58.6	60.6	51.7	45.0	**50.6**	9.8
4 mm	93.0	51.4*	96.2	94.1*	96.5	95.2	95.8*	**88.9**	16.6
5 mm	95.3*	99.3	99.4	98.3	99.5	98.0	99.3	**98.4**	1.5

**Extrapolated value*

Some laboratories show that the operating time (on the mechanical shaker) tends to increase the percentage of material passing, especially on the first sieves. Percentages of material passing tend also to decrease when the mass of tested sample increases. Below 100 g with 315 mm diameter sieves and after 30 minutes shaking, no significant trends have been observed.

An extended shaking time may modify the size of granular due to the erosion of shiv on the metallic mesh and/or due to the separation of the remaining fibres attached to the shiv.

8.1.1 Fibre content

Inspection of the fractions shows that most of fibres are attached to the shiv particles, especially the coarser ones. Sometimes, material retained in the sieves included a significant quantity of fibre mixed with shiv which formed a loose roll or fiber balls. The unattached fibres can be then be manually separated with greater efficiency. It should be noted that various lengths of fibre can be removed. Mass fractions of fibre are reported in Table 10.

Table 10 *Mass content of fibre*

Lab B	Lab C	Lab G
3.25%	4.5%	5.9%
±0.3%	±0.5%	±0.52%

8.1.2 Dust content

Dust is defined as particles with length or width lower than a given size. This size has not been clearly set. If we consider particles passing the 0.5 mm nominal aperture sieve to be dust, the mass content of dust is reported in Table 9. However, dry sieving of finest particles can lead to scattered results. Moreover, finest fibres are able to pass the 0.5 mm sieve while dust should mainly consist in round particles, or mineral particles, that could remain stuck to the shiv particles.

8.2 Image processing

8.2.1 Picture acquisition

Particles are spread in a manner that they do not touch or overlap one another, in a single layer arrangement [7].

Particles have a light color. In order to obtain the maximum contrast, a black background is usually selected. Pictures are taken by the means of camera or flatbed scanner. In both cases, a picture calibration is required. Scanning method allows pre-selected resolution (DPI) leading to homogeneous and accurate calibration. Measurement and accuracy improves with higher resolution, but restrictions come from handling the file size of recorded images. 8-bit gray scale images are usually needed to be processed with image analysis software (i.e: pixel intensity levels ranging from 0 to 255). Other format files of acquired images have to be converted.

8.2.2 Image analysis

Among the image analysis software used, two are freely available: ImageJ (V. 1.45) developed at the National Institute of Health, USA, and ImageTool (V. 3.0) developed at the University of Texas Health Science Center in San Antonio, USA. The Table 11 presents the protocols of Laboratories.

Image processing algorithms require a binary image that can be produced by converting the image into binary image by proper thresholding. Since hemp shiv could include dust, i.e, particles lower than 0.5 mm, and attached fibres to shiv particles, thresholding operations may be critical.

Examination of ground biomass material reveals that they are of irregular shape with rough edges due to the shredding action of the de-fibering process. As observed in many studies with various materials [3], the shape of particles tends to be convex polygonal but some particles tend to deviate and are non convex polygonal.

Morphological operations can be applied such as "opening" routines in order to remove the finest fibres attached to particles and avoid bias in particle measurements, but results do not show significant effect on the global size distributions.

Table 11 ***Image processing and analysis protocols***

Lab.	Mass	Min. area	Analysed Particles	Image processing/Scale Picture physical size	Software used Analysis method
B	1 g among particles retained on 0.5 mm sieve	1 mm^2	830	Camera Pic.: 420 × 297 mm	ImageTool: Féret*
C	4 g	0.08 mm^2	2600	Scan 600 DPI 23.6 pixels/mm Pic.: 9 (210 × 297 mm^2)	ImageJ: Ellipse + Féret ImageTool: Féret*
D	5 g	0.44 mm^2	3100	Camera Pic.: 400 × 300 mm^2	ImageJ: Ellipse
E	*	0.5 mm^2	3200	Camera Pic.: 5 (400 × 300 mm^2)	ImageJ: Ellipse + Féret
I	8 g	1 mm^2	5350	Scan 300 DPI 11.8 pixels/mm	ImagePro: Féret
G	6 g	0.03 mm^2	3700	Scan 600 DPI 23.6 pixels/mm	ImageTool: Féret*

8.2.3 Measurements

For each detected particle the following parameters are stored in data files: projected area and projected perimeter. The mean equivalent diameter (i.e. 4 × Area/Perimeter) leads to an average size assuming that particles have spherical shape. However, length and width of particles can be assessed using image analysis.

The direct measurement of particle length is usually considered as the major axis length (i.e. the length of the longest line that can be drawn between any two points along the object boundary or the diameter of the circumscribed circle) also known as maximum Feret's diameter.

Many methods exist to evaluate the particle width: it can be considered as the length of the longest line that can be drawn through the object perpendicular to the major axis (e.g: ImageTool method) or it can also be considered as the minimum calliper, also called minimum Feret's diameter (e.g. ImageJ method).

Other methods consisting of fitting basic shapes onto identified particles exist [8] and could be used to estimate length and/or width. Ellipse is a common shape suitable to fit the shiv particles and available in many software. The ellipses have same area and centroid (i.e. center of mass) as the original objects. In such case length and width can also be done along the major and minor axis from the fitted ellipse.

8.2.4 Shape descriptors

Among various shape descriptors, the aspect ratio (i.e. length to width ratio) is relevant since it is an indicator of the anisotropic properties induced in case of particles arrangement with a preferential orientation.

8.2.5 Characterisation of size distributions

Many PSD could be done from image analysis data depending on the considered quantification of the material: number, projected area or volume of particles. It should be noted that in some studies distribution based on particle number (i.e. frequency) is only considered. But this distribution is very sensitive to numerous finest particles and is not representative in case of hemp shiv. The minimum area of detected particles can not be accurately set.

A grading curve from area-based results can also be done directly from recorded data. The weighted arithmetic mean of size x or expected value, E_{am}, and its associated standard deviation Sd_{am} are basic parameters to compute. If the dimension x_i and area A_i of each particle i are known, E_{am} and Sd_{am} can be written as follow:

$$E_{am}(x) = \frac{\sum A_i(x_i)}{\sum A_i} \qquad \text{(eq.2)}$$

$$Sd_{am} = \sqrt{\frac{\sum A_i(x_i - E_{am}(x))^2}{\sum A_i}} \qquad \text{(eq.3)}$$

When incorporating logarithm of size, the weighted geometric mean X_{gm} and its associated standard deviation Sd_{gm}, see eq.(3) and eq.(4), are useful parameters to compute since many standards use this parameter [9][10]. It is always lower than arithmetic mean, and it is less sensitive to coarsest particles. It could be considered more representative in case of PSD of ground biomass [5][11][12].

$$X_{gm} = \exp\left(\frac{\sum A_i \ln(x_i)}{\sum A_i}\right) \qquad \text{(eq.4)}$$

$$Sd_{gm} = \exp\left(\sqrt{\frac{\sum A_i(\ln(x_i) - \ln(X_{gm}))^2}{\sum A_i}}\right) \qquad \text{(eq.5)}$$

In the case of individual particle measurement, the considered size x is whether the width or length of particles, or the aspect ratio (i.e. length to width ratio) in Table 11.

In case of sieving, the dimension of the normative aperture d_i and the mass M_i of the particles retained on sieve can be considered instead of x_i and A_i in eq. 1, 2, 3 and 4.

In Tables 12, 13 and 14, representative cumulative dimensions such as D_k, for k = 95, 75, 50, 25, and 5 corresponds to particle dimension in mm at respective k% cumulative undersize area.

Table 12 ***Effect of Length and Width estimation method on distribution parameters***

	Width			Length		
	C, Ellipse	C, Féret*	C, Féret	C, Ellipse	C, Féret*	C, Féret
D_5	0.90	0.94	1.03	3.01	3.21	3.17
D_{25}	1.47	1.54	1.65	5.48	5.66	5.66
D_{50}	2.05	2.17	2.29	7.80	7.96	7.93
D_{75}	2.87	3.03	3.21	11.12	11.69	11.61
D_{95}	4.32	4.89	5.28	17.57	18.38	18.50
X_{gm}	2.00	2.16	2.31	7.62	7.94	7.91
Sd_{gm}	1.64	1.70	1.67	1.72	1.71	1.71
ε				4.18	4.05	3.69
$Sd(\varepsilon)$				1.89	1.92	1.47

Table 13 ***Results of particle length distributions***

	B ellipse	C ellipse	D ellipse	E ellipse	G Féret*	I Féret	Average length (mm)	Standard deviation (mm)	Relative standard deviation
D_5	3.00	3.01	3.23	2.68	3.59	3.13	3.10	0.30	9.7%
D_{25}	4.97	5.48	5.73	4.77	6.02	5.30	5.38	0.47	8.7%
D_{50}	7.54	7.80	8.08	6.92	8.28	7.46	7.68	0.49	6.3%
D_{75}	11.77	11.12	11.19	9.92	11.37	10.48	10.97	0.66	6.0%
D_{95}	19.20	17.57	17.30	15.75	18.01	16.30	17.35	1.23	7.1%
n	827	2600	3088	3204	3709	5299			
R^2	0.9989	0.9993	0.9985	0.9991	0.9995	0.9996			
μ	2.02	2.05	2.09	1.93	2.11	2.01			
σ	0.60	0.54	0.51	0.55	0.49	0.51			
$e^{\mu}{}_{\cong D50}$	7.55	7.75	8.06	6.88	8.28	7.45	7.66	0.49	6.4%
X_{gm}	7.54	7.62	7.86	6.77	8.17	7.37	7.55	0.47	6.3%
e^{σ}	1.83	1.71	1.67	1.74	1.63	1.67	1.71	0.07	4.0%
Sd_{gm}	1.78	1.72	1.68	1.73	1.63	1.64	1.70	0.06	3.3%

Table 14 ***Results of particle width distributions***

	B ellipse	C ellipse	D ellipse	E ellipse	G Féret*	I Féret	Average length (mm)	Standard deviation (mm)	Relative standard deviation
D_5	0.92	0.90	0.88	0.95	1.20	0.97	0.97	0.12	12.1%
D_{25}	1.54	1.47	1.52	1.55	1.80	1.54	1.57	0.12	7.5%
D_{50}	2.19	2.05	2.05	2.06	2.33	2.10	2.13	0.11	5.1%
D_{75}	3.17	2.87	2.69	2.71	3.01	2.79	2.87	0.19	6.5%
D_{95}	5.91	4.32	3.92	4.14	4.21	4.32	4.47	0.72	16.1%
n	827	2600	3088	3204	3709	5299			
R^2	0.9985	0.9996	0.9978	0.9990	0.9997	0.9997			
μ	0.80	0.71	0.71	0.72	0.85	0.74			
σ	0.54	0.49	0.44	0.44	0.39	0.45			
$e^{\mu}{}_{\cong D50}$	2.22	2.04	2.04	2.06	2.33	2.09	2.13	0.13	6.1%
X_{gm}	2.24	2.00	1.98	2.04	2.31	2.08	2.11	0.13	6.4%
e^{σ}	1.71	1.64	1.55	1.55	1.47	1.57	1.58	0.08	5.1%
Sd_{gm}	1.78	1.64	1.57	1.55	1.48	1.58	1.60	0.10	6.5%
ε	3.73	4.18	4.39	3.64	3.79	3.68			
$Sd(\varepsilon)$	2.00	1.89	2.14	1.68	1.49	1.61			

Particle size distribution can be represented closely by mathematical expressions. Basic models, based on a continuous and semi infinite variable, use equations with two parameters, one relative to the mean size and the other one indicating the width of distribution. The log-normal distribution can be a powerful approximation to characterise size of particles generated in dry or wet milling process [13].

A random variable X is said to be log-normally distributed if log(X) is normally distributed. The log-normal cumulative distribution function can be then written as follow:

$$P_{Log.N}(X \leq x) = \frac{1}{2}\left[1 + \text{erf}\left(\frac{\ln(x) - \mu}{\sigma\sqrt{2}}\right)\right] \quad \text{(eq.6)}$$

Where X is the particle size, and $P_{Log.N}(X \leq x)$ is the percentage based on the cumulative projected area of particles smaller than x. The median size $D_{50(Log.N)}$ corresponds to e^{μ}. Due to the symmetrical curve produced in log-scale, the geometric mean size of the distribution is equivalent to the median diameter for log-normal graphs, i.e. equal to e^{μ}, and the associated geometric standard deviation is equal to e^{σ}. The values of the weighted geometric mean size X_{gm} of the experimental data and their associated standard deviation Sd_{gm} values could be then respectively compared to e^{μ} and e^{σ} in order to consider the relevance of the log-normal fit.

8.2.6 Comparison of image analysis methods

In this section, the same set of binary images of hemp shiv, including 2600 particles with a projected area greater than 0.08 mm^2, has been used in order to evaluate the influence of analysis parameters such as area threshold of particles, width and length determination methods.

8.2.6.1 Effect of particle area threshold

The minimum area of considered particles ranges from 0.03 to 1 mm^2. This detection threshold can be set at various values according to the image resolution since smaller objects need to contain a minimum number of pixels to be representatively analysed. More than 40 pixels per particle should be recommended to obtain representative objects in binary images. Below this limit, the sizes and area of finest particles is more sensitive to thresholding operations due to the "halo effect" around the object boundary occurring in color or gray level images [7]. Shape indicators are also not relevant in such cases.

Results show that variation of the area threshold, between 0.08 and 0.9 mm^2 (i.e: between 45 and 500 pixels) induces only small change in the cumulative distributions, especially for the lower sizes (width and length). Beyond 0.9 mm^2, the fraction of ignored particles becomes not negligible, and the cumulative distributions deviate significantly.

8.2.6.2 Effect of Length and Width estimation method

The image analysis method used in this study has an appreciable influence on width distribution of particles while it has no significant influence on length distribution (see Figure 6).

Methods consisting of fitting an ellipse on identified particles lead to lower width than methods using direct measurement such as minimum Féret's diameter or minor axis length named as Féret* in Figure 6 and Figure 7. The data computed using ImageTool globally produce size distributions bounded by the others. The distribution parameters of width and length are presented in Table 12.

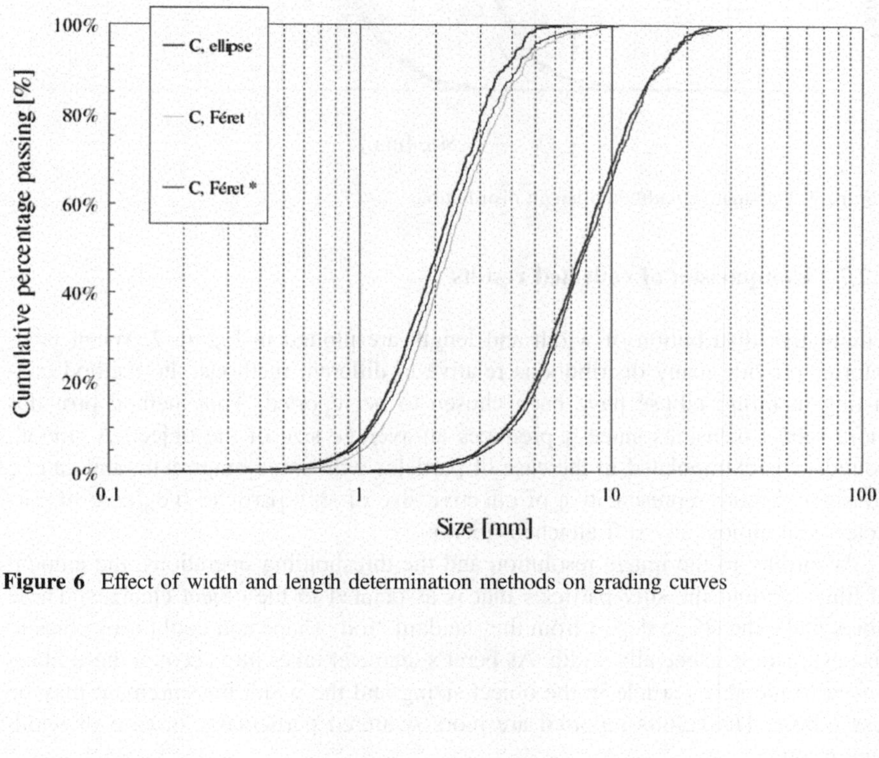

Figure 6 Effect of width and length determination methods on grading curves

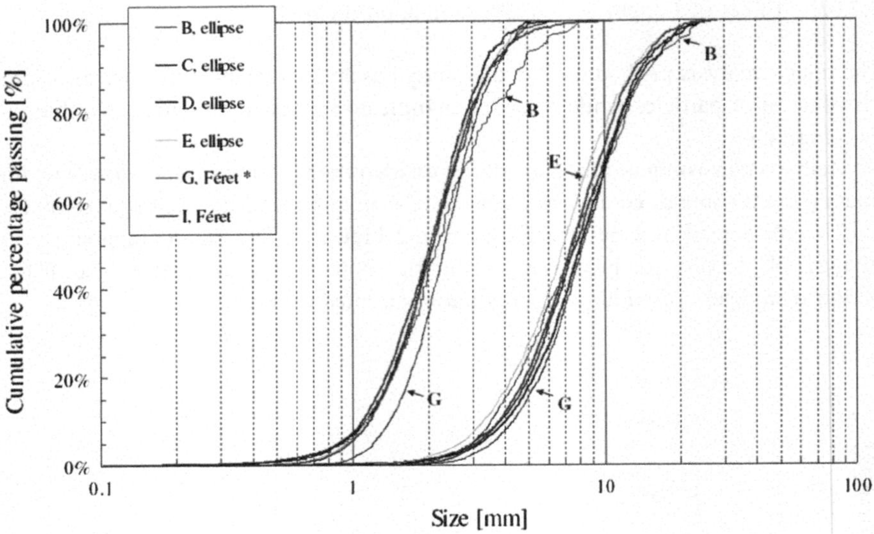

Figure 7 Cumulative width and length distributions

8.2.7 Comparison of collected results

Cumulative distribution of width and length are plotted in Figure 7. When laboratories provide many distributions relative to different methods, the method consisting in fitting ellipse have been chosen to be reported. This method provides more robust measures since it produces an average size of the object. A smooth boundary is extrapolated in the case of particles with some outgrowth, and can be considered more representative of effective size of shiv particle, (i.e., core of particles with almost any soft attached fibers).

According to the image resolution and the thresholding operations, the amount of fibres around the shiv particles that is associated to the object changes. These fibres make the shape depart from the standard "rod" shape and could cause bias in size estimation, especially width. As Féret's diameter takes into account these fibres linked to the shiv particle in the object sizing and the width measurement may be less robust. The results reported are more scattered particularly in case of width repartition.

Generally, the cumulative distributions exhibit less scattering results around the median values, D_{50}, than for the tails of distributions (D_5 and D_{95}). Investigation of finest particles needs specific image processing. Discrepancy observed with coarser particles can be due to the sampling quality and the analysis method used.

Log.normal functions (see eq. 5) provide a good fit for most of cumulative size distribution. The coefficients of determination, R^2, are reported in Tables 13 and 14. The parameters μ and σ deduced from fitting are in accordance with the weighted geometric mean and the associated standard deviation.

The elongation (or aspect ratio) ε, reported in Table 14 is deduced from the weighted arithmetic mean of the length to width ratio of each analysed particle. The associated arithmetic standard deviation Sd(ε) is also reported.

Since the width distribution depends on analysis method, the aspect ratio is greater when major and minor axis lengths are considered as length and width. In that case the elongation ratio is also called elliptical ratio.

It should be noted that, even if some coarser particles tend to be more elongated, in case of the studied hemp shiv, the mean elongation is almost constant whatever the class of area considered, and the analysis method used. Globally, the mean elongation ratio is around 4 with significant standard deviation about 2.

Less than one thousand particles are analysed in B. The reduced sample size is probably less representative and may explain the difference. In case of B, the sampling has also been made from particles retained on the 0.5 mm sieve while the other samples come from the whole material, without previous sieving.

In G, all detected particles have been considered in image analysis. The area threshold is very low 0.03 mm^2, and an overestimation of finest particles may occur.

In E, a calibration bias may explain the difference observed with particle length. But width distribution is close to the others.

8.3 Comparison between sieving and image analysis results

This study focuses on comparisons of the data from several laboratories. In order to compare measurements with usual sieving methods, results from image analysis have to be based on mass fractions. This issue needs more measurements or hypothesis on density and shape of particles in different size classes that have not been considered here.

In this study, the sieving data, using a mass based distribution, relate closely to the width based distribution calculated using an area based distribution (see Figure 8). In theory, these two distribution types should not be directly comparable. However, this assumes that the applied sieving process separates particles mainly based on their width. Indeed, this is what is found when examining the material retained on the sieves, where cross sectional dimensions of particles are very homogeneous and close to the mesh size.

Comparison of Tables 9 and 14 shows that image analysis provides results less spread than sieving does with retained fraction on sieves with 2 mm nominal opening size where approximately 50% of the mass fraction is supposed to be retained. Nevertheless sieving methods seems to provide better results than image analysis does in the tails of distributions, especially with finest particles.

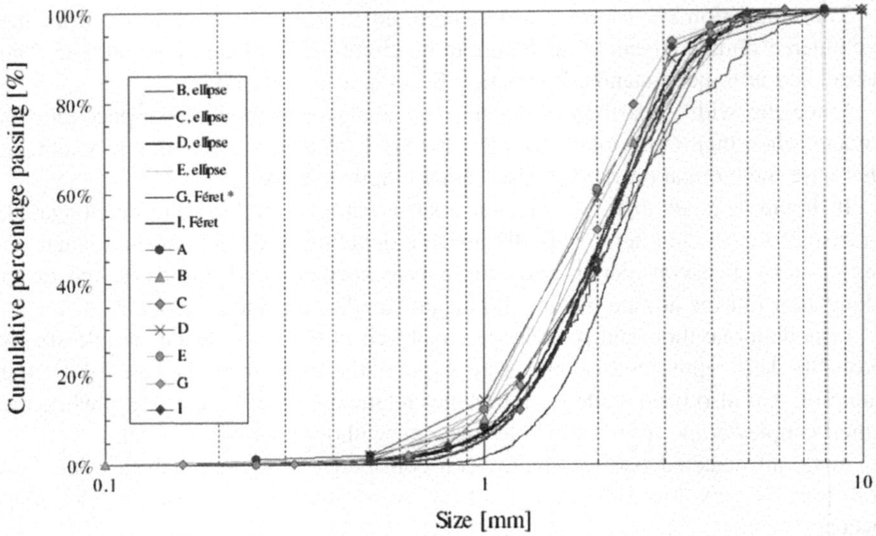

Figure 8 Cumulative width distributions obtained by sieving and image analysis

Table 15 ***Area fraction (%) of particles with width lower than 0.5, 1, 2, 4 and 5 mm***

Width size	B Ellipse	C Ellipse	D ellipse	E Ellipse	G Féret*	I Féret	Average	Standard deviation
0.5 mm	1.2	0.9	1.0	0.0	0.1	0.1	0.5	0.5
1 mm	7.0	7.5	7.0	5.9	1.6	5.6	5.8	2.2
2 mm	43.7	48.0	47.1	47.1	33.8	45.9	44.3	5.3
4 mm	82.3	92.1	95.7	93.7	91.7	92.9	91.4	4.7
5 mm	93.1	99.1	99.6	98.2	98.4	97.2	97.6	2.4

8.4 Concluding remarks on Particle Size Distribution

- Results are globally consistent, especially for length distributions obtained using image processing.
- Mechanical sieving leads to width separation of particles. Both methods, sieving and image analysis, could be complimentarily used to study width distribution. Sieving method remains more appropriate to study the finest particle contents such as dust.
- Most of the fibres could be removed by sieving methods on the first screens when sieving method is carried out. Hemp shiv with high cortical fibre content is more difficult to characterise with both methods.

- Results show uni-modal distributions of both particle width and length of the tested hemp shiv. The log-normal distribution models provide adequate and robust fits. Geometric standard deviations, median and mean lengths match well those computed from the raw data. The size distribution can be then closely described using only the two parameters: μ and σ.
- The image analysis is reliable to characterise shiv particles but the selected threshold size of detection, and the analysis methods have to be specified. It assesses many particle shape factors that are needed to study anisotropic properties of the composite materials. Some particles have a shape that deviates from "rod" shape due to the shredding action of hemp straw and remaining attached fibres, but ellipse fitting allows the determination of an average shape that seems to be more representative and robust.
- Sampling quality is a key point to consider with digital image processing, since this method can not be commonly applied to samples exceeding few grams.

SUMMARY OF FINDINGS

To obtain the principal physical characterisation of plant aggregate as hemp shiv, the literature shows that several protocols are in use. Some of them are directly derived from the methodology to describe the properties of mineral aggregates or other materials. Hemp shiv exhibits some very specific properties in terms of geometry, water absorption, bulk density etc. which are very different from normal mineral aggregates. As a result, the objective of the inter-laboratory tests presented in this paper is to analyse the data obtained from several protocols in order to propose recommendations which are able to reduce the discrepancies resulting from the limitations and lack of adaptation of the protocols to the specificity of hemp shiv.

At this stage five characteristics are investigated. The discussion of the results allows the proposal of the following principal recommendations:

Dry the material at 60 °C until constant mass is reached (variation less than 0.1% between two readings over a 24 hour period). Sampling quality is a key point to consider before any determination od properties.

3. Bulk density: The container size has to be adapted to the maximal length of the particle. In addition, before the test insufficient loosening of the packed material induces an error in the determination of the bulk density. The quantity of the material has to be adjusted to be half the volume of the container.
4. Particle Size Distribution: 100 gr for sieving method and 3 to 6 gr for image analysis method are necessary with hemp shiv. Representative analyses require more than 3000 identified particles represented by more than 40 pixels. As a consequence, with a given resolution (i.e. pixel size) the sample mass has to adjusted according to the mean size of the particles. Both methods, sieving and image analysis, could be complimentary used to study width distribution. Sieving method remains more appropriate to study the finest particle contents such as

dust. The image analysis is reliable to characterise shiv particles but the selected threshold size of detection, and the analysis methods have to be specified. Sampling quality is a key point to consider with digital image processing, since this method can not be commonly applied to samples exceeding few grams.

ACKNOWLEDGEMENTS

References

[1] Amziane S., Collet F., Lawrence M., Magniont C., Picandet V., Round robin test for hemp shiv characterisation: committee report of tc - 236 bio-based building materials part 1: evaluation of initial water content and water absorption, Industrial Crop And Product, submitted, May 2016.

[2] Construire en chanvre: Règles professionnelles d'exécution, Fédération Française du Bâtiment, Collection recherche, développement métier, 2009.

[3] Igathinathane C., Pordesimo L.O., Columbus E.P., Batchelor W.D., Sokhansanj S. Sieveless particle size distribution analysis of particulate materials through computer vision. Computers and Electronics in Agriculture 2009;66:147–58.

[4] Mani S, Tabil LG, Sokhansanj S. Grinding performance and physical properties of wheat and barley straws, corn stover and switchgrass. Biomass and Bioenergy 2004;27:339–52.

[5] Bitra V S P; Womac A R; Chevanan N; Miu P I; Igathinathane C; Sokhansanj S; Smith D R. Direct mechanical energy measures of hammer mill comminution of switchgrass, wheat straw, and corn stover and analysis of their particle size distributions. Powder Technology 2009;193:32–45.

[6] Allaire SE, Parent LE. Physical Properties of Granular Organic-based Fertilisers, Part 1: Static Properties. Biosystems Engineering 2004;87:79–87.

[7] Igathinathane C., Melin S., Sokhansanj S., Bi X., Lim C.J., Pordesimo L.O., Columbus E.P., Machine vision based particle size and size distribution determination of airborne dust particles of wood and bark pellets. Powder Technology 2009;196:202–212.

[8] Igathinathane C., Pordesimo L.O., Columbus E.P., Batchelor W.D., Methuku S.R. Shape identification and particles size distribution from basic shape parameters using ImageJ, Computers and Electronics in Agriculture 2008;63:168–82.

[9] ASABE Standards. Method of determining and expressing particle size of chopped forage materials by screening ANSI/ASAE S424.1. St. Joseph, MI: ASABE; 2006: 619–21.

[10] Allaire SE, Parent LE. Size Guide Number and Rosin–Rammler approaches to describe particle size distribution of granular organic-based fertilisers. Biosystems Engineering 2003;86:503–9.

[11] Bitra V S P., Womac A R., Chevanan N., Yang Y.T., Miu P I., Igathinathane C., Sokhansanj S. Mathematical model parameters for describing the particle size spectra of knife-milled corn stover. Biosystems Engineering 2009;104:369–83.

[12] Miao Z, Grift T.E, Hansen A.C, Ting K.C. Energy requirement for comminution of biomass in relation to particle physical properties. Industrial Crops and Products 2011;33:504–513.

[13] Limpert E, Stahel W.A, Abbt M. Log-normal Distributions across the sciences: keys and clues. BioScience 2001;51,5:341–52.

ROUND ROBIN TEST FOR HEMP SHIV CHARACTERISATION

PART 3: THERMAL CONDUCTIVITY

Authored by: Sofiane Amziane[1], Florence Collet[2], Mike Lawrence[3], Camille Magniont[4], Vincent Picandet[5]

1: Chair of RILEM TC-BBM 236, Institut Pascal, Clermont Université, France, sofiane.amziane@univ-bpclermont.fr
2: LGCGM, Université Rennes 1, Rennes, France, florence.collet@univ-rennes1.fr
3: BRE Centre for Innovative Construction Materials, University of Bath, Bath, UK m.lawrence@bath.ac.uk
4: Université de Toulouse, UPS, INSA, LMDC (Laboratoire Matériaux et Durabilité des Constructions), camille.magniont@insa-toulouse.fr
5: Université Bretagne Sud, IRDL - CNRS FRE 3744, F-56100 Lorient, France, vincent.picandet@univ-ubs.fr

With the contribution of:

6: Laurent Arnaud, Art et Métiers, ENSAM Cluny, laurent.arnaud@ensam.eu
7: Sylvie Prétot, LGCGM, Université de Rennes 1, France, sylvie.pretot@univ-rennes1.fr
8: Christophe Lanos, LGCGM, Université de Rennes 1, France, christophe.lanos@univ-rennes1.fr
9: Thibaut Collinart and Pierre Tronet, Université Bretagne Sud, IRDL - CNRS FRE 3744, F-56100 Lorient, France, thibaut.collinart@univ-ubs.fr
10: Vincent Nozahic, Institut Pascal, Clermont Université, France, vincent.nozahic@hotmail.fr
11: Etienne Gourlay, CETE de l'Est - Laboratoire Régional de Strasbourg, etienne.gourlay@developpement-durable.gouv.fr
12: Sandrine Marceau, Paris-Est, IFSTTAR, MAST, Marne-La-Vallée, France, sandrine.marceau@ifsttar.fr
13: Gilles Escadillas, Université de Toulouse, UPS, INSA, LMDC (Laboratoire Matériaux et Durabilité des Constructions), Gilles.Escadillas@u

S. Amziane and F. Collet (eds.), *Bio-aggregates Based Building Materials*, RILEM State-of-the-Art Reports 23, DOI 10.1007/978-94-024-1031-0

14: S. Dubois, Université de Liège (Ulg), Gembloux Agro-Bio Tech, Belgique, s.dubois@student.ulg.ac.be
15: Laetitia Bessette, Centre Technique Louis VICAT, L'Isle d'Abeau, France, L'Isle d'Abeau, l.bessette@vicat.fr
16: Paulien Debruijn, Lund University, Faculty of Engineering (LTH), Dept of Building Materials, Lund, Sweden, pbdebruijn@gmail.com
17: Ulrike Peter, Lhoist, Lhoist Recherche et Développement S.A., Belgium, ulrike.peter@lhoist.com
18: Sara Pavia, Dept of Civil Engineering, Trinity College Dublin, Ireland, PAVIAS@tcd.ie

Abstract
The paper presents the experience of a working group within the RILEM Technical Committee 236-BBM 'Bio-aggregate based building Materials'. The work of the Technical Committee (TC) will be to study construction materials made from plant particles. These materials are obtained from the processing of hemp, flax, miscanthus, pine, maize, sunflower, bamboo and others. The first round robin test of the TC-BBM was carried out to compare the protocols in use by the different laboratories (labs) to measure initial water content, bulk density, water absorption, particle grading and thermal conductivity. The aim is to define a characterisation protocol derived from those used by the different labs. This first round robin test was carried out on one variety of hemp shiv. Nine laboratories from European universities and research centers were involved (Table 1). The test results of 7 laboratories constitute a set of statistically representative data in order to propose recommendations to characterise hemp shiv after analysing the different methodologies in use in these labs. This paper presents the results on thermal conductivity.

Table 1 Participating Labs

Letter	*City*	*Labs*
A	*Bath (UK)*	*BRE Centre for Innovative Construction Materials/University of Bath*
B	*Clermont Ferrand (France)*	*Institut Pascal*
C	*Lorient (France)*	*LIMatB/Université de Bretagne Sud*
D	*Lyon (France)*	*DGCB/ENTPE*
E	*Paris (France)*	*IFSTTAR*
F	*Rennes (France)*	*LGCGM/Rennes 1*
G	*Toulouse (France)*	*LMDC/Université de Toulouse/UPS/INSA*
I	*Combloux (Belgium)*	*Combloux-Agro ressource – Université de Liège*

9 INTRODUCTION

This study focus on bio-based aggregate coming from the stem of plants cultivated either for their fibers (hemp, flax, etc.) or for their seeds (oleaginous flax, sunflower, etc.). Owing to the structure of the stem of the plant they are made from, such aggregates are generally malleable, elongated and highly porous with a low apparent density. They are very different from the mineral aggregates typically used in concretes, for which there are standardised tools and techniques for characterisation. Amongst these, hemp shiv (the woody core of the stem of the hemp plant) is probably the most widely used in alternative or eco-friendly building materials in Europe and is also representative of most of the aggregate coming from the stem of an annual crop. This is usually mixed with a lime-based binder and the resultant 'bio-concrete' is known as 'hemp-lime'.

This kind of aggregate is a co-product of hemp industry that is renewable and produced in an annual cycle while the price of mineral aggregates is steadily increasing as resources become less readily available. The characterisation of these aggregates, however, which is crucial to a proper understanding of the quality of the materials in which they are incorporated, requires adaptations to be made to the techniques usually employed for mineral aggregates, or the devising of new characterisation procedures.

The first round robin test of the Rilem TC-BBM was carried out to compare the protocols in use by the different labs. The aim was to define a test method to measure bulk density, particle size distribution, water absorption and thermal conductivity of bio-aggregates. This first round robin test was carried out with one variety of hemp shiv (the woody core of the plant stalk chopped into lengths of a few centimeters) coming from the same production of a processing factory located in France. Seven labs conducted the measurements (Table 2). This paper presents the results on thermal conductivity.

Table 2 ***Description of the interlaboratory test***

Characteristics	Participating Labs
Thermal conductivity	B, C, D, F, G

10 MATERIAL

One variety of hemp shiv was selected for this inter-laboratory test. It comes from the same processing factory where the bast fibers are stripped off (de-cortication), leaving the shiv behind. This shiv was provided by LCDA producer under the commercial name "KANABAT" (Tables 3 and 4). This shiv is in line with the French national recommendation provided by "Construire en Chanvre" association (Table 5) [1].

Table 3 Physical properties of shiv as supplied by the producer

Characteristic	Measurement
Density	100 to 110 kg/m^3 (depending on ambient relative humidity) Loosely packed, not compressed
Water absorption	198% (NFV 19 002)
Water absorption of mineral elements	24 meq per 100 g of raw material
Calorific value	3804 cal/g (NF M 07-030 12/9G)
Thermal Conductivity (10 °C in a dry state)	0.0486 W/m.K (NF EN12667)

Table 4 ***Chemical composition of the shiv*** as supplied by the producer*

Water	9 to 13%
Dry material	85 to 90% of which
Total organic material	97.5% on a dry basis of which:
Net cellulose:	52%
Lignin:	18%
Hemicellulose:	9%
Minerals:	Calcium: 1% on a dry basis Magnesium: 0.03% on a dry basis Phosphorus: 9 mg/100 g Potassium: 0.8% on a dry basis Total Nitrogen: 0.4 to 1% on a dry basis Total carbon: 496 g/kg on a dry basis C/N: 87 Ash: 2% pH in suspension at 10%: 6.7

*These data are supplied as indicative, certain values can vary depending on the year, the variety, the cultivation location etc.

Table 5 French national recommendations for the shiv characteristics

Characteristics	**Recommendation**
Apparent density of particles in kg/m^3	Quoted apparent density +/−15%
Maximal length (L_{dmax} in %)	Quoted L_{dmax} +/−10%
Initial moisture content	<19%
Dust content	<2% passing a 0.25 mm sieve
Colour	% of particles not conforming <5%

11 THERMAL CONDUCTIVITY

In association with their high porosity, hemp shiv show low thermal conductivity. In addition, this porosity is the location of moisture storage and transfer that impacts on the thermal properties of the material. In this study, two kinds of experimental methods are used to measure the thermal conductivity of hemp shiv. With the guarded hot plate, measurements are held under steady state. If the specimens are wet, this induces a moisture flux that hampers the reliability of thermal conductivity value. With the hot wire, measurements are held under transient state. This does not induce (or does limit) moisture flux, allowing the characterisation of the variation of thermal conductivity against humidity.

11.1 Experimental methods

11.1.1 Guarded hot plate

The guarded hot plate method was used by two labs (see Table 6).

Table 6 Experimental devices – guarded hot plates

	Lab C	Lab G
Bench	Home-made with lateral polyurethane guard	λ-Meter EP 500
Size of the heating unit/of the sample	10 0 × 10 cm^2/9.5 × 9.55 × 4.5 cm^3	50 × 50 cm^2/15 × 15 × e cm^3 1 cm < e < 20 cm
Temperatures	23 °C; T_h = 28 °C, T_c = 18 °C	25 °C; $T_h - T_c$ = 5 K
Steady state criteria		Change in conductivity lower than 1% in 60 minutes
Accuracy	0.05 °C for temperatures, 2 mW for power and 0.5 mm for thickness; Precision range 5 to 7%	0.1% for temperature <1% for thickness < 90 mm

(continued)

Table 6 (continued)

	Lab C	Lab G
Number of tests	2 to 4	3
Sample conditioning	cling wrap	Hemp shiv are placed in a PVC box (λ = 0.11934 W/m.K) (Figure 2)
Hydric state of hemp shiv	Dry state (dried in an oven at 60 °C); ambient temperature/50%RH (stabilised with salt solution); 95% RH (stabilised in a climatic chamber)	Dry state (dried in an oven at 50°C until the change in mass of the sample was less than 0.1% over 24 hours)
Bulk density of hemp shiv	140 to 200 kg.m^{-3}	About 135 kg.m^{-3}

Figure 1 Hot plate (Lab C)

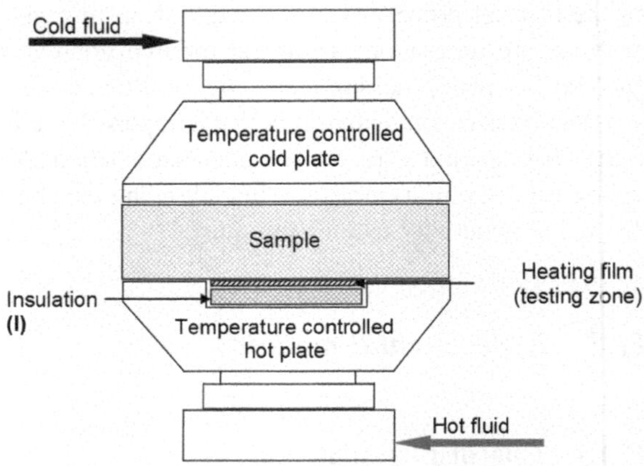

The sample is placed between two temperature-controlled plates. One plate is heated while the other plate is cooled. Their temperatures (resp. T_h and T_c) are monitored until they are constant (Figure 1). The steady state temperatures, the thickness of the sample (e) and the heat input (q) are used to calculate the thermal conductivity (eq. 9).

$$\lambda = \frac{q.e}{S(T_h - T_c)} \qquad \text{(eq.9)}$$

Figure 2 Hemp shiv placed in a PVC box

11.1.2 Hot wire

The Hot wire method was used by three labs: Lab B (NEOTIM FP2C apparatus), Lab D (home made apparatus) and Lab F (CT Meter apparatus).

The wire is a heating probe and its temperature rise is measured versus heating time t (Figure 3 and Table 7). There is a proportional relationship between temperature rise $\Delta T$and logarithmic heating time (t) (eq. 10), where q is the heat flux per meter. The thermal conductivity λ [mW.m^{-1}.K^{-1}] is obtained from the slope of this curve.

$$\Delta T = \frac{q}{4\pi\lambda}(\ln(t) + cste) \qquad \text{(eq.10)}$$

The main advantage of this method, compared to hot plate method, is that it is a transient method that does not induce (or that does limit) the water migration during test [2][3].

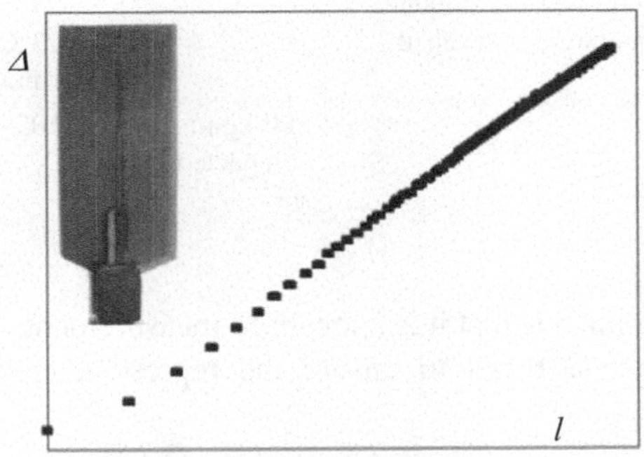

Figure 3 Hot wire: temperature rise versus logarithmic heating time

Figure 4 Measurement of thermal conductivity of hemp shiv (Lab F)

Table 7 Experimental devices - Hot wires

	Lab B	Lab D	Lab F
Length of the hot wire		5 cm	5 cm
Electric resistance	12.8 Ω		6.14 Ω
Power/Duration of measurement	0.1 W/100 s	0.1 W/120 s, 140 s, 160 s, 180 s 0.16 W/160 s	0.159 W/60 s (best setting within 8 tested)
Number of tests	3	1	5
Sample conditioning	Probe plunged in the middle of 500 g of aggregates which is a consistent volume to approximate infinite volume conditions	Bulk aggregates and compacted hemp shiv (under a compression stress 0.05 MPa)	Controlled thanks to moulds (Figure 4)
Hydric state of hemp shiv	Undried and dried aggregates (dried in an oven at 60 °C for 48 h and stored in a hermetic and airtight bag till ambient temperature is reached	–	Dry state (dried in an oven at 50 °C - measurement held at 23 °C in a dried chamber), stabilised at 23°C, 50%RH and at 23°C, 80% RH in a climatic chamber
Bulk density of hemp shiv	–	180 kg.m^{-3} for compacted hemp shiv	105–116 kg.m^{-3}

The main disadvantage is that it is a localised measurement. Thus, measurements should be taken several times to ensure the representativeness of the thermal conductivity value.

11.2 *Results*

11.2.1 Hot plate

The variation of thermal conductivity against relative humidity and against bulk density for given relative humidity is given Figure 5.

The range of bulk density goes from 140 to 200 $kg.m^{-3}$.

For dry state, the average value of thermal conductivity is 54.8 $mW.m^{-1}.K^{-1}$, and the standard deviation is 2.8 $mW.m^{-1}.K^{-1}$. For ambient relative humidity of 50%, the average value of thermal conductivity is 62.6 $mW.m^{-1}.K^{-1}$, the results are more scattered with a standard deviation of 8.9 $mW.m^{-1}.K^{-1}$. This variation is much higher than the given precision of the measure. This may be due to moisture migration during the measurement. Lastly, for relative humidity of 95%, the average value of thermal conductivity is 75.8 $mW.m^{-1}.K^{-1}$, and the standard deviation is 2.6 $mW.m^{-1}.K^{-1}$.

The variation of thermal conductivity of hemp shiv versus bulk density is given Figure 6.

The range of bulk density goes from 134 to 138 $kg.m^{-3}$. The average value of thermal conductivity at dry state is 55.3 $mW.m^{-1}.K^{-1}$, and the standard deviation is 1.7 $mW.m^{-1}.K^{-1}$.

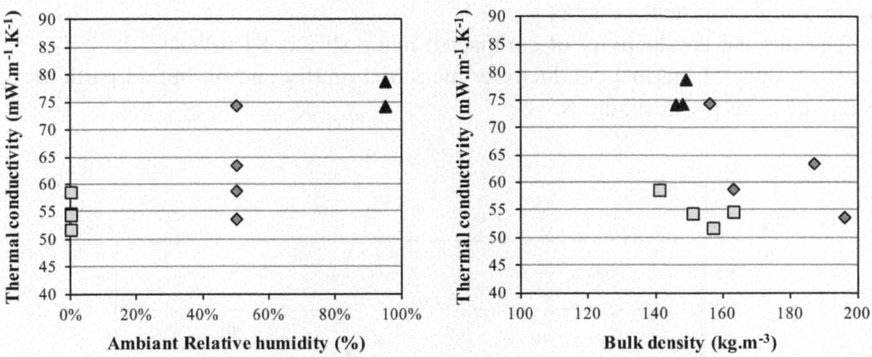

Figure 5 Variation of thermal conductivity of hemp shiv - left: versus ambient relative humidity - right: versus bulk density for given relative humidity - Lab C

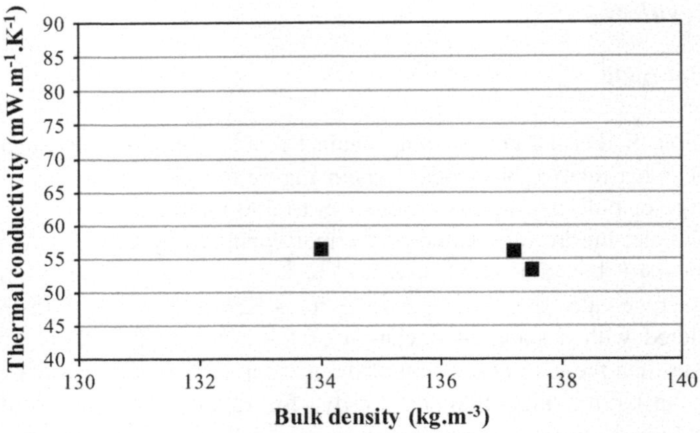

Figure 6 Variation of thermal conductivity versus bulk density - Lab G

11.2.2 Hot wire

The variation of thermal conductivity of hemp shiv versus heating time is given Figure 7. The average value of thermal conductivity is 73.5 $mW.m^{-1}.K^{-1}$, and the standard deviation is 2.5 $mW.m^{-1}.K^{-1}$.

Thermal conductivity for dry state is thus estimated about 66.3 $mW.m^{-1}.K^{-1}$ using a self-consistent scheme.

The thermal conductivity of compacted hemp shiv is 84 $mW.m^{-1}.K^{-1}$.

The values of thermal conductivity measured on dry and on humid particles of hemp shiv are given Figure 8.

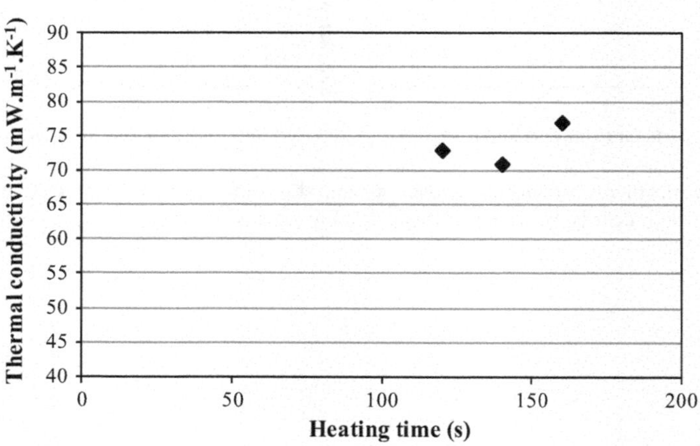

Figure 7 Thermal conductivity of bulk hemp shiv versus heating time of hot wire – Lab D

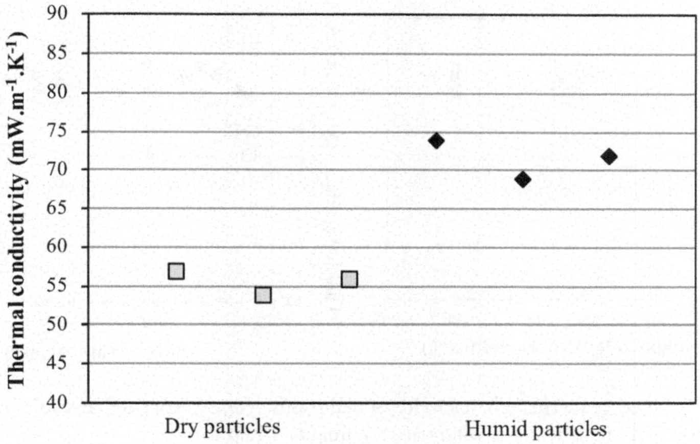

Figure 8 Thermal conductivity of hemp shiv – Lab B

For dry particles, the average value of thermal conductivity is 56 $mW.m^{-1}.K^{-1}$, and the standard deviation is 2 $mW.m^{-1}.K^{-1}$. While for humid particles, the average value of thermal conductivity is 72 $mW.m^{-1}.K^{-1}$, and the standard deviation is 3 $mW.m^{-1}.K^{-1}$. The correlation coefficient is 0.999.

The variation of thermal conductivity of bulk hemp shiv versus setting of power and heating time is given Table 8. The setting n°3 is considered to be invalid as the temperature rise is less than 10 °C. In a similar way, the setting n°5 shows a slightly high temperature rise. For these two settings, the correlation coefficient is lower than the ones obtained with the other settings. The average value of thermal conductivity (without settings 3 and 5) is 79.1 $mW.m^{-1}.K^{-1}$, and the standard deviation is 4 $mW.m^{-1}.K^{-1}$.

The setting used for the following measurement is $P = 0.159$ W, $t_c = 60$ s. The variation of thermal conductivity versus relative humidity and versus bulk density for given relative humidity is shown Figure 9.

Table 8 Thermal conductivity versus settings of hot wire

		Heating time	λ	R^2	ΔT
	P [W]	tc [s]	[$W.m^{-1}.K^{-1}$]		[°C]
1	0.23	80	0.083	0.9999	19.1
2	0.159	80	0.0753	0.9993	15.32
3	0.102	80	0.1067	0.9965	8.15 < 10 °C
4	0.159	60	0.0797	0.9997	12.64
5	0.23	60	0.0906	0.997	17.7
6	0.159	90	0.084	0.9996	13.07
7	0.159	60	0.0785	0.9999	12.73
8	0.159	60	0.0745	0.9998	12.8

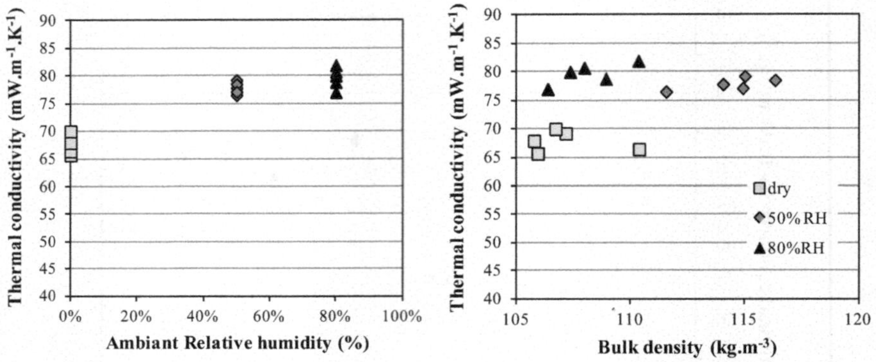

Figure 9 Variation of thermal conductivity of hemp shiv – *left*: versus ambient relative humidity – *right*: versus bulk density for given relative humidity - Lab F

For dry state, the average value of thermal conductivity is 68 $mW.m^{-1}.K^{-1}$, and the standard deviation is 2 $mW.m^{-1}.K^{-1}$. The correlation coefficients are all higher than 0.9996. For ambient relative humidity of 50%, the average value of thermal conductivity is 78 $mW.m^{-1}.K^{-1}$, and the standard deviation is 1 $mW.m^{-1}.K^{-1}$. The correlation coefficients are all higher than 0.9998. Lastly, for relative humidity of 80%, the average value of thermal conductivity is 80 $mW.m^{-1}.K^{-1}$, and the standard deviation is 2 $mW.m^{-1}.K^{-1}$. The correlation coefficients are all higher than 0.9998 as well.

11.2.3 Synthesis

Figure 10 shows a synthesis of the results obtained from all the participant labs (average values with standard deviation bars when available).

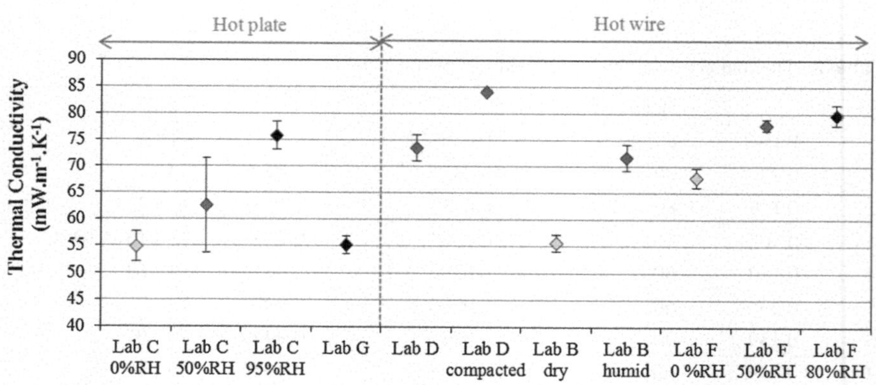

Figure 10 Synthesis of results from all participant labs (red: dry state, blue: humid state)

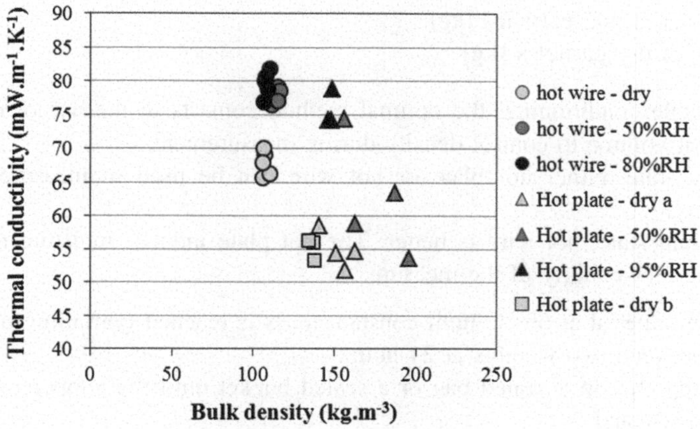

Figure 11 Synthesis of variation of thermal conductivity versus bulk density for given relative humidity

The dry state was studied by four labs. The average value is 58 $mW.m^{-1}.K^{-1}$ with a standard deviation of 6 $mW.m^{-1}.K^{-1}$. That induces a variation of 11%. Similar average values are obtained with hot plate and hot wire.

For humid particles, the study was held by four labs. The average value is 74 $mW.m^{-1}.K^{-1}$ with a standard deviation of 6 $mW.m^{-1}.K^{-1}$. That induces a variation of 8%. The measurements held with the hot plate show higher discrepancy than those held with hot wire. This may be due to moisture migration during the test.

Figure 11 gives the variation of thermal conductivity versus bulk density for given relative humidity and for measurements made with the two methods. This figure shows a cloud of points that remains difficult to analyse. Finally, thermal conductivity is more closely linked to humidity than to bulk density.

11.3 Concluding remarks

Hemp shiv must be stabilised at dry state or at given relative humidity. The drying of hemp shiv is done in an oven at 60 °C. The stabilisation (at dry state or given RH) is reached when two successive daily measures of weight agree within 0.1% of mass of the sample. The stabilisation kinetics should be given. The water content should be calculated from humid and dry mass according to [Eq. 2].

$$w = \frac{m - m_0}{m_0} \text{ [kg.kg}^{-1}\text{]} \qquad \text{(eq.11)}$$

m mass of humid particles [kg]
m_0 mass of dry particles [kg]

For sample conditioning, the optimal method consists in the use of container with known volume to control density during measurement.

For dry state, either hot plate or hot wire can be used to measure thermal conductivity.

For humid state, hot wire is better. The hot plate induces moisture migration altering the repeatability of the measure.

1. Dry the material at 60 °C until constant mass is reached (variation lower than 0.1% between two weights at 24 hours).
2. Put hemp shiv in a sealed bag or a sealed bucket until the temperature of the room is reached.

Hot plate method

3. To measure thermal conductivity with hot plate, the minimal thickness is 4.5 cm and the minimal lateral length is twice the thickness.
4. The bulk density must be measured: hemp shiv are placed in a known volume. Hemp shiv are weighed before and after the measurement.
5. The measurement is held once the steady state is assumed to be reached (when the change in conductivity was less than 1% in 60 minutes)
6. The gradient of temperature should be given

Hot wire

3. Hemp shiv are placed in containers to measure bulk density.
4. The power of hot wire should be between 0.1 and 0.2 W. The duration of the measurement should be between 60 and 120 s.

SUMMARY OF FINDINGS

To obtain the principal physical characterisation of plant aggregate as hemp shiv, the literature shows that several protocols are in use. Some of them are directly derived from the methodology to describe the properties of mineral aggregates or other materials. Hemp shiv exhibits some very specific properties in terms of geometry, water absorption, bulk density etc. which are very different from normal mineral aggregates. As a result, the objective of the inter-laboratory tests presented in this paper is to analyse the data obtained from several protocols in order to propose recommendations which are able to reduce the discrepancies resulting from the limitations and lack of adaptation of the protocols to the specificity of hemp shiv.

At this stage five characteristics are investigated. The discussion of the results allows the proposal of the following principal recommendations:

Dry the material at 60 °C until constant mass is reached (variation less than 0.1% between two readings over a 24 hour period). Sampling quality is a key point to consider before any determination od properties.

5. Thermal Conductivity: The hot plate and the hot wire methods are both well adapted to measure bulk thermal conductivity. The initial dry state of the sample is a critical parameter in order to obtain valid data.

ACKNOWLEDGEMENTS

References

[1] Construire en chanvre: Règles professionnelles d'exécution, Fédération Française du Bâtiment, Collection recherche, développement métier, 2009.

[2] Glouannec P., Collet F., Lanos C., Mounanga P., Pierre T., Poullain P., Pretot S., Chamoin J. et Zaknoune A.: Propriétés physiques de bétons de chanvre. Matériaux & Techniques, EDP Sciences, Volume 99, Numéro 6, 657-665 (2011).

[3] J Hladik, "Métrologie des propriétés thermophysiques des matériaux", Masson collection mesures physiques, 1990

